WIND TURBINES

In the multi-disciplinary field of wind energy, students and professionals can often be uncomfortable outside their own specialist areas. This essential textbook explains, in a single readable text, the key aspects of wind turbine technology and its application. Covering a broad range of multi-disciplinary topics, including everything from aerodynamics to electrical and control theory, structures, planning, economics, and policy, this reference is an excellent toolkit for undergraduate students, postgraduate students, and professionals in the field of wind energy. Key concepts, including more challenging ones, such as rotational sampling of turbulence, vortex wake structures, and reactive power management, are explained using clear language and simplifying illustrations, including experimental graphs, photographs, and line drawings.

Colin Anderson is a consulting engineer specialising in renewable energy technology and an adjunct lecturer at Edinburgh University School of Engineering, where he teaches a course entitled Principles of Wind Energy.

WIND TURBINES

Theory and Practice

COLIN ANDERSON

CAMBRIDGE
UNIVERSITY PRESS

CAMBRIDGE
UNIVERSITY PRESS

University Printing House, Cambridge CB2 8BS, United Kingdom

One Liberty Plaza, 20th Floor, New York, NY 10006, USA

477 Williamstown Road, Port Melbourne, VIC 3207, Australia

314–321, 3rd Floor, Plot 3, Splendor Forum, Jasola District Centre, New Delhi – 110025, India

79 Anson Road, #06–04/06, Singapore 079906

Cambridge University Press is part of the University of Cambridge.

It furthers the University's mission by disseminating knowledge in the pursuit of education, learning, and research at the highest international levels of excellence.

www.cambridge.org
Information on this title: www.cambridge.org/9781108478328
DOI: 10.1017/9781108777469

First published 2020

Printed in the United Kingdom by TJ International Ltd, Padstow Cornwall

A catalogue record for this publication is available from the British Library.

Library of Congress Cataloging-in-Publication Data
Names: Anderson, Colin, 1957– author.
Title: Wind turbines : theory and practice / Colin Anderson.
Description: New York : Cambridge University Press, 2020. | Includes index.
Identifiers: LCCN 2019029326 (print) | LCCN 2019029327 (ebook) | ISBN 9781108478328 (hardback) | ISBN 9781108777469 (epub)
Subjects: LCSH: Wind turbines.
Classification: LCC TJ828 .A53 2020 (print) | LCC TJ828 (ebook) | DDC 621.31/2136–dc23
LC record available at https://lccn.loc.gov/2019029326
LC ebook record available at https://lccn.loc.gov/2019029327

ISBN 978-1-108-47832-8 Hardback

CONTENTS

ACKNOWLEDGEMENTS

A number of individuals and organisations made valuable contributions to the material in this book, and I would like to offer my thanks in particular to the following: Peter Jamieson and Irvin Redmond, for discussions of the pioneering Howden technology; Penny Dunbabin and Jonathan Whale, who generously allowed me to reproduce some of their early PhD research work in Chapters 3 and 10; my Aerpac colleague Jaap de Boer, for feedback on the material in Chapters 7 and 8; and Roger Borre, for generously approving Figure 7.16. The contributions of many other ex-Aerpac colleagues in the Netherlands and the UK are implicitly acknowledged, with a special thanks to Rob Roelofs and Henk Heerkes. I am grateful to Richard Yemm for providing useful comments on Chapter 7 and to Walt Musial and Scott Hughes at NREL for feedback on blade testing and the use of photos and test data in Chapter 8. The dramatic photo in Figure 5.19 is reproduced with kind permission of KEMA Labs, with thanks to Bas Verhoeven.

The family firm of Mackies Ltd pioneered the use of wind turbines in the UK, and their contribution is reflected in several chapters; I have used data from Mackies windfarm to illustrate wind characteristics in Chapter 2 and some economic points in Chapter 11. Many thanks are due to Mac and, in absentia, his late father, Maitland Mackie, CBE. Still in Aberdeenshire, thanks also to Grant Mackie at Greenspan Energy for the use of SCADA data from Balquhindachy to illustrate wake effects and for providing the perfectly timed photo in Figure 3.21; thanks also to Neil and Sara Macrae, owners of Mains of Dudwick farm when it was used for a wind measurement campaign referenced in Chapter 2.

I have used the Isle of Gigha community windfarm to illustrate several technical and economic points: thanks are due to past and present members of the Gigha Heritage Trust for their assistance, not least to Andy Oliver for feedback on Chapter 11. Donald Mackay kindly supplied the photo in Figure 11.11. A little farther north, the Isle of Luing hosted an earlier demonstration project, and my thanks are to Shane Cadzow for his assistance and for the photo in Figure 9.19; thanks also to fellow islander Anja Lamont for material relating to the Luing bird surveys and the photo in Figure 10.2. Paul Pynn kindly supplied the photo in Figure 9.27. Island wind projects are a recurring theme, and extreme wind data from the Outer Hebrides are reproduced in Chapter 2 with kind permission from David Cameron and David Wake at the North Harris Trust. Charlie Robb generously provided some excellent photos of wind turbines large and small and offered valuable comments on the text. Thanks also to Anne Phillips of Highlands and Islands Airports for permission to reproduce the Tiree Airport obstacle avoidance map in Chapter 8.

In addition, I am grateful to Jens Peter Hansen of the Danish Wind Turbine Owners' Association and to Anders Lønne of Vindenergi Danmark for their information on Danish ownership; Peter Thisted and Tor Helge Kjellby of Windtechnique A/S for details of their proprietary rock anchor system, Vestas Wind Systems A/S (with particular thanks to Mark Powell and Christina Schmidt for photographs); and Scottish and Southern Electricity Networks (Rob Broughton and Greg Clarke) for use of the Gigha grid map. Thanks also to Dave Collett of Collett Transport for his ever-enthusiastic support and for generously providing illustrations for Chapter 9.

An especial thanks is due to the team at Cambridge University Press, including Steven Elliott, Dominic Stock, and Amy Mower, and the many unseen hands involved in bringing this book to press. Finally, I must thank Professor Stephen Salter, MBE, for giving me – and many others – their opening into the field of renewable energy. And, for her unfailing help (not least with Chapter 5), support, and companionship during the preparation of this book, I thank my wife, Steffi Anderson.

PREFACE

This book offers a broad overview of wind energy technology, explaining the principles underlying the design, manufacture, and operation of modern wind turbines. The scope and content are based on a master's course I have taught at Edinburgh University since 2009, which has been continually updated in an attempt to keep pace with the evolution of the technology. To write a book about 'contemporary' wind turbines is, however, to offer a hostage to fortune. By the time it is published, it will be out of date, so references to the most powerful turbine, the longest blade, or the largest offshore array must carry the proviso 'at the time of writing'. So be it; but over the past few decades, the evolution of wind turbine technology and the increase in its worldwide reach have been quite breathtaking, and they show little signs of slowing. Wind power has firmly passed from alternative to mainstream.

The target readership for this book is the undergraduate or postgraduate student, professionals new to the field of wind energy, or anyone already working in it who seeks background reading outside his or her own specialism. Although a reasonable understanding of physics or engineering is required for some of the material, the level of mathematics is not severe, with illustrations and graphics used to clarify difficult concepts. I have tried where possible to illustrate key topics with examples from my own experience in the wind industry, in which I am fortunate to have been working since the mid 1980s. In this time I am privileged to have worked for the first UK manufacturer of large wind turbines, one of the earliest businesses to install a turbine for its own use, and the first community in Scotland to develop and own a grid-connected windfarm. These experiences have all been drawn on in the book, and I hope that in each chapter, the reader will either learn something new or understand better something already known.

Chapter Overview

Note. If this book is to be used as a course text, then the chapters should ideally be read in order, although the first and last can be read in isolation. There are example exercises at the end of each chapter; most are of exam standard, but a few (based on real situations) are included as examples in problem solving. For general readers or researchers who wish to explore more deeply the topics touched on in the book, there is a comprehensive reference list at the end. The following is a chapter summary.

Chapter 1 is a brief review of the trajectory of wind power from the pre-electric era through to the modern day. A recurring theme is that development of the technology has historically been driven by energy security. The chapter also relates how sophisticated measurements of rotor aerodynamic loads were being made 70 years before Michael Faraday built the first electric generator. Chapter 2 describes the origin and characteristics of the wind. Topics such as wind shear and turbulence are illustrated with site measurements, and the chapter includes an example of some thought provoking wind conditions recorded during a severe storm in the Outer Hebrides.

Chapter 3 is a refresher on aerodynamic theory. It includes the basic mathematical development of blade element momentum theory and a qualitative description of the more complex but also more physically realistic vortex wake theory. There is a comparison of wake measurements from full-scale and model wind turbines; vorticity data from the latter verify the predictions of vortex wake theory with images that would not be out of place in a gallery of modern art. Chapter 4 extends the aerodynamic discussion to show how net rotor loads (thrust, torque, and power) and the dimensionless C_p, λ curve are derived. The relationship between optimum blade solidity and tip speed ratio and the influence of blade pitch are explained, leading to the broader discussion of rotor aerodynamic control in Chapter 6.

Chapter 5 meantime deals with electrical issues and is broadly divided in two. The first half explains the operating principles of the different types of generator (there are several) found on wind turbines and their influence on dynamics and electrical power quality. The second half deals with electrical networks and further examines the issue of power quality; the role of reactive power, and how generators can manipulate it (some better than others) to aid voltage stability, is explained. The role of statcoms, SVCs, and pre-insertion resistors is discussed. Chapter 6 then examines the subject of wind turbine control, drawing together material from the preceding three chapters. The main topic is real-time power limiting, with explanations of stall regulation, constant-speed variable pitch (CSVP), and variable-speed variable pitch (VSVP) control. A comparison of the control accuracy of CSVP and VSVP strategies is made using power measurements from full-scale wind turbines.

The subject of Chapter 7 is structural loading and response. A recap on the dynamics of a single degree of freedom system leads into a discussion of multi-DOF systems and modal analysis. The cyclic loads affecting a wind turbine structure are described, with explanations of stochastic and deterministic loading and the principle of aerodynamic damping. The last part of the chapter draws on an early experimental campaign in which the dynamic loading on a full-scale wind turbine was measured and compared with a modal simulation. Results from these trials also demonstrated the difference in rotor loading arising from positive and negative pitch control. The chapter concludes with a brief summary of fatigue prediction.

Chapter 8 is an overview of rotor blade technology, covering structural design, manufacture, and testing. The material properties of glass fibre– and carbon fibre–reinforced plastics and wood-epoxy laminate are compared, and their superiority to metals is explained. Blade stresses are analysed using a simple cantilever beam model, with bending moment theory modified for composite structures. Blade manufacture using the vacuum resin infusion (VRIM) process is illustrated and described, and different blade root attachment methods are compared. The chapter concludes with a look back at the ten-fold scaling of wind turbine rotors that has occurred in the modern era and how it was achieved without subverting the fundamental laws of physics.

Chapter 9 steps away from the wind turbine to consider the external factors involved in siting and construction for onshore wind projects. The measure-correlate-predict (MCP) procedure for site wind assessment is described, and analytic models for wake loss and added turbulence are illustrated with experimental data from large and small arrays. The second half of the chapter looks at construction: topics include foundations, transport and access, and wind turbine erection. Some novel examples include rock anchor foundations that require almost no concrete and turbines that have been winched into place without using cranes.

Chapter 10 is an overview of the planning and environmental issues that attend onshore wind turbine developments, including ecological factors (birds and mammals), public acceptance (noise, visual impact), and safety. Examples are given of the type of information required in UK planning submissions but of generally wider application. The origins and treatment of wind turbine noise are dealt with in detail, and some simple rules are given for noise prediction. The chapter also considers impacts on other human activities, including radio-frequency communications, aviation, and radar. The principle of a radar-absorbing 'stealth' blade is described: such blades have recently been developed and put into windfarm service.

The final chapter (Chapter 11) addresses economic and political aspects; it is not overtly technical and may hold some interest for those working in the field of energy policy. The standard formula for the cost of generation, the levelised cost of energy (LCoE), is given in full and simplified forms. Historic installation costs from the UK onshore and offshore sectors are presented from 1990 to the present day and used to calculate LCoE at representative capacity factor and discount rates. A section on ownership considers two case studies. One examines the economics of self-supply for a business whose electricity demand is largely met by its own windfarm; the statistics for consumption, export, and balancing for this case usefully predict the impact of wind power at a national level. The second case study considers the Isle of Gigha community windfarm, which is an interesting technical story in itself, but informs a wider debate about ownership and public acceptance. The concluding section is on UK wind energy policy, past and (at the time of writing!) present.

Chapter 1 INTRODUCTION

1.1 A Wake-Up Call

On 8 November 1977, President Jimmy Carter made a televised address to the US nation on the subject of energy. There was a crisis. Geopolitical tensions had resulted in an embargo on oil exports from the Middle East, on whose output much of the industrial world then relied. The price of oil, which had been unchanged in real terms since the Second World War, trebled in little over a year. Carter acknowledged America's dependency on oil and outlined a range of measures to encourage energy saving, new fuels, and the production of electricity from renewable sources: for the first time these would include solar and wind power on a national scale. The 'energy crisis' of the 1970s was felt worldwide and similar policies were implemented in many countries; but it was in America, where the use of oil was most prevalent, that the pace of deployment of renewables was set. Within a few years the first large-scale windfarms started to appear in California and soon there were thousands of wind turbines feeding into public electricity networks. These early machines were relatively small, and appeared in a variety of designs and configurations with two, three, and even four blades. Interestingly, more than half of them came from Denmark, a country with a population less than a fifth that of California, but one which had played a pivotal role in the history of wind energy, and was about to play an even greater one.

A further hike in oil prices occurred in 1979 with the price almost doubling again; see Figure 1.1. This was again triggered by politics rather than fundamental limits to the oil resource. Over the next decade or so the price subsided to more agreeable levels but its historic stability did not return; and although there is plenty of oil being pumped today, its price and availability is no longer taken for granted. Also, notably absent from Carter's speech was any mention of the environment. In 1977 the overarching concern was short-term economic security, but since then man-made climate change has become increasingly recognised as a threat, and is now a major driver for renewable energy policies. Ultimately this too is an argument about security, but on a long-term basis and a global scale. It is no surprise, then, that sources of power generation requiring no fuel and producing minimal pollution are becoming mainstream; among them wind power has emerged as the economic front-runner, and the extent to which the technology has developed can fairly be called spectacular. In the mid 1980s commercial wind turbines had 15 m rotors and 50 kW output; at the time of writing they have evolved into giants with 160 m rotors

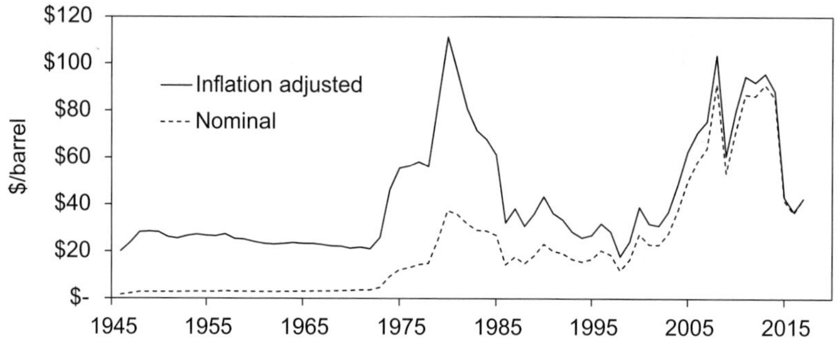

Figure 1.1 Historic price of crude oil 1945–2017. The 'energy crisis' began in 1973 and has arguably never ended. President Carter's landmark speech was in 1977. (*Source*: Inflationdata.com)

and 8 MW output (see Figure 1.2) and the latest offshore arrays rival the capacity of nuclear power stations.

Electricity-generating wind turbines were by no means a new idea in 1977, however, and their evolution can be traced back for almost a century before. The Danes were employing wind turbines in significant numbers well before the First World War, while the use of windmills and wind pumps in the pre-electric era goes back over a thousand years. The following is a brief historical review.

1.2 Early History

Archaeologists have found evidence that sailing boats were in use in the Middle East as long as 7000 years ago (Carter, 2006) and in an era when the only other sources of mechanical power were human and animal labour it seems reasonable to speculate that at some point, someone had the idea to make sails do useful work on land. No date exists for the first practical application of wind power, but the earliest windmills appear to have been used in the Middle East (Mesopotamia and Persia[1]) according to documentary evidence dating from around 500–900 AD (Dodge, 2014). From the start the principal uses of wind power were water pumping and grain milling. The early Persian machines were *panemones*, with sails of cloth or reed matting rotating about a vertical axis. Horizontal-axis windmills and pumps appeared in Europe in mediaeval times, thought to have been introduced from the Middle East by the Crusaders (BWEA, 1982), and wind pumps of a type known in the Aegean around 1300 AD can still be seen on the island of Crete.

These early horizontal-axis windmills employed sailcloth blades, which (like sailboats) benefit from aerodynamic lift, giving them a clear efficiency advantage over the vertical-axis panemone, which depends purely on drag. As such, horizontal-axis windmills were much more

[1] Modern-day Iraq and Iran, respectively.

Figure 1.2 Thirty years' growth: (a) the 'Californian wind rush', San Gorgonio, 1986; (b) Vestas 8 MW wind turbine in the Aberdeen Offshore Array, 2018: this single machine has a greater power output than all the wind turbines in the top picture combined.

representative of modern wind turbines: the importance of lift is discussed in Chapter 3. Thus began the evolution of the classic European windmill, usually based on a four-blade rotor with sailcloth or wooden-slatted blades. A major step in its development was to enable the rotor to face the wind from any direction: this was first achieved on *post mills* by manually rotating the whole tower (or getting a donkey to do it), while on the later *smock mills* only the top section rotated, similar to the nacelle of a modern wind turbine. Eventually this feature was mechanically automated by the use of tail vanes or fantail drives. In this and other ways European windmills developed quite sophisticated control systems, precursors of those in modern horizontal-axis wind turbines (HAWTs), but achieved without electric or hydraulic power. A good example is Herne Mill in Kent: built in 1789 this smock mill features geared drive shafts, a fantail yaw drive, flyball governors to vary the millstone clearance in response to rotor speed (a form of load control), and spring-loaded slats to vary the rotor aerodynamic loading (Eggleston, 1987). Similar functions are nowadays achieved electromechanically (though fantail yaw drives persisted into the electric era; see Figure 1.4).

The Anglo-Scots engineer John Smeaton (1724–92) is credited with the first scientific measurements of aerodynamic forces on a rotating windmill model. Smeaton was already famous as a structural engineer and the designer of the third Eddystone Lighthouse (the first two were destroyed in storms – Smeaton's was still in use a century later) when in 1759 he published a treatise on the use of water and wind power to turn mills and other machines (Smeaton, 1759). Smeaton devised an ingenious 'whirling arm' with weights and cables to enable a model windmill rotor to be rotated about its shaft axis, at the same time being driven forward to simulate the oncoming wind; see Figure 1.3. His mechanism was the effective precursor of the wind tunnel, and Smeaton's force measurements were used by aerodynamic researchers throughout the nineteenth century, right up to the time of the Wright Brothers, who used (and subsequently corrected) his force coefficients in their early aircraft designs; for a description of Smeaton's work and influence on aerodynamic research see (Anderson, 1997).

The development of large windmills in Europe declined from the late eighteenth century as the Industrial Revolution took hold, with the rise of fossil fuels (initially coal, then oil) marking a wholesale shift away from renewable energy sources. This fundamental change was driven by economics: coal was cheap, easily transportable, and with the high energy density demanded by the new processes of iron and steel making; coal was also capable of heating the homes of the rapidly growing urban populations of northern Europe. Windmill technology nevertheless continued to evolve in rural locations and by the mid nineteenth century small multi-blade wind pumps had become common on farms in the USA, where they provided a cheap and reliable means of water pumping in the pre-electric era; and it was the marriage of US wind pump rotor designs and the nascent technology of electrical generators that gave the world its first horizontal-axis wind turbine. It was, however, the world's second wind turbine to run, the first being a vertical-axis machine.

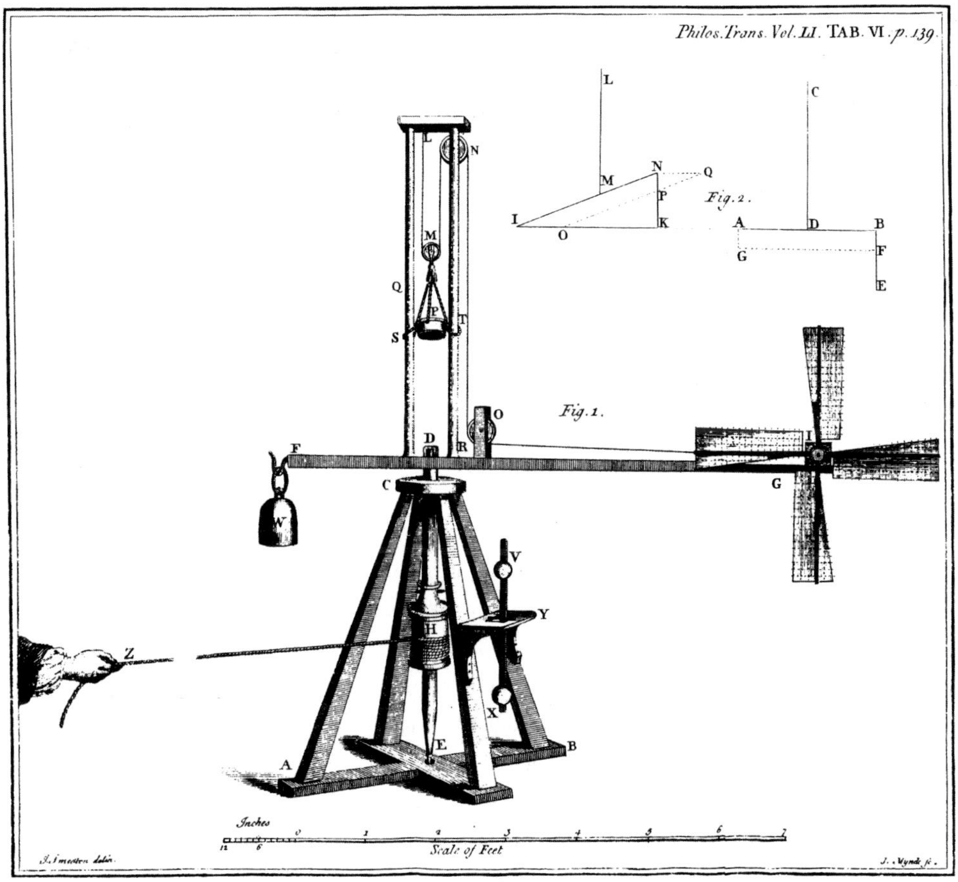

Figure 1.3 The first measurements. John Smeaton's apparatus for measuring the aerodynamic forces on rotating windmill blades (Smeaton, 1759). His experimental force coefficients were used by aerodynamic researchers for nearly 150 years. (*Source*: Smeaton (1759–60, 51))

1.3 The First Wind Turbines

The first recorded use of wind-generated electricity was at Marykirk in northeast Scotland, where in July 1887 Professor James Blyth built a large vertical-axis turbine, reminiscent of the early panemone designs, based on a 10 m rotor with cloth sails. This machine drove a generator to charge accumulators (batteries) to power the lighting at Blyth's home, the first to have electricity supplied by wind power (Price, 2005); Blyth's offer to supply surplus electricity to his Marykirk neighbours was declined: apparently they thought it to be 'the work of the devil'. The wind turbine operated for 25 years but did not lead to commercial developments, probably due to its low efficiency and unwieldy design. In recognition of his pioneering work, though, James Blyth was awarded a prize medal by the Royal Scottish Society of Arts. Just a few months after Blyth's first success, Charles F Brush of Cleveland, Ohio, demonstrated a horizontal-axis wind turbine, whose origins in the US multi-blade water pumps are evident. Brush, however, took the key step of incorporating a 50:1 speed-increasing gearbox to achieve the high rotation speed (500 rpm) needed for electric generation, while maintaining low rotation speed at the 17-m-diameter rotor. His low-speed, high-solidity design had a modest 12 kW rating (a modern rotor this size could produce 100 kW) but the Brush wind turbine ran for an impressive 20 years.

The early twentieth century saw the emergence of Denmark as a key player in the development of wind power, since when this country's contribution to the technology has been spectacularly greater than its size. The Danish scientist Poul la Cour (1846–1908) is regarded as one of the great pioneers of the modern wind turbine, and in 1891 he constructed his first electricity generator by merging features of contemporary European and US windmill designs. He went on to produce many machines with outputs in the range 20–35 kW, and aerodynamic efficiencies that are respectable by today's standards. La Cour's designs were rapidly taken up in Denmark, which by 1918 had an installed wind energy capacity of 3 MW, corresponding to some 3% of national electricity consumption. La Cour himself was an innovator, teacher, and social reformer, who advocated the use of wind generation for rural electrification (Poul la Cour Foundation, 2005); he also recognised the issue of intermittency and researched ways to store surplus wind generation, including hydrogen production.

After the First World War development of large wind turbines lapsed for almost half a century, due to the increasing use of fossil fuels and later the rise of nuclear power. Mention should, however, be made of the 1941 Smith–Putnam wind turbine built in Vermont, USA: this was the world's first megawatt-size machine, with rotor diameter of 53 m and rated output of 1.25 MW. It ran grid-connected for a total of 1100 hours between 1941 and 1945 when it suffered a major blade failure. Until 1979 it remained the largest wind turbine ever built; photos of its construction can be seen on the 'Wind Works' website (Gipe, 2017). The 1950s saw a rekindling of interest in wind energy in Denmark, however, where electricity generation had become heavily dependent on imported coal and oil, and the country's topography offered little opportunity for hydroelectric power. The 1956 Gedser wind turbine was designed by another Danish pioneer, Johannes Juul (1887–1969), and was arguably the first modern HAWT. Rated at 200 kW it featured an asynchronous grid-connected generator, stall-regulated rotor with air brakes, and electric yaw drive; its peak power density of 442 W m^{-2} is similar to some present-day machines. Like la Cour, Juul based his design largely on existing technology, exemplifying the incremental approach to development that came to characterise the Danish wind industry. A pictorial history of Danish wind power, including details of the pioneering work of la Cour and Juul, can be found on the Winds of Change website (Nielsen, 2000).

The first grid-connected wind turbine in the UK was built by Glasgow-based engineering company John Brown and installed at Costa Head in the Orkney Islands in 1955. It was rated at

100 kW with a 15 m rotor comprising three slender steel blades with variable pitch (based on contemporary helicopter technology). The wind turbine suffered from a variety of mechanical problems, largely due to high blade stresses (albeit the site was one of the windiest in Europe) before eventually being destroyed in a winter storm. Film footage of the Costa Head prototype survives at the National Library of Scotland (Archive, 1955). The choice of Orkney for the project was significant in that the islands had no link to the UK electricity grid at that time, and relied on a diesel power station. Fuel costs were high, and wind energy was seen as a potential means to reduce the cost of generation. In this respect the Costa Head wind turbine was ahead of its time, pre-empting by 20 years the worldwide increase in oil prices that triggered the exponential rise of wind power.

1.4 The Wind Revolution

As noted earlier, wind energy blossomed internationally in the decade following the 1977 Carter legislation. The USA introduced a market for wind-generated electricity via a tax credit system, which led to the first windfarm developments in California and acted as a stimulus for mass production and development of small US and European wind turbine designs; initially the average size of these machines was around 30–50 kW, but it rapidly grew. At the same time a number of very large prototypes were developed under various national initiatives aimed at utility-scale generation. Megawatt-scale machines included the German Growian, Danish Nibe, British LS-1 (see Figure 8.2), and US MOD series. While these large prototypes served as valuable research tools and helped introduce a generation of engineers into the field of wind energy, the turbines did not evolve into commercial designs, being characterised by high weight (steel blade spars were common) and suffering from issues of fatigue, noise, and vibration. They were typically built by large aerospace or heavy engineering companies who perhaps did not fully appreciate the differing structural requirements of aircraft and wind turbines in terms of loading and fatigue. In fairness no-one did, but the lessons would be more easily learned at small scale, and the huge wind turbines we now see evolved via a continual process of technical development and up-scaling from the small mass-produced machines that populated the early windfarms – evolution rather than revolution.

A key factor in this process was the establishment of stable markets for renewably generated electricity, guaranteeing a long-term return to wind turbine manufacturers to develop their products at an appropriate pace (the early US tax credit market having led to a somewhat boom-and-bust approach). The countries that first introduced steady incentives, most notably Denmark and Germany, became home to the main European wind energy manufacturing and R&D effort (for a good history of the rise of modern wind power, see Maegaard, 2013). Among the first wind turbines to achieve series production, a special place may be reserved for the 1975 design of Danish carpenter Christian Riisager, which was manufactured with off-the-shelf components

Figure 1.4 A Danish classic. Christian Riisager's 1975 design was the first series-produced wind turbine, rated at 22 kW and manufactured with off-the-shelf components. The example shown was installed on South Ronaldsay on Orkney in 1982. Note the fantail yaw drive, echoing traditional windmill designs.

and sold to Danish farmers; see Figure 1.4. Now regarded as a classic design the Riisager turbine was initially rated at 22 kW with 10 m rotor, though larger versions followed. It was essentially a scaled-down version of the Gedser prototype of 20 years earlier, embodying the hallmark 'Danish concept' of a three-blade fixed-pitch rotor, grid-connected induction generator, and power control via stall regulation. Though the numbers produced were modest, the Riisager machines have the distinction of being the first commercial grid-connected wind turbines in Denmark.

1.5 Scaling Up

In the time since Riisager's first design appeared the dimensions of new wind turbines have increased by an average of around 7% per year, with corresponding power increase of 16% per annum. The result of 40 years of this compound interest can be seen in Figure 1.2: at the time of writing the largest commercial wind turbine has a rotor diameter of over 160 m and output rating of 8 MW. Over the same period, the cost of wind energy has progressively reduced and offshore windfarms in Europe are currently securing generation contracts at near-marginal electricity prices (this topic is discussed in Section 11.2.4). The development and scaling up of large wind turbines is a technological success story with few modern parallels, but there is a paradox – according to the laws of physics wind turbines should become less, not more, economic the larger they become. This is an outcome of the 'square-cube law', whereby the energy yield of a wind turbine is proportional to rotor swept area but the mass of the structure, and hence its cost, is proportional to volume. More formally, if R is rotor radius then energy yield scales as R^2 but structural mass scales as R^3, so material efficiency is proportional to $1/R$. Have wind turbines somehow defied the natural laws of scaling? The truth is more subtle: the square-cube law is sound[2] but through a succession of innovative steps the designers of wind turbines have managed to stay one step ahead of it. Although today's multi-megawatt wind turbines may look like massively scaled versions of the early machines, they incorporate many advances that have enabled them to grow in size without becoming unfeasibly heavy and uneconomic.

To illustrate the point, if we hypothetically scaled a 1984 Vestas V17 by a factor of 10 in every detail its diameter would rise from 17 to 170 m, and its rating from 75 kW to 7.5 MW. The head weight (rotor plus nacelle) would in theory scale by a factor of 1000. In contrast the real-world Vestas V164 (seen in Figure 1.2) has a rated output of 8.0 MW and is just under 10 times the diameter of the V17, but its head weight is only 80 times that of its predecessor. This dramatic improvement in material efficiency is the result of advances in many fields, including aerodynamics, structures, materials, generator design, control strategy, and power electronics.

Some of the key advances made in the last four decades of wind engineering are

- improved blade design. Modern aerofoil sections have superior aerodynamic properties to earlier designs, with thicker profiles giving higher structural rigidity. Blade roots are based on large-diameter shells rather than flanged designs, with higher material efficiency and improved fatigue resistance.

[2] The rule is approximate but essentially correct; for more detailed discussion of scaling, see Jamieson (2011).

- use of composite materials. Glass and carbon fibre–reinforced composites (GFRP and CFRP) are now used in place of traditional metals in rotor blades, bringing significantly greater structural efficiency and fatigue resistance. Selective fibre orientation optimises blade stiffness and strength, and advanced moulding techniques help control blade weight.
- blade pitch control. Early rotors were stall regulated, with fixed-pitch blades and power output limited by aerodynamic stall. Although simple and effective, this concept did not scale efficiently to MW scale. Variable pitch improves energy yield and reduces rotor weight.
- rotor speed control. Early wind turbines were directly grid-connected and ran at fixed speed, so aerodynamic performance was optimal for only a narrow range of wind speeds. Variable-speed generators enable rotor speed to be matched to wind speed over a broad operating range, maximising efficiency. Variable speed also enables smoother power control in high winds, so drivetrain components can be lighter.
- power electronics. Advances in power semiconductors have been key to the development of variable-speed generators, enabling synchronous machines to operate at variable frequency, or doubly fed induction generators (DFIGs) to operate with variable-frequency excitation. Power semiconductors also facilitate voltage control, improving power quality and maximising export capacity on weak grids.[3]
- detailed design changes. Numerous small and less obvious design changes have improved wind turbine performance. Attention to blade trailing edge thickness and tip design has reduced rotor aerodynamic noise, allowing higher speed operation and again greater efficiency; integral lightning protection eliminates blade damage; intelligent control algorithms allow the newest wind turbines to ride out severe storms without switching off; and so on.

As significant as any of the above has been the improvement in our understanding of the forces acting on a wind turbine. Components can now be designed with conservative strength margins, but avoiding the gross oversizing and weight penalty typical of early designs. In this, wind turbine technology has benefited from knowledge gained from research prototypes and commercial machines, and many advances in theoretical understanding have been the outcome of problems experienced on new production types. The following chapters provide more detail on this remarkable and continuing story.

1.6 Some Definitions

Some technical terms and abbreviations recur throughout this book, and the following is a summary of some of the more important ones.

[3] Equally, the wind turbine industry has helped to push the development of power semiconductor technology.

Aerofoil. The streamlined shape that forms the cross section of a rigid wing or blade; characterised by a rounded leading edge and sharp trailing edge. In cross section a wind turbine blade looks very much like an aircraft wing.

Binning. A method for finding the average power curve from a large data set, by averaging all the measured data within a narrow wind speed range or 'bin'. Binning is also used to obtain wind frequency distributions and other experimental data.

Capacity factor. The ratio of a wind turbine's annual energy output to that assuming it ran continuously at full power, i.e. the ratio of average to rated power. On good sites a figure of 30% may be expected; the record is about 60%.

DFIG. Doubly-fed induction generator. A type of generator that combines features of the simple and traditional induction machine with modern power electronics.

HAWT. Horizontal-axis wind turbine. A wind turbine is strictly an electricity generator, as opposed to a windmill or wind pump (informal use of the term 'windmill' is, however, almost impossible to eradicate in the wind energy community).

Nacelle. The part of the wind turbine housing the main shaft, gearbox (if present), and generator. Located on top of the tower, the nacelle is turned to face the wind at all times during operation.

PMG. Permanent magnet generator.

Power coefficient (C_p). The proportion of the power extracted from the air flowing through a wind turbine rotor. It is impossible to extract all the energy without causing the air to stop moving, so C_p is less than unity, and the theoretical limit is 0.59 (the Betz limit); well-designed rotors can achieve better than 0.50.

Power factor. An electrical term, meaning the ratio of real to apparent (or total) power flow at the point of measurement (explained in Section 5.2). Not to be confused with power output, capacity factor, or efficiency, and often best ignored by those with no interest in electrical matters.

Rated power. The maximum electrical power output at which a wind turbine generator can continuously operate without damage. Often referred to simply as 'the rating'. Not always the best indicator of a wind turbine's likely energy output in a given wind regime (the rotor diameter is a better one).

Solidity. The ratio of the area occupied by the rotor blades to the overall swept (disc) area, and a key design parameter. Solidity may be as low as 4% on a modern HAWT, which may nevertheless operate at close to the Betz limit. Explained in Chapter 3.

Tip speed ratio. The ratio of the rotor blade tip speed to the ambient wind speed; a value usually in the range 3–8, and another key parameter in rotor design. The blade moves faster than the wind driving it, while at the same time extracting power. This is due to the action of aerodynamic lift: again, Chapter 3. Formally, tip speed ratio $\lambda = \omega R/V$, where ω is rotor angular speed, R is blade tip radius, and V is the freestream wind speed.

VAWT. Vertical-axis wind turbine. At the current time this turbine configuration is uncommon, and for reasons of brevity discussion of VAWTs has not been included in this book.

VRIM. Vacuum resin infusion moulding. The most widely used process for manufacturing large composite blades. Sometimes known as RIM or SCRIMP.

WEC, WTG. Wind energy converter, wind turbine generator, aka wind turbine.

Wind shear. The variation of wind speed with height above the ground. See Section 2.2.

Chapter 2 The Wind and Its Characteristics

2.1 Introduction

Wind energy is a concentrated form of solar energy, continually being replenished. At a basic level, heat from the sun causes the air at the earth's equatorial latitudes to expand and rise, while at the colder poles it contracts and sinks, setting up recirculating currents between these regions. This simple convective model is, however, complicated by the rotation of the earth, which causes the moving airflow to deflect relative to the earth's surface to an extent depending on latitude and local air velocity. Inertial and viscous effects impose additional forces, and the wind patterns we experience at the surface of the earth are ultimately quite complex and site specific. High above the surface, however, wind patterns are more predictable and the principal atmospheric forces acting here are

- pressure gradient
- Coriolis force
- centrifugal force

These large-scale forces combine to form the *geostrophic wind*, which represents the wind speed at the top of the earth's boundary layer, typically 1–2 km above the ground.

2.1.1 The Geostrophic and Gradient Winds

Solar heating gives rise to pressure gradients that drive atmospheric air from high to low pressure, in a direction normal to the isobars (the lines of constant pressure familiar from weather charts). When observed from the earth's rotating frame of reference, however, moving air experiences an apparent deflection due to the Coriolis force,[1] whose strength depends on the air velocity, the rate of rotation of the earth, and the latitude at the point of observation. Air is deflected to the right of its forward motion in the northern hemisphere (see Figure 2.1) and to the left in the southern. The

[1] Coriolis and centrifugal force are sometimes referred to as 'fictitious', as they do not strictly exist in a Newtonian (non-accelerating) frame of reference. Objects that are stationary or moving with constant velocity behave as though these forces do exist, however, to an observer who is accelerating. So it is with the air above the surface of the earth.

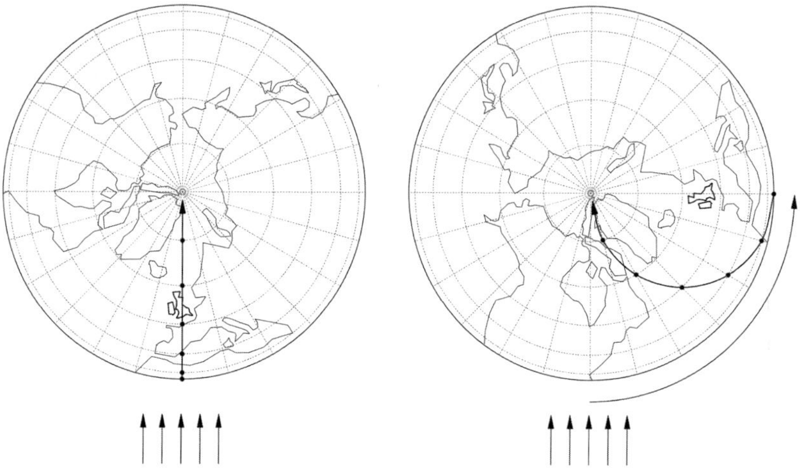

Figure 2.1 The Coriolis effect. Solar heating causes air to flow from equator to pole; its trajectory is seen (*left*) in an external frame of reference and (*right*) by an observer on the earth's surface. Air currents in the northern hemisphere appear deflected to the right of their forward motion.

Coriolis force thus accelerates the air at right angles to its forward motion until it comes into equilibrium with the pressure gradient force, when the resulting atmospheric motion is known as the *geostrophic* wind, with velocity given by

$$U_g = \frac{\partial p/\partial n}{-2\rho\omega \sin\phi} \tag{2.1}$$

where ρ is air density, ω the earth's angular rotation speed, ϕ latitude, and $\partial p/\partial n$ the normal pressure gradient. The equilibrium direction of flow is *parallel* to the isobars, which may seem puzzling: in a stationary frame of reference the wind would blow strictly at right angles to the pressure gradient. This, however, is due to the influence of the Coriolis force; it explains, for example, why in the northern hemisphere air circulates anticlockwise round areas of low pressure, and clockwise round high (vice versa in the southern hemisphere).

Real isobars are generally curved, and the moving air is additionally subject to centrifugal force; a modified equation is then

$$U_{gr} = U_g - \frac{U_{gr}^2}{2\omega R \sin\phi} \tag{2.2}$$

where U_{gr} is the *gradient* wind speed, U_g is the geostrophic speed from Equation (2.1), and R is the isobar radius of curvature. The magnitude of U_{gr} is slightly, though not significantly, modified relative to the geostrophic wind, but the equilibrium direction remains parallel to the isobars.

The gradient wind is more complex to model than the geostrophic, and an advantage of the latter is that wind speeds can be estimated solely on the basis of pressure measurements. As the results of the two models are broadly similar, the geostrophic wind is often the preferred representation of the air velocity at the top of the earth's boundary layer, around 2000 m above the surface. Wind turbines are, however, located deep in the boundary layer, with rotor tip height generally less than 200 m above the surface, where conditions are significantly different from the geostrophic wind. An analysis of surface wind speeds across the British Isles, measured at 10 m above ground level, found them to be 2–3 times lower than the geostrophic speed, with the greatest difference found inland (Moores, 1988). This reduction in wind speed with proximity to the ground is due mainly to two factors, namely (a) surface roughness and (b) topography. Surface roughness is intimately associated with the phenomena of wind shear and turbulence, both of which are of importance in wind engineering, as described below.

2.2 Wind Shear and Turbulence

2.2.1 Shear Profiles

Wind shear refers to the variation in horizontal wind speed with height above the earth's surface. From boundary layer theory, the relationship between speed U and height z is given by the Prandtl log law:

$$U = (u^*/K)\ln(z/z_0) \tag{2.3}$$

in which u^* is the friction velocity, K the von Karman constant ($K \cong 0.4$), and z_0 the local roughness length, which is related to the vertical dimensions of surface features, but is not a direct measurement of them. For large features, e.g. trees or buildings, the value of z_0 is around 1/10 the true height, whereas over smooth terrain (e.g. sand or ice) the applicable scaling ratio is 1/30 (Nikuradse, 1950). The correspondence between terrain type and roughness is somewhat empirical, and typical values of z_0 are given in Table 2.1. The friction velocity u^* and constant K are fundamental to meteorological analyses, but less used in day-to-day wind engineering, and the following form of the Prandtl boundary layer equation enables wind speed to be extrapolated from height z_1 to z_2 without reference to either term

$$U_2/U_1 = \ln(z_2/z_0)/\ln(z_1/z_0) \tag{2.4}$$

Equation (2.4) is widely used in wind turbine siting and engineering practice, for instance to extrapolate a hub height wind speed from measurements taken nearer the ground, or to estimate the variation of the incident wind speed on a large blade as it rotates.

The influence of surface roughness is illustrated in Figure 2.2, with shear profiles corresponding to low, intermediate, and high roughness (z_0 = 0.002 m, 0.40 m, and 4.0 m, respectively).

Table 2.1 Landscape and Terrain Characteristics Used for Wind Shear Estimation

Landscape type	Roughness length z_0(m)	Shear index α
Water surface (smooth)	0.0002	0.08
Short grass, flat open terrain, smooth concrete	0.002	0.11
Open agricultural land, gently rolling	0.02	0.14
Level country, occasional small trees	0.04	0.15
Agricultural land, crops, hedges, some trees	0.10	0.18
Agricultural land, distributed buildings and trees	0.2	0.20
Villages or small towns, wooded countryside	0.4	0.24
Larger towns, tall buildings	0.8	0.29
Highly urban landscape, skyscrapers	1.6	0.36

Note. The equivalence of z_0 and α is here based on shear evaluated between 10 m and 80 m height.

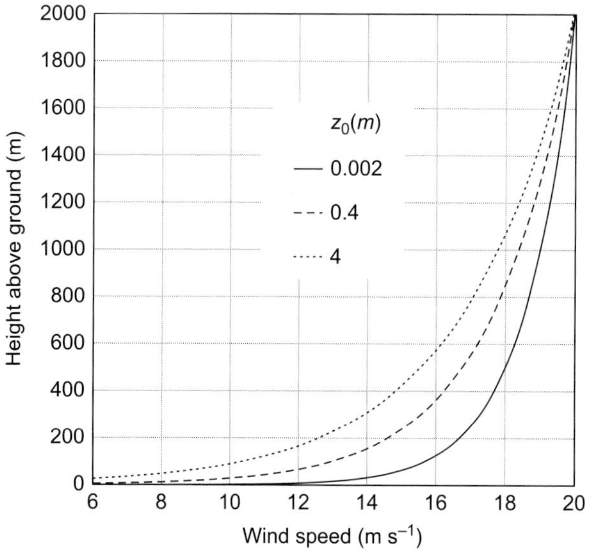

Figure 2.2 Wind shear profiles for different roughness length z_0. The greater the roughness, the more the air is retarded at the earth's surface. Gradient wind speed of 20 m s^{-1} is here assumed at 2 km height.

The wind speed at 2000 m represents the gradient wind (in this example, 20 m s^{-1}) while the surface speed is in all cases zero, according to the 'zero slip' condition of boundary layer theory. The greater the roughness, the more the air is retarded, resulting in reduced wind velocities at lower height. Smooth sites are characterised by a steeper shear profile, with high winds extending farther towards the ground. An alternative wind shear formula is the power law

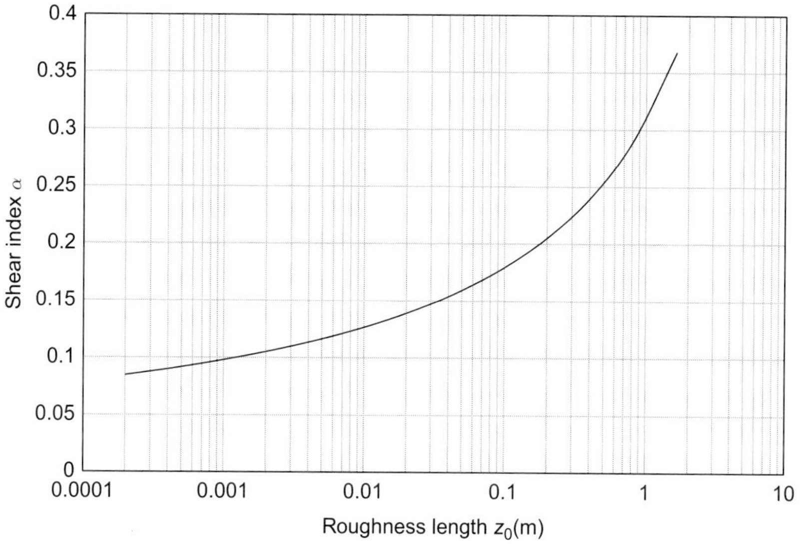

Figure 2.3 Equivalence of wind shear power law index α and roughness length z_0. The curve here assumes a shear profile evaluated between 10 m and 80 m elevation; Equation (2.7) applies.

$$U = kz^{\alpha} \tag{2.5}$$

where k is a constant and α an empirical coefficient, again based on terrain properties. The power law is more often seen in the following form, which dispenses with k:

$$U_2/U_1 = (z_2/z_1)^{\alpha} \tag{2.6}$$

Typical values for the coefficient α are included in Table 2.1. The log and power laws are both extensively used for practical wind shear modelling; care is needed, however, as the two descriptions are not mathematically synonymous and equivalence between z_0 and α must be referred to a specific height range: for instance the values given in Table 2.1 would be valid for shear evaluated between 10 m and 80 m height, a range typically of interest for wind engineering or siting studies. With this caveat the following equation may be used to relate α and z_0:

$$\alpha = \frac{\ln\{\ln(z_2/z_0)/\ln(z_1/z_0)\}}{\ln(z_2/z_1)} \tag{2.7}$$

The equivalence between α and z_0 is plotted in Figure 2.3, again assuming shear evaluated between 10 m and 80 m height. The wind shear profile at a site can be measured directly using multiple anemometers located at several heights on a suitably tall mast, or, alternatively, estimated from landscape inspection in conjunction with terrain descriptions like those in Table 2.1. The latter method is necessarily approximate, and judgement is needed when assessing landscape characteristics. In addition surface roughness often varies around a given location so that z_0 and α depend on

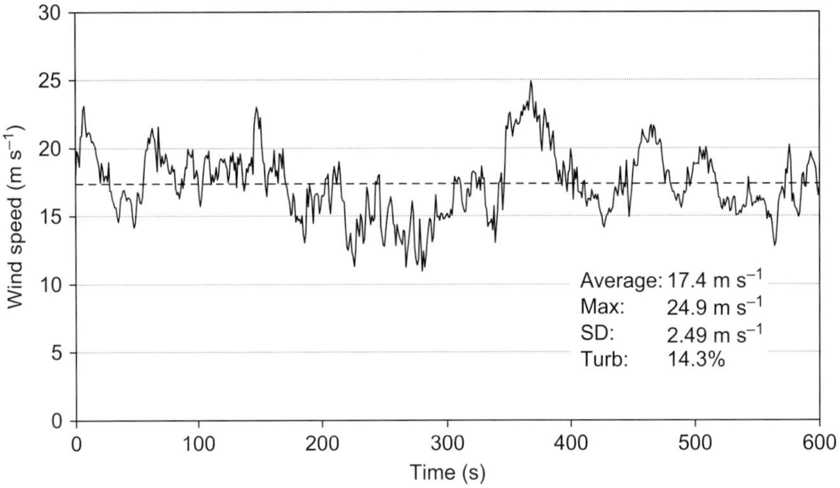

Figure 2.4 Turbulent wind record. This 10 min trace from a high-wind site illustrates turbulent variation of the horizontal wind speed; turbulence intensity is defined as the ratio of the standard deviation to the mean. (Data provided by kind permission of the North Harris Trust)

the incident wind direction, and directionally weighted average values would normally be used to characterise a site. Surface roughness is also an important factor in determining the level of turbulence (see Section 2.2.2).

2.2.2 Turbulence

The wind is never entirely steady, but continually subject to short-term fluctuations, or turbulence. This quasi-random variation is caused by shear stresses in the boundary layer, giving rise to recirculating eddies; the eddies are three-dimensional, but it is the longitudinal or axial speed variation that has the greatest influence on aerodynamic loading, and unless otherwise specified turbulence measurements normally refer to the longitudinal component. Turbulence intensity I is then defined by

$$I = \sigma/U \tag{2.8}$$

where U is the average wind speed and σ the variance (standard deviation) over a specified measurement period. A 10 min record length is standard in wind engineering, though 1 hour measurements are generally used for meteorological records. Both these periods lie within the 'spectral gap' of the typical wind distribution, being longer than the timescales associated with turbulent variation but shorter than diurnal or synoptic (weather front-related) changes in wind speed.[2] A typical 10 min record from a high-wind site is shown in Figure 2.4 in which the mean wind speed is 17.4 m s^{-1} with

[2] For more detail on this topic, see e.g. Burton (2011, Chapter 2).

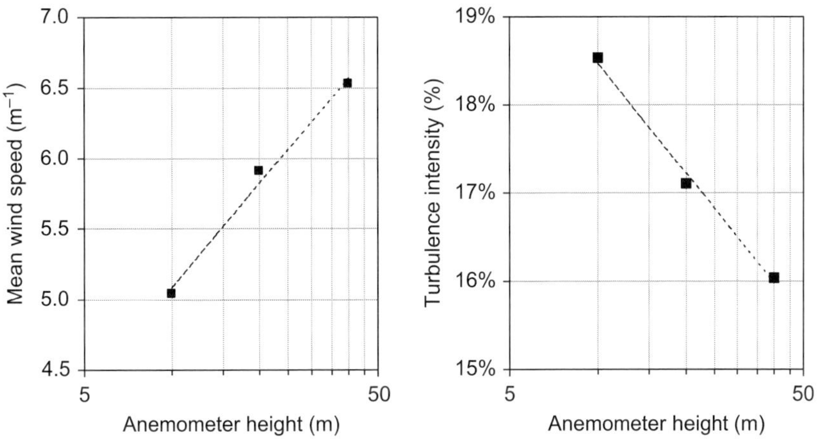

Figure 2.5 Boundary layer characteristics: (*left*) wind speed and (*right*) turbulence intensity measured at heights of 10, 20, and 40 m on a smooth upland site; the profile is a good logarithmic fit in both cases. The turbulence figures are averages for wind speeds exceeding 5 m s^{-1}.

standard deviation of 2.49 m s^{-1}, giving turbulence intensity of 14.3% (turbulence intensity may be expressed as a percentage or a dimensionless ratio). There is a linear relationship between peak gust speed and turbulence intensity, and this is discussed in Section 2.4.

From boundary layer theory the magnitude of turbulence intensity is related to roughness length z_0 and height z above the surface according to

$$I(z) = \frac{1}{\ln(z/z_0)} \tag{2.9}$$

Equation (2.9) has similar logarithmic form to the wind shear relationship of Equation (2.3); an outcome of these laws is that turbulence intensity generally reduces with height, while the mean wind speed increases. These characteristics are borne out on most sites and an example of wind speed and turbulence profiles is shown in Figure 2.5, from measurements taken at 10, 20, and 40 m elevation on an exposed upland site; a logarithmic profile is seen to be a good fit in both cases. Where hub height wind measurements are not available, the following equation allows turbulence intensity to be extrapolated from one height to another:

$$I(z_2) = \frac{1}{\{1/I(z_1) + \ln(z_2/z_1)\}} \tag{2.10}$$

where $I(z_1)$ is turbulence intensity at height z_1 and $I(z_2)$ is the extrapolated value at z_2. Strictly speaking Equation (2.10) is applicable only where the shear profile is logarithmic, but this is the common case. The logarithmic relationships break down, however, on sites exhibiting severe boundary layer flow separation, for instance in complex terrain. In such cases shear and turbulence characteristics must be assessed in more detail, using full-height anemometry and/or computational

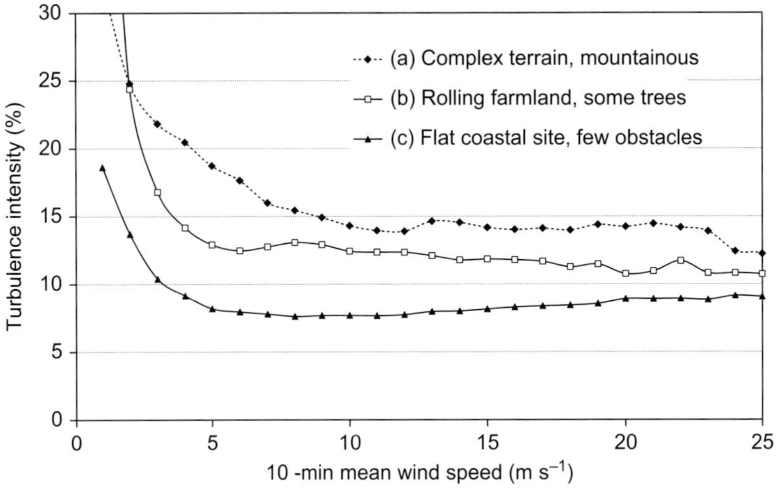

Figure 2.6 Measured turbulence intensity at 50 m height for three sites with different terrain roughness characteristics. The higher the surface roughness, the greater the turbulence.

fluid dynamic (CFD) modelling. Such sites are, however, frequently unsuitable for siting wind turbines.

Turbulence intensity varies with mean wind speed, with the general trend towards lower turbulence in high winds. This may be seen in Figure 2.6, in a comparison of measurements at 50 m height taken at three sites with different landscape characteristics. In all cases turbulence intensity initially reduces sharply with mean wind speed, then flattens out or rises gently towards high wind speeds; the overall turbulence levels vary significantly, however, with the terrain characteristics. For site characterisation purposes the turbulence intensity in a mean wind speed of 15 m s^{-1} is often used, and the following values apply to the three sites shown in Figure 2.6:

Site	*Description*	*Turbulence at* $V = 15$ *m* s^{-1}
a	Complex terrain in mountainous region, subject to flow separation	14%
b	Agricultural landscape, rolling fields with some trees	12%
c	Flat coastal location with few terrain obstacles, no trees	8%

The IEC 61400-1 standard defines characteristic turbulence intensity, denoted I_{15}, as the average turbulence intensity in a mean wind speed of 15 m s^{-1} (found by binning) plus an allowance based on the level of scatter in the data set (IEC, 2005); the scatter correction results in I_{15} being slightly greater than the simple average (used in the example above for simplicity). Sites with characteristic turbulence not exceeding 18% are denoted Class A, and not exceeding 16%, Class B. The IEC standard also includes formulae to calculate turbulence as a function of mean wind speed for different wind classes, for use in wind turbine design and site characterisation.

2.3 Time and Space Scales

There is a strong relationship between the duration of atmospheric events and their physical scale. Wind patterns that develop over a long timescale affect large areas; for example, in winter the entire British Isles can experience the same south-westerly gales for days at a time. In contrast, short-term gusts measured in seconds are caused by turbulent eddies with dimensions of only a few tens of metres. The temporal variation of wind speed is conveniently subdivided into the following categories:

- inter-annual
- annual (seasonal)
- time of day (hourly)
- short-term (gusts and turbulence)

These timescales have differing significance for the siting and operation of wind turbines. Inter-annual variation is important for economic reasons, with accurate long-term wind forecasting essential for windfarms designed to operate for 20 years or more. As a rule of thumb at least 10 years' data are considered necessary to characterise the average wind speed for a site. A measurement taken over 12 months might be accurate only to within ±10% of the long-term

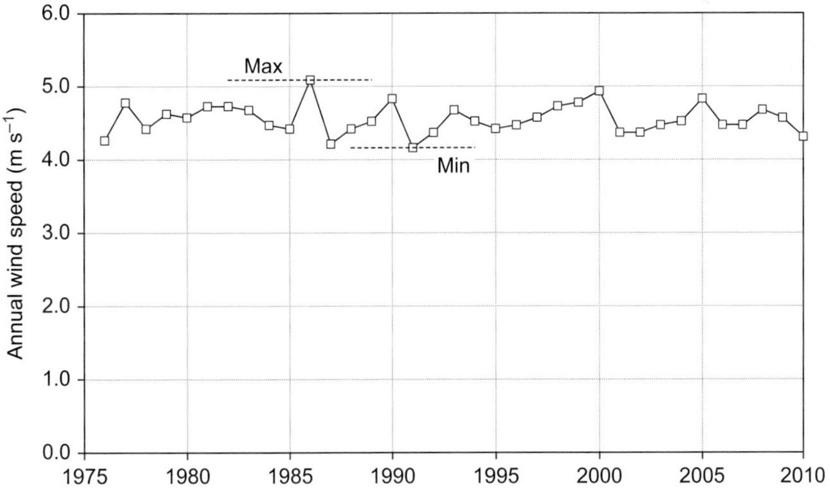

Figure 2.7 Inter-annual wind speed variation at Aberdeen Airport over 35 years. Extreme values deviate from the mean by only ±10.5%. (*Data source*: Tutiempo.net)

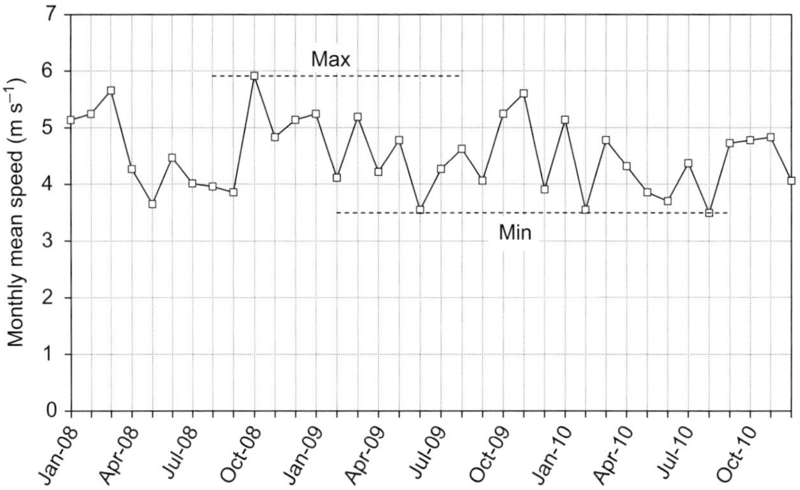

Figure 2.8 Annual (monthly) wind speeds at Aberdeen Airport over a 3 year period. Variation is ±27% about the long-term mean: compare with the inter-annual variation seen in Figure 2.7. (*Data source*: Tutiempo.net)

mean, and a business model relying on such an estimate could be highly unreliable. There is, however, a well-established method to obtain accurate long-term wind prediction on the basis of relatively short site measurement campaigns: this is the MCP procedure, as described in Chapter 9.

Inter-annual variation can be illustrated with reference to Figure 2.7, which shows records from Aberdeen Airport in north-east Scotland over a 35 year period. During this time the long-term average was 4.6 m s^{-1}, with extreme values deviating by approximately ±10.5%. Annual (or seasonal) variation refers to the difference in speeds observed within a year: seasonal variation is illustrated in Figure 2.8, which shows monthly averages at Aberdeen over a 3 year period, with variation of ±27% about the long-term mean.

In north-west Europe the trend is for the highest winds to occur in the winter months and the lowest in summer. The variation shown in Figure 2.7 and Figure 2.8 is fairly typical for a UK site, and the impact on wind energy production is illustrated in Figure 2.9: this shows figures for a wind turbine at Hill of Easterton in Aberdeenshire over a 10 year period. The annual output varied by only ±7% about the mean, whereas monthly output varied by a factor greater than 4. These data illustrate the sensitivity of wind turbine output to short-term wind speed variation, emphasising the need for accurate long-term assessment to reduce economic risk: this topic is discussed further in Section 9.2. The data also show, however, that wind energy is inherently predictable on a long-term basis.

Hourly, or time of day, wind speed variation is dominated by quasi-random influences, and one day's wind speeds can be entirely different to the next. Averaging can however reveal underlying trends, the most common of which is diurnal variation caused by the temperature change from day to night; daytime winds tend to be higher, and the lowest speeds occur

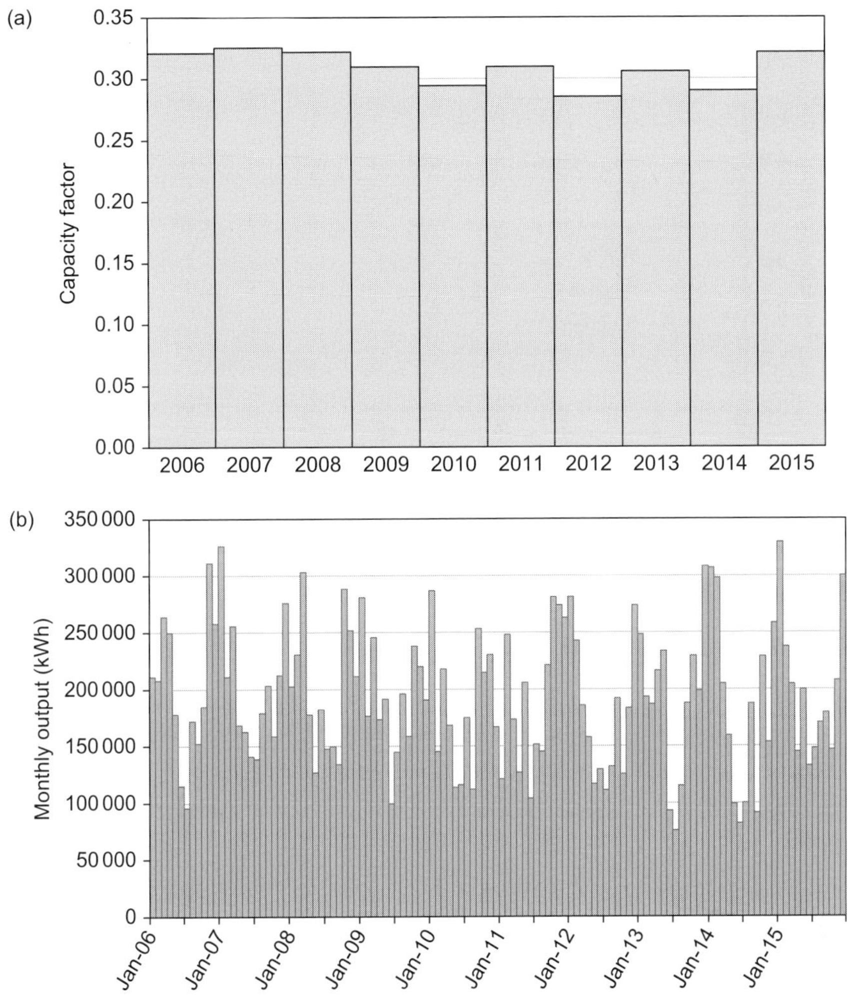

Figure 2.9 Comparison of (a) annual and (b) seasonal output of an 850 kW wind turbine at Hill of Easterton in Aberdeenshire, NE Scotland, over a 10 year period. Variation in annual output was around ±7%, but monthly output varied by a factor of 4. (Data provided by kind permission of Mackies Ltd)

between midnight and sunrise. This trend is illustrated in Figure 2.10, which shows hourly wind speeds measured over a 5 year period at a coastal site, again in Aberdeenshire. In this case the peak wind speed occurs around mid afternoon; the trend is more heavily pronounced in summer than in winter due to the greater diurnal temperature variation. This trend is noted in both temperate and tropical latitudes.

Diurnal variation may also be apparent in wind shear measurements, and is again a temperature-related phenomenon. At night the air is colder and there is consequently less turbulent mixing: the air remains more stratified so that a stronger wind shear profile is evident

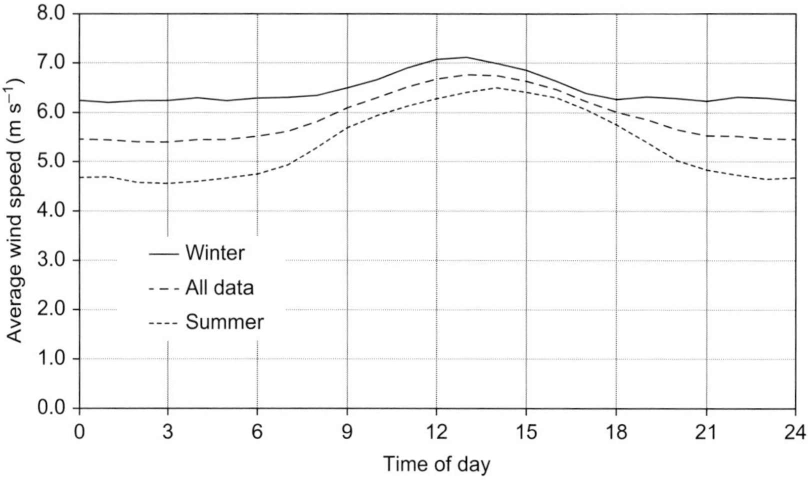

Figure 2.10 Diurnal wind speed variation. Hourly average wind speeds measured over a 5 year period at Hill of Dudwick in NE Scotland. The trend is much stronger in the summer months (April–September).

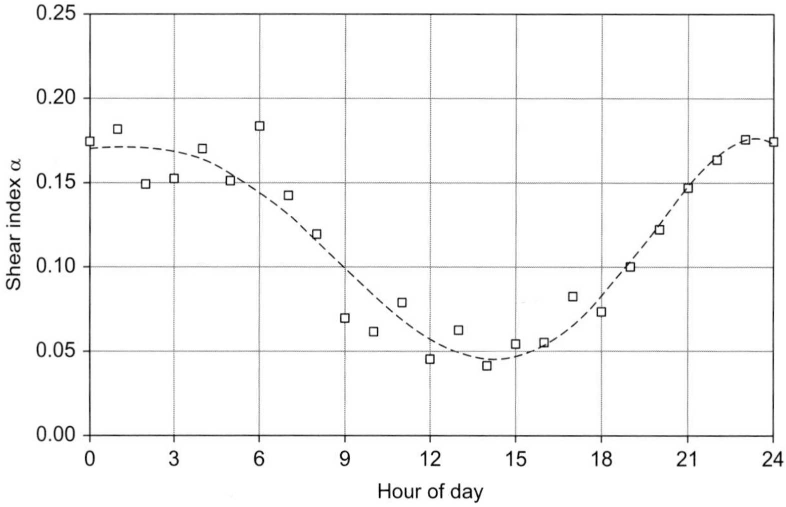

Figure 2.11 Diurnal wind shear variation. Power law shear profile measured between 10 m and 50 m elevation at Hill of Easterton, NE Scotland, during the summer months.

by night than by day. This trend is illustrated in Figure 2.11 using measurements from Hill of Easterton over a 3 month summer period, when diurnal temperature variation is strongest. In this case the power law shear index α varies from around 0.18 by day to as low as 0.04 by night, the latter indicating an almost flat velocity profile.

2.4 Gust and Extreme Wind Speeds

Gusts are transient wind speed variations lasting only a few seconds, and are important in the context of extreme structural loading on wind turbines, and rapid changes in their output power. The gust factor G is defined as the ratio of the maximum (V_g) to average (V_m) wind speed during a given wind record:

$$G = V_g / V_m \tag{2.11}$$

When making comparisons the record length and gust duration must both be quoted. The standard length for meteorological records is 1 hour, and for wind engineering purposes 10 min; within these periods gusts are commonly measured over a 3 s interval. Gust factor is directly proportional to turbulence intensity, and inversely related to gust duration. In UK building codes (BS6399-2, 1997) extreme gust estimates are based on the following equation, which assumes hourly wind records with arbitrary gust duration of t seconds (Cook, 1985):

$$G(t) = 1 + 0.42I\ln(3600/t) \tag{2.12}$$

The IEC wind classification system specifies a 3 s gust as standard (IEC, 2005); setting $t = 3$ s in Equation (2.12) then gives the following simple relationship between gust factor G and turbulence intensity I:

$$G = 1 + 3.0I \tag{2.13}$$

Note that Equation (2.13) is based on hourly wind records, and the use of 10 min data results in slightly lower values for G (Wieringa, 1973). This is seen in Figure 2.12, which shows measured gust factor based on 10 min wind records (with 3 s gust) for a number of sites in wind speeds ranging up to 30 m s^{-1}. The linear dependence on turbulence intensity is clearly seen, but the magnitude of G is a few per cent less than the predicted hourly peak gust. Similarly if gusts are measured over periods other than 3 s (this may be function of the type of anemometer or data logger used) they can be scaled using Equation (2.12); for example, a 1 s gust is typically 3%–4% higher than the corresponding 3 s gust, for the same record length.

A common requirement in wind engineering studies is to estimate the extreme gust speed at a site for which long-term wind measurements are not available. The theory of extreme wind speeds is rather complex and in the UK an empirical procedure, originally developed for wind loading calculations on buildings, is often used (BS6399-2, 1997). The basis of the method is to take regional wind speeds from long-term isovents and adjust to local site conditions via a number of factors that account for terrain elevation, topography, surface roughness, and wind direction. Different return periods can be modelled, e.g. to estimate a 25 year or 50 year extreme gust. The results are considered to be conservative.

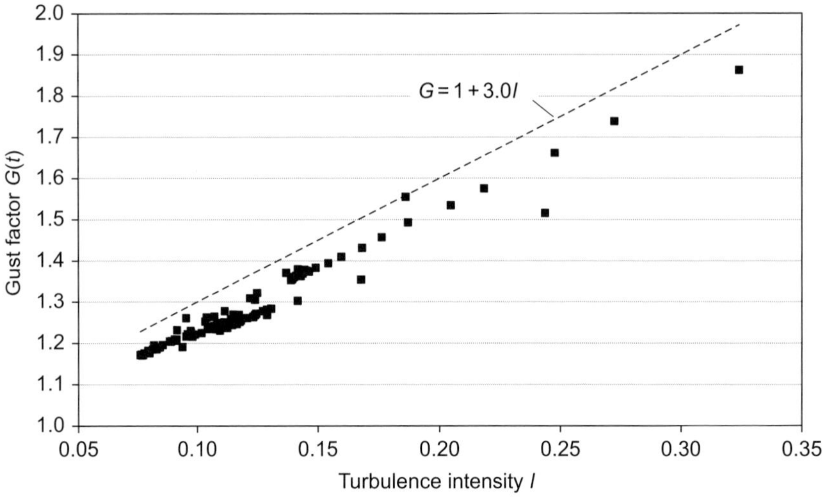

Figure 2.12 Gust factor as a function of turbulence intensity. The data points are based on 10 min records and 3 s gust duration at a number of sites (average speeds in the range 1–30 m s^{-1}). The dashed line is the IEC recommended factor based on hourly wind records and 3 s gust.

2.5 The Weibull Wind Distribution

While the long-term average wind speed indicates the attractiveness of a prospective site, equally important is the wind frequency distribution, or the number of hours (or proportion of time) spent annually at a given speed. Most (though not all) sites can be characterised by a Weibull distribution, which gives the probability of occurrence P of an hourly mean wind speed U from

$$P(U) = (k/C)(U/C)^{k-1} exp\{-(U/C)^k\} \tag{2.14}$$

in which C is the *characteristic* wind speed, and k the Weibull shape parameter. The characteristic wind speed is related to the average speed U_{av} according to

$$C = \frac{U_{av}}{\Gamma(1 + 1/k)} \tag{2.15}$$

where Γ is the gamma function. The value of k reflects the amount of variation about the mean, with higher values indicating a narrower distribution of wind speeds.[3] A special case is the Rayleigh distribution, for which $k = 2$, and $C = 1.13U_{\mathrm{av}}$; this is often used as an initial estimate in site wind assessments where only the mean wind speed is known. A useful form of the Weibull equation is

[3] The characteristic wind speed C is formally defined as the speed exceeded for a proportion of time equal to $1/e$, or about 37% of the time.

the cumulative exceedance Q, which gives the probability that the wind speed will exceed a particular value V:

$$Q(V) = \exp\{-(V/C)^k\} \quad (2.16)$$

Using Equation (2.16) we can for instance calculate the number of hours per year the wind speed lies between an arbitrary lower and upper limit, respectively V_1 and V_2, according to

$$Q(V_2 > V > V_1) = 8760[\exp\{-(V_1/C)^k\} - \exp\{-(V_2/C)^k\}](\text{hours/year}) \quad (2.17)$$

Equation (2.17) is at the heart of wind turbine yield predictions; it is used in conjunction with the turbine's published power curve to calculate the energy yield in each wind speed range and hence the total annual output (see Chapter 8). An example of a measured wind speed distribution is shown in Figure 2.13, where a year's worth of hourly wind speeds is 'binned' at 1 m s^{-1} intervals to produce a histogram of occurrence. The best-fit Weibull curve is found by firstly integrating the frequency distribution to obtain the exceedance distribution $Q(V)$, then applying a double-logarithm transfer to Equation (2.16) to yield

$$\ln\left(-\ln(Q)\right) = k\ln(V) - k\ln(C) \quad (2.18)$$

A plot of $\ln\left(-\ln(Q)\right)$ against $\ln(V)$ should then yield a straight line with slope k and intercept $-k\ln(C)$, from which k and C are found. To illustrate, Figure 2.14 shows the exceedance curve $Q(V)$ derived from the distribution in Figure 2.13; application of Equation (2.18) to the data yields the best-fit line shown in Figure 2.15, from which k and C are derived (in this case $k = 2.2$ and $C = 11.3$ m s^{-1}). If high-quality wind measurements are used a straight line plot of this form is usually obtained.

2.6 Topographic Factors

As noted earlier, the wind speed near the earth's surface differs significantly from the geostrophic conditions at the top of the atmospheric boundary layer. One of the main reasons for this is the influence of topography, or the variation in size and shape of the terrain.[4] Topographic effects include

- sheltering behind hills or in valleys
- creation of turbulence due to large-scale flow separation
- speed-up of the airflow over smooth hills (streamline compression)
- deflection of the wind from its prevailing direction

[4] The term *orography* is also used when referring to large-scale landscape features such as mountains.

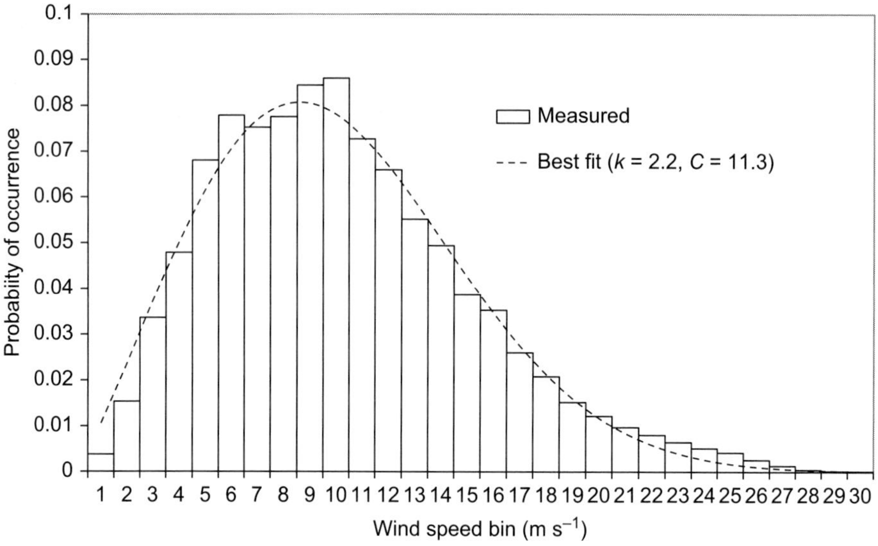

Figure 2.13 Measured wind distribution with best-fit Weibull curve.

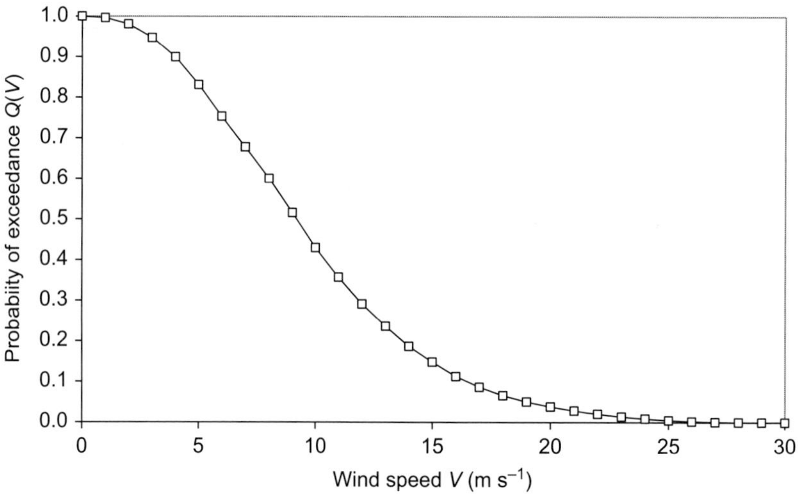

Figure 2.14 Exceedance curve $Q(V)$ derived from the measured wind distribution in Figure 2.13.

Prospective wind sites may experience some or all of these effects, and local wind measurements are essential where large variations in wind conditions occur within a relatively small area, such as in mountains, or on sites with extensive forestry or urban development. In complex terrain the airflow over steep-sided features can cause flow separation, where the boundary layer breaks away from the surface, giving rise to severe turbulence and large-scale eddies; the simple laws governing shear and turbulence no longer apply. These conditions are found in the lee of steep-sided hills, as

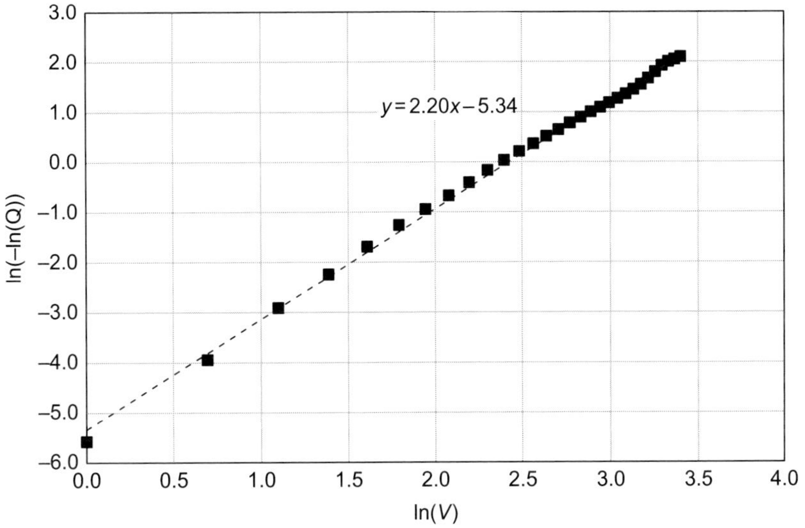

Figure 2.15 Extraction of best-fit Weibull characteristics from the exceedance distribution $Q(V)$ shown in Figure 2.14. Equation (2.18) applies; the slope of the line gives k and the intercept $-k\ln C$.

illustrated in Figure 2.16: the topography shown is that of a site at Monan in North Harris, in the Outer Hebrides, where detailed wind measurements for a prospective wind project indicated that large-scale flow separation occurred when the wind came from certain directions.

Figure 2.17 shows a 10 min trace of wind speed and direction from North Harris, measured at 50 m agl during a winter storm. The wind was blowing from the south (180°) with a mean speed of around 20 m s^{-1}, which is well within the operational envelope of most wind turbines. Overlying the mean, however, are cyclical gusts up to 55 m s^{-1} occurring with a repeat period of 90 s, and during which the wind direction simultaneously swings through ±45° about the mean direction. Such gusts would exceed the normal operating limits for most wind turbines.

The observed velocity patterns are similar to the conditions adjacent to a von Karman vortex street, and a qualitative explanation is offered in Figure 2.18. The separated flow is characterised by strong circulatory eddies overlaid on the freestream wind; as each eddy passes the measurement position it causes a strong variation in local velocity, with the highest speed recorded when the vortex is at its closest separation from the mast. The maximum directional change, however, occurs in the interval midway between eddies, and at the moment of maximum wind velocity the measured direction corresponds to the freestream value. These are similar to the conditions observed in Figure 2.17.

Short-term wind conditions in complex terrain are hard to predict, and computational fluid dynamic (CFD) modelling is increasingly used; in a study of the North Harris site, high-frequency wind measurements were compared with a CFD model based on an empirical orthogonal function technique, with good results (Abiven et al., 2011). It should be said that siting wind turbines in

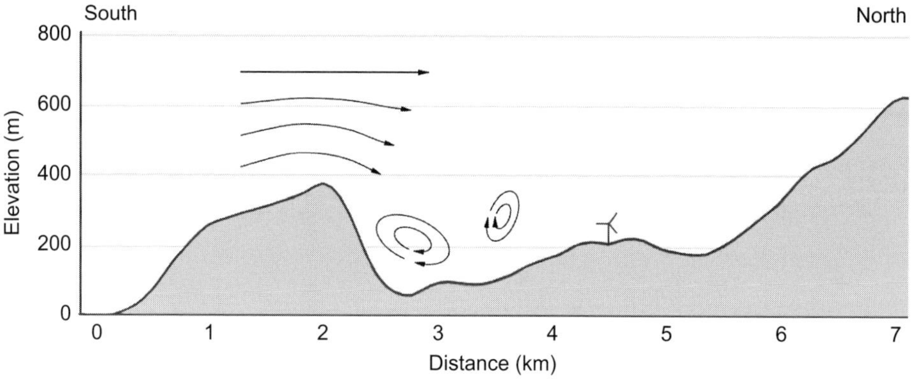

Figure 2.16 Terrain profile (the vertical scale is exaggerated) illustrating flow separation and vortex shedding in the lee of a steep-sided hill. The topography is based on Monan in North Harris, where wind measurements indicated extreme turbulence from certain directions. See Figure 2.17.

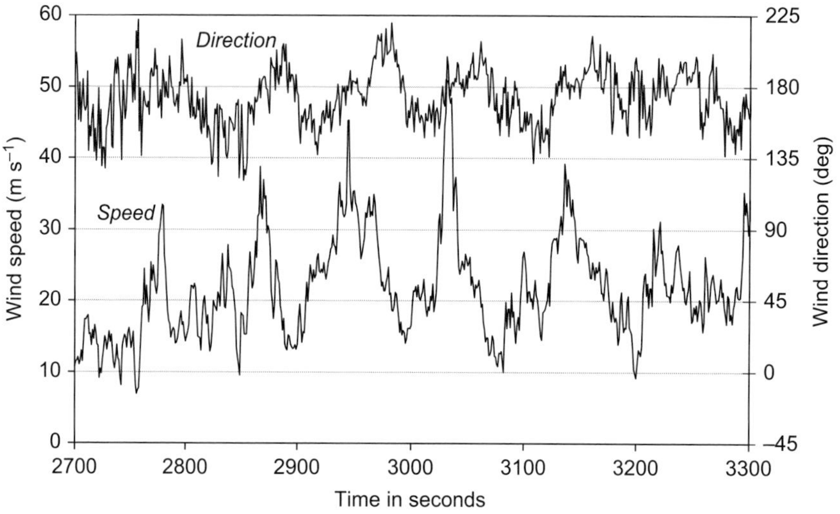

Figure 2.17 Extreme wind conditions in complex terrain. Measurements at 50 m agl from Monan, on North Harris, during a storm in January 2006. This level of turbulence is characteristic of large-scale flow separation. (From data provided by kind permission of the North Harris Trust)

complex terrain is normally to be avoided: as a rule of thumb the downwind distance between a wind turbine and a sheltering obstruction should be at least 20 times the height of the latter (Gipe, 1993), whether it is a building, vegetation, or a large terrain feature as in Figure 2.16. This cannot always be achieved in practice, and modern wind turbines can be programmed with directional sector management (see Section 6.5) to curtail their output, or shut them down altogether, to avoid critical wind conditions.

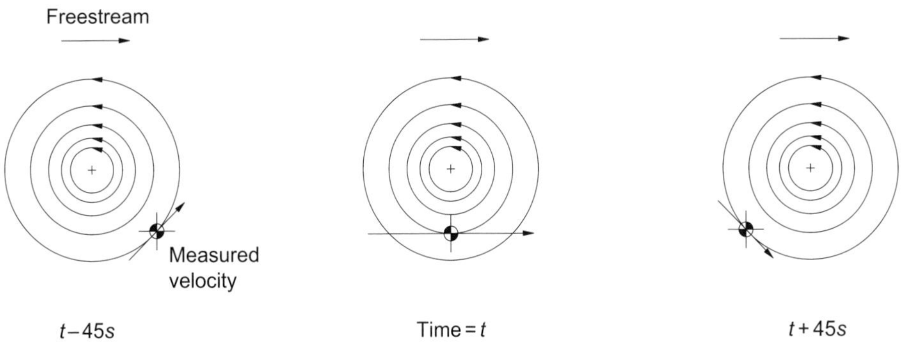

Figure 2.18 A qualitative explanation of the wind record in Figure 2.17. As a vortex (travelling left to right) passes the measurement mast, superposition of freestream and circulatory airflows causes a change in local velocity and direction. The highest velocity (middle) coincides with zero directional change.

2.7 Exercises

2.7.1 Geostrophic Wind

Use the information shown on the isobar chart below (Figure 2.19) to estimate the geostrophic wind velocity (speed and direction) at Pitlochry (latitude 56.7°, longitude −3.7°). Assume air density of 1.225 kg m^{-3} and equilibrium atmospheric conditions.

2.7.2 Roughness Length

If the surface roughness length z_0 in the vicinity of a windfarm site is 0.04 m estimate the turbulence intensity I at a height of 50 m above ground level.

2.7.3 Turbulence Intensity

Measurements from an anemometer at 10 m above ground level indicate average turbulence intensity I of 15.0%. Using this information extrapolate the turbulence intensity to heights of (a) 50 m and (b) 80 m.

2.7.4 Wind Shear Estimates

A prospective wind site is in open agricultural land, in gently rolling terrain. The wind speed at 10 m above ground level is 6.60 m s^{-1}. Using empirical terrain data estimate the corresponding wind speed at 65 m above ground level using (a) the power law and (b) the log law.

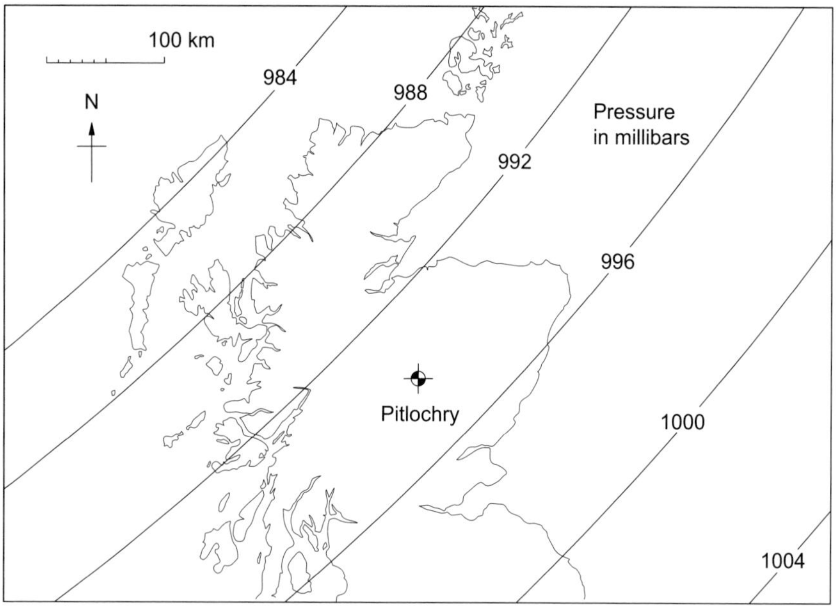

Figure 2.19 See Exercise 2.7.1.

2.7.5 Shear Law Comparison

Measurements from an anemometer mast indicate an average wind speed of 6.40 m s^{-1} at 10 m height, and 7.50 m s^{-1} at 30 m height. Estimate the corresponding wind speed at 80 m assuming (a) the power law and (b) the log law.

2.7.6 Weibull Function

If the characteristic wind speed is 9.0 m s^{-1} at a measurement site and the corresponding Weibull parameter $k = 1.9$, then (a) what is the probability of occurrence of an hourly speed of 15.0 m s^{-1}, and (b) what is the probability that the wind speed will exceed 15.0 m s^{-1}?

2.7.7 Rayleigh Distribution

The measured wind speed at a prospective windfarm site conforms to a Rayleigh distribution. If the mean speed is 7.80 m s^{-1}, calculate (a) the characteristic wind speed C and (b) the number of hours per year that the wind speed will exceed 12 m s^{-1}.

2.7.8 Gust Speed

In a site wind survey an hourly average speed of 12.5 m s^{-1} is recorded during which the peak 3 s gust is measured at 17.7 m s^{-1}. From these data estimate the turbulence intensity. What would the corresponding gust speed have been if the gust duration was 1 s?

CHAPTER 3 AERODYNAMIC THEORY

3.1 INTRODUCTION

Although the aerodynamic principles of wind turbine rotors are now well understood, some aspects can be subtle, and perhaps even surprising. On a modern multi-MW wind turbine (e.g. Figure 1.2) the blades are very slender, and in fact the rotor solidity may be as little as 4%: with so little material in direct contact with the air we might expect that most of the wind would blow through without giving up its energy; yet the reverse is true, and modern wind turbines operate quite close to the theoretical limit of energy extraction (the Betz limit; see below) and are substantially more efficient than the old water-pumping windmills whose blades occupied a much greater proportion of the swept area. Rotor aerodynamic theory explains this apparent contradiction, and can now be used to predict the power output and aerodynamic loading on the blades to a high degree of accuracy.

Investigations of rotor aerodynamic performance began in earnest in the eighteenth century with John Smeaton's work on windmills (see Section 1.2), and reached maturity in the twentieth century in the work of Prandtl, Glauert, Goldstein, and others. Much of this work was concerned with propeller theory,[1] but whereas the power output of a propeller can in principle be increased indefinitely (by driving it with a more powerful motor) the power generated by a wind turbine rotor is limited by the amount of energy in the airflow – and not all can be extracted. A key aspect of rotor aerodynamic theory is therefore to calculate the amount by which the freestream is slowed down, and how much power is generated, for a given combination of rotor geometry and external conditions. Mathematical models range from the relatively simple 'actuator disc', through more complex vortex wake descriptions, and ultimately to full-blown computational fluid dynamic (CFD) treatments.

Much useful information can be obtained with blade element momentum (BEM) theory, which lies somewhere in the mid range of mathematical sophistication. The basis of BEM theory is to equate the momentum change in the flow through the rotor plane with the thrust loading on the blades, and a good BEM code will yield rotor loading and power output accurate enough for industrial use, at low computational expense. The theory is, however, less useful for modelling the rotor wake and for this purpose more sophisticated vortex wake analyses have been developed, embodying mathematically accurate descriptions of the three-dimensional wake airflow.

[1] See e.g. Chapter 12 of von Mises (1959).

Furthermore, with recent developments in flow imaging techniques the air is no longer invisible, and mathematical descriptions of the flow through a wind turbine rotor can be verified in detail. These topics are discussed in the present chapter.

3.2 The Actuator Disc

The actuator disc[2] is a hypothetical surface that occupies the swept area of the wind turbine rotor and extracts energy continuously from the flow; see Figure 3.1. The disc is oriented normal to the flow, has no physical thickness, and is assumed to comprise an infinite number of blades whose influence is averaged over its surface. A streamtube of air flows continuously through the disc, from far upstream to far downstream; in doing so it decelerates and loses part of its kinetic energy to the actuator disc. In a real wind turbine the energy is converted to electricity, but with the actuator disc we do not need to specify where the energy goes. The following assumptions also apply:

- The flow is incompressible.
- The air pressure far upstream and downstream is ambient.
- Thrust loading over the disc area is uniform.
- Frictional drag and wake rotation are neglected.
- Flow deceleration is smooth and steady with no velocity discontinuity at the rotor plane.

The last point is important: the air velocity immediately upstream and downstream of the actuator disc is almost the same, hence there is no kinetic energy transfer at the rotor plane itself. The variation in streamtube properties is shown in Figure 3.2. As the flow smoothly decelerates

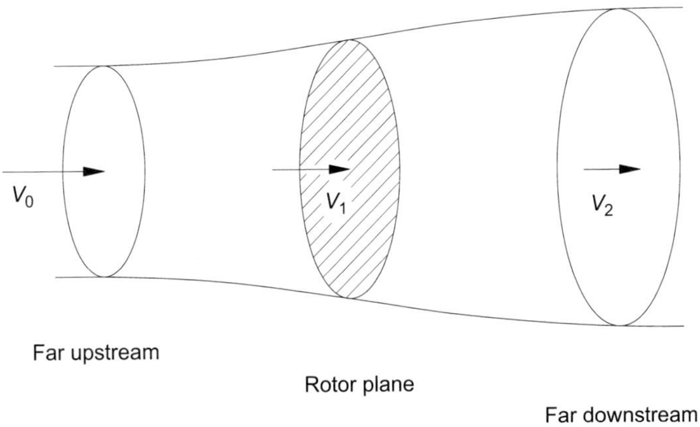

Figure 3.1 The actuator disc. Axial velocity decelerates continuously from the freestream value V_0 far upstream to the final value V_2 far downstream of the rotor plane. See also Figure 3.2.

[2] Actuator disc theory is also applicable to vertical-axis wind turbines, in which case the 'disc' is not circular but rectangular, or of more complex shape, depending on the VAWT configuration.

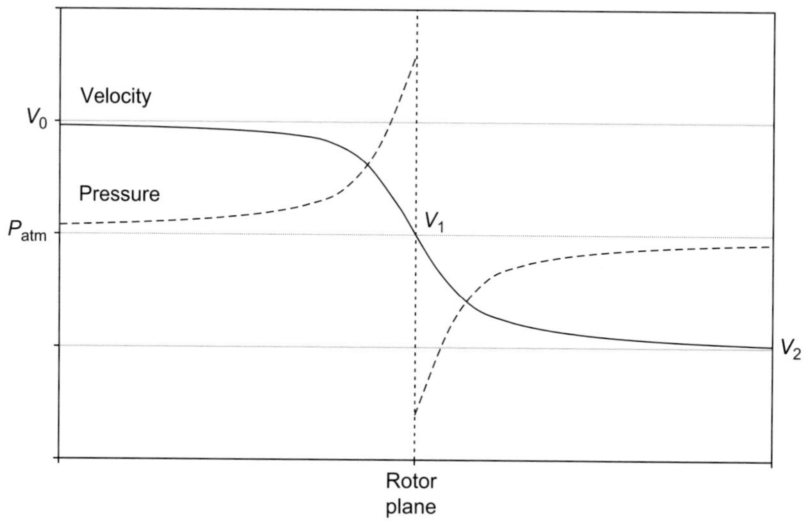

Figure 3.2 Variation in velocity and pressure along the actuator disc streamtube.

kinetic energy is progressively converted to pressure energy; an area of high pressure is developed on the upstream surface of the disc with correspondingly low pressure on the downstream surface. This pressure discontinuity results in a net thrust force acting on the actuator disc in the downwind direction, and it is pressure energy rather than kinetic energy that is extracted. Flow behind the rotor plane continues to decelerate, and experiences a gradual pressure recovery back to ambient in the far wake. Conservation of mass flow dictates that the decelerating airflow expands, such that the streamtube diameter in the downstream wake is greater than that of the rotor disc (conversely the streamtube diameter is smaller upstream of the disc).

Mathematical treatment of the actuator disc can be found in most fluids textbooks and only the key outcomes are quoted here. Application of Bernoulli's theorem to streamlines upwind and downwind of the disc enables the pressure drop to be equated to the rate of change of fluid momentum. The axial induction (or interference) factor a is then defined as the fractional decrease in freestream velocity at the plane of the disc with

$$V_1 = V_0(1 - a) \tag{3.1}$$

where V_0 is the freestream velocity and V_1 is the velocity at the rotor plane. Momentum theory shows that the flow velocity in the far wake is decelerated by twice the amount seen at the rotor plane, or

$$V_2 = V_0(1 - 2a) \tag{3.2}$$

The theory ultimately yields expressions for the thrust loading on the actuator disc and corresponding power extracted as follows:

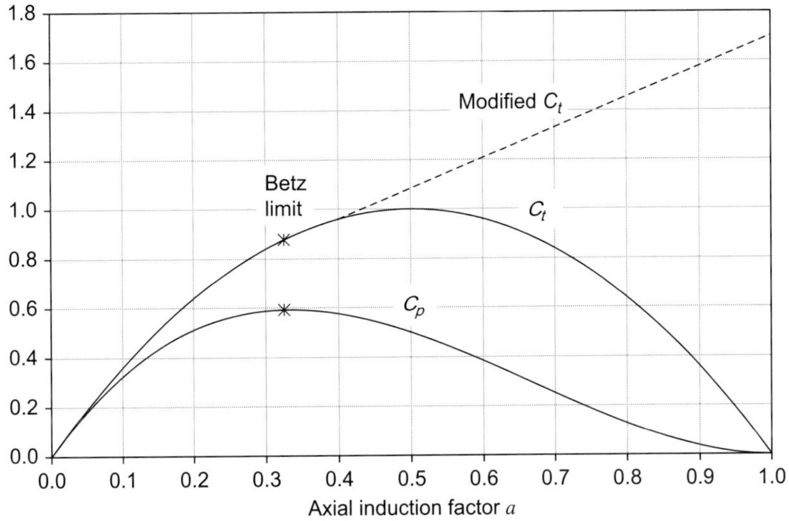

Figure 3.3 Thrust and power coefficients as functions of axial induction factor. The dots indicate the Betz limit (optimum efficiency) when $a = 0.33$. For a above ~0.4, high blockage exists and momentum theory breaks down; C_t has then to be found empirically (see Section 3.6.3).

$$\text{Thrust :} \quad T = \frac{1}{2}\rho V_0^2 A C_t \tag{3.3}$$

$$\text{Power :} \quad P = \frac{1}{2}\rho V_0^3 A C_p \tag{3.4}$$

where ρ is air density, A is the disc area and C_t and C_p are the dimensionless thrust and power coefficients, respectively; both coefficients are functions of axial induction factor a with

$$C_t = 4a(1 - a) \tag{3.5}$$

$$C_p = 4a(1 - a)^2 \tag{3.6}$$

Variation of C_t and C_p with a is shown in Figure 3.3. When $a = 0$ there is no velocity reduction, hence no thrust or power developed. Both coefficients increase with a until C_p reaches a maximum of 0.59 (16/27) when $a = 0.33$. This is the Betz limit, the theoretical condition for maximum power extraction (it is impossible to extract all the energy without stopping the flow); under this condition the velocity at the rotor plane is two-thirds of the undisturbed freestream velocity V_0, and thrust coefficient $C_t = 0.89(8/9)$. Theory therefore predicts high thrust loading under conditions of optimum power extraction: the above value of C_t is not far off the drag coefficient of a circular flat plate ($C_d = 1.1$ at high Reynolds number).[3]

Despite the simplicity of the actuator disc model its predictions are largely borne out on real wind turbines. Figure 3.4 shows measured thrust and power coefficients for a medium-size HAWT

[3] C_t is equivalent to a drag coefficient based on the rotor disc area.

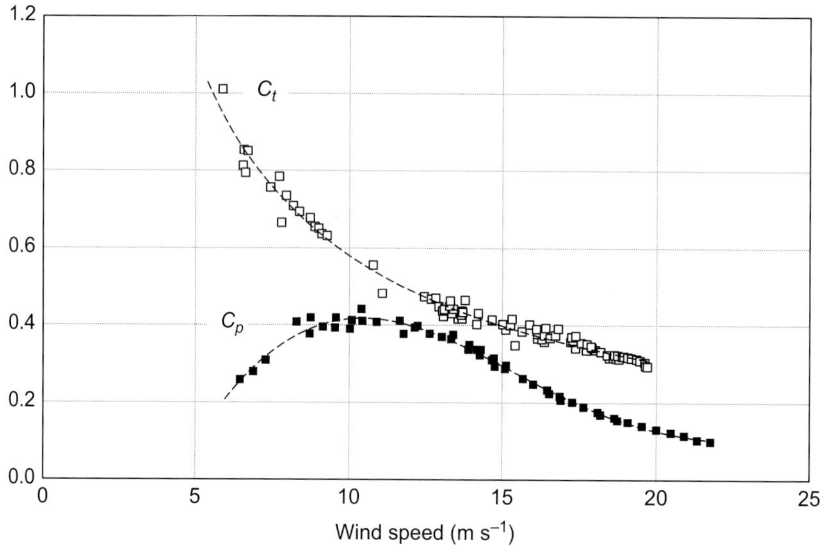

Figure 3.4 Measured thrust and power coefficients for a constant-speed wind turbine (thrust derived from blade root axial loading). The high C_t values in low winds correspond to increased flow blockage, when simple momentum theory breaks down.

as functions of wind speed: the peak power coefficient is just over 0.4 in a wind speed of 10 m s^{-1}, with corresponding thrust coefficient just under 0.6. The graph is essentially an image of Figure 3.3 reversed left to right: high wind speeds correspond to low induction factor a when the rotor extracts only a small part of the available energy. Note that in low wind speeds C_t continues to rise: this behaviour is not predicted from simple momentum theory, which breaks down at high values of a and empirical modification to the basic thrust equation is necessary; the modified thrust relationship is indicated by the dashed line in Figure 3.3. This topic is discussed further in Section 3.6.3.

Another limitation of the simple actuator disc model is the assumption of uniform pressure distribution over the disc area; conditions on a real wind turbine rotor are generally non-uniform, and the theory must be extended to allow for radial variation in thrust loading. This is achieved by replacing the simple actuator disc flow with a multiple streamtube model, as now described.

3.3 Multiple Streamtube Theory

In the multiple streamtube model momentum theory applies as before, but the actuator disc is replaced by a set of concentric rings as shown in Figure 3.5, each corresponding to the intersection of an annular streamtube with the rotor plane. The streamtubes are assumed to be radially independent with no flow between them, so that different interference factor and thrust coefficient can exist at different radii. Equations (3.5) and (3.6) still apply, but now the thrust and power coefficients are defined locally, and for a given streamtube the axial thrust dF_{ax} at the rotor plane is given by

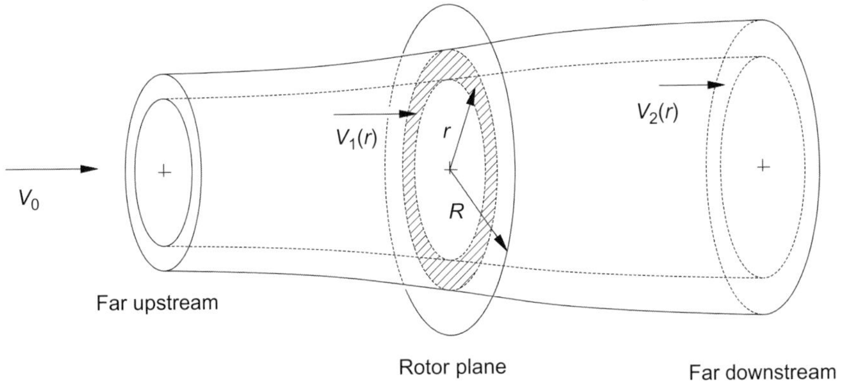

Figure 3.5 Multiple streamtube model. The annular streamtubes (one shown) are radially independent, but each obeys the same momentum theory principles as a simple actuator disc. Thrust loading can now vary with radial position.

$$dF_{\text{ax}} = \frac{1}{2}\rho V_0^2(2\pi r dr)C_t \tag{3.7}$$

where the annular area ($2\pi rdr$) replaces that of the simple disc, and C_t is now the local thrust coefficient. Substituting for C_t from Equation (3.5) then gives

$$dF_{\text{ax}} = 4\pi r\rho V_0^2 a(1-a)dr \tag{3.8}$$

This model is more representative of the airflow through a real HAWT rotor, though in itself not sufficient to enable aerodynamic load predictions: for a given freestream velocity V_0, Equation (3.8) contains two unknowns, namely the annular thrust force and the interference factor a. The thrust force is assumed to be uniformly distributed over the annular fluid surface, whereas on a real rotor the blades occupy only a small fraction of the annulus. To reconcile these points we turn to blade element theory.

3.4 Blade Element (2D Wing) Theory

In blade element theory the HAWT blade is divided into radial segments, each of which can be treated as an independent two-dimensional wing. The general case is illustrated in Figure 3.6, and the lift force on an element is given by

$$L = \frac{1}{2}\rho V^2 C_l c dr \tag{3.9}$$

where lift L acts at right angles to the incident air velocity V. The element area is the product of chord length c and span dr, and the non-dimensional lift coefficient C_l is (for small angles) proportional to incidence α. Drag is for the moment neglected: this is a reasonable assumption as the fundamental operation of a wind turbine is governed by lift. Figure 3.6 is, however, more representative of an aircraft wing, for which the incident velocity V is dictated only by the forward

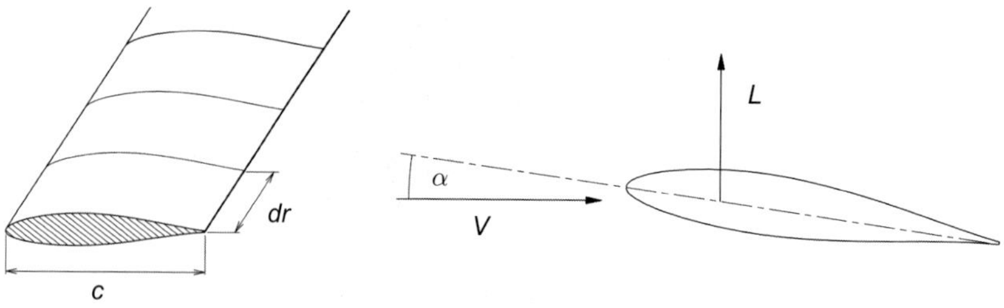

Figure 3.6 Blade element theory. Each discrete element behaves as a 2D wing: aerodynamic lift L is proportional to incidence α and acts at right angles to incident velocity V. Elemental area is cdr.

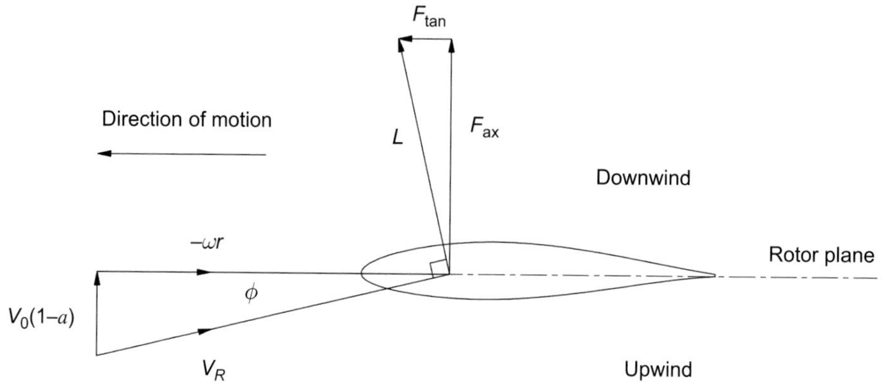

Figure 3.7 Aerodynamic conditions at a rotating blade element. The section experiences an effective air velocity V_R, which is the resultant of axial and tangential velocity components. Drag and tangential induction are for the moment neglected.

motion of the aircraft and is uniform along the span. On a rotating wind turbine blade the effective air velocity is governed by the combination of the blade's motion and the incident wind speed, and varies with radial position.

The aerodynamic conditions on a rotating blade element are shown in Figure 3.7: the section chord line is here shown parallel to the rotor plane (this corresponds to the tip section of most blades, or to an arbitrary section of an untwisted blade: twist is discussed in Section 4.5). The axial wind speed at the rotor plane is $V_0(1-a)$ and the tangential velocity of the element is ωr where ω is the angular rate of rotation; the resultant inflow velocity V_R is then given by

$$V_R = \sqrt{(V_0[1-a])^2 + (-\omega r)^2} \tag{3.10}$$

(For simplicity we neglect the tangential induction factor, which makes a modest correction to the flow geometry; its relevance is discussed in Section 3.6.2.) The resultant velocity V_R is then the airflow 'seen' by the blade element, taking the place of V in the generic 2D wing, and determining the lift force.

The inflow angle ϕ is made between the rotor plane and V_R with

Figure 3.8 Highly loaded rotor. In normal operation, the axial thrust loading is much greater than the tangential and causes the visible downwind bending of the blades. (Courtesy Vestas Wind Systems A/S)

$$\phi = \tan^{-1}\left\{\frac{V_0(1-a)}{-\omega r}\right\} \tag{3.11}$$

For an untwisted blade element as shown ϕ is synonymous with incidence α. Lift force L is orthogonal to V_R, and may be resolved into an axial thrust component F_{ax} acting normal to the rotor plane, and a tangential component $F_{\tan}$ acting parallel to it, with

$$F_{\mathrm{ax}} = L\cos\phi \tag{3.12}$$

$$F_{\tan} = L\sin\phi \tag{3.13}$$

In normal operation ϕ is a small angle (<10°) over most of the blade and to a first approximation F_{ax} equals the lift force. Note that the axial force component is generally much greater than the tangential, even though it is the latter that generates the torque to drive the rotor. The axial force does no useful work on the blade, but is responsible for the majority of the stress imposed on it, and in a strong wind the blades of a large HAWT can be clearly seen bending in the downwind direction under the influence of axial thrust (see Figure 3.8).

On the outer blade elements, where most of the power is generated, the resultant velocity V_R is dominated by the tangential velocity component $(-\omega r)$ rather than the axial, and at the blade tip V_R is many times greater than the freestream wind speed. This is a characteristic of lift driven rotors: the blade moves much faster than the wind driving it. The inflow angle ϕ dictates the direction of the lift vector, and in Figure 3.7 lift L is effectively 'pulling' the blade in the direction of motion. The vector relationships shown in Figure 3.7 are worth taking time to understand: they will be familiar to windsurfers or sailors as the conditions for 'reaching', or sailing at right angles to the wind, when the boat moves much faster than the wind propelling it; likewise, the tip of a HAWT blade may have a tangential velocity 8 times higher than the freestream wind speed.

Following from Equation (3.12), the axial thrust force on a blade element is given by

$$F_{\mathrm{ax}} = \frac{1}{2}\rho V_R^2 c C_l dr \cos\phi \tag{3.14}$$

And the corresponding tangential force by

$$F_{\tan} = \frac{1}{2}\rho V_R^2 c C_l dr \sin\phi \tag{3.15}$$

3.5 BEM: The Combined Theory

Following from the above, the key step in BEM theory is to equate the axial thrust forces independently derived from blade element and momentum (streamtube) theories. The analytic geometry is shown in Figure 3.9. The streamtube thrust force is given by Equation (3.8) and the thrust on a single blade element by Equation (3.14). Assuming an N-bladed rotor and equating the two expressions gives

$$4\pi r \rho V_0^2 a(1-a)dr = \frac{N}{2}\rho V_R^2 c C_l dr \cos\phi \tag{3.16}$$

Equation (3.16) can be simplified, firstly by defining the local solidity σ as the fraction of the annulus occupied by the blade elements

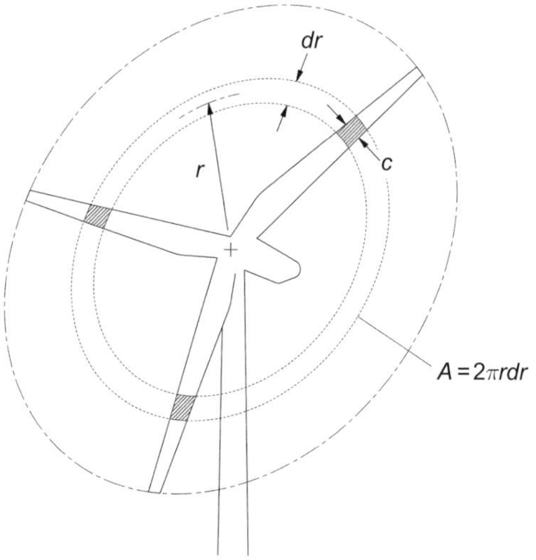

Figure 3.9 Equating blade element and momentum forces. The momentum thrust acting on annular area $2\pi r dr$ is equated to the thrust on the discrete blade elements (shaded) of combined area $Ncdr$, where N is blade number.

$$\sigma = \frac{Nc}{2\pi r} \tag{3.17}$$

and secondly by noting the following relationship:

$$\sin\phi = \frac{V_0(1-a)}{V_R} \tag{3.18}$$

Making the above substitutions Equation (3.16) yields the following expression:

$$\frac{a}{(1-a)} = \frac{\sigma C_l}{4\,\sin\phi\tan\phi} \tag{3.19}$$

Equation (3.19) reconciles the thrust loading from momentum and blade element theories: this is the key relationship in BEM theory. The equation is not, however, explicit and must be solved by an iterative procedure. To simplify, we first define the local velocity ratio λ' for a blade element at radius r:

$$\lambda' = \frac{\omega r}{V_0} \tag{3.20}$$

Then, from Equations (3.11) and (3.20),

$$\phi = \tan^{-1}\left[\frac{(1-a)}{\lambda'}\right] \tag{3.21}$$

An iterative procedure for solving Equation (3.19) is then as follows:

1. Make an initial estimate of ϕ by assuming $a = 0.33$ and applying Equation (3.21).
2. Evaluate lift coefficient C_l using tabulated (C_l, α) data for the relevant aerofoil. For an untwisted section as here, $\alpha = \phi$; otherwise twist must be accounted for, but the calculation is essentially unchanged (see Section 4.5).
3. Substitute C_l and ϕ into the right hand side of Equation (3.19) and solve for a new value of a. Recalculate ϕ using Equation (3.21).
4. Compare ϕ with the previous estimate, and repeat steps 2–4 until convergence.
5. Once the calculation has converged the final values of C_l and ϕ are used to calculate blade element axial and tangential forces using Equations (3.14) and (3.15), respectively. The resultant inflow velocity V_R is found from Equation (3.10) using the converged value of induction factor a.

In a typical BEM code the blade will be divided into 30–50 elements; the iteration procedure is carried out for each element to yield radial distributions of axial and tangential force, which are then integrated to give the net loads transmitted to the rotor hub. Assuming a uniform windfield all the blades experience the same radial load distribution, and the calculations only require to be carried out for one blade. The net loads for an N-bladed rotor are then found by numerical integration with

$$\text{Axial thrust}: T = N\sum F_{\text{ax}} \tag{3.22}$$

$$\text{Shaft torque}: Q = N\sum F_{\tan} r \tag{3.23}$$

$$\text{Power}: P = N\omega\sum F_{\tan} r \tag{3.24}$$

where r is the radius of a blade element and ω the rotor angular velocity. The bending moments at the blade root are found from

$$\text{Blade root axial moment}: M_{\text{ax}} = \sum F_{\text{ax}} r \tag{3.25}$$

$$\text{Blade root tangential moment}: M_{\tan} = \sum F_{\tan} r \tag{3.26}$$

Similar integrations yield the radial distributions of bending moment and shear force along the blade, which are key inputs to blade structural design (see Chapter 8). Under optimal (Betz limit) conditions the axial induction factor a would be 0.33 at all blade elements, and the local thrust coefficient $C_t = 8/9$. The axial and tangential force per blade element can then be found from the streamtube momentum equations with

$$F_{\text{ax}} = K(r/R) \tag{3.27}$$

$$F_{\tan} = \frac{K}{1.5\lambda} \tag{3.28}$$

where R is tip radius, λ is tip speed ratio, and K is given by

$$K = \frac{8\pi\rho V^2 R^2}{9Ne} \tag{3.29}$$

where ρ is density, V freestream velocity, N blade number, and e the number of elements into which the blade is divided. Under optimum conditions the axial force F_{ax} per blade element is then proportional to radius, while the tangential force $F_{\tan}$ per element is constant along the blade. These relationships are illustrated in Figure 3.10, which compares elemental load distributions for a real blade design (dashed lines) and an ideal blade operating at the Betz limit (solid lines). The forces are normalised with respect to K, and the tangential force is calculated for tip speed ratio $\lambda = 6$. The differences between the optimum force distributions and those calculated for the real blade are attributable to the influences of drag, and tip loss: these factors are discussed further below.

Lift and drag coefficients for HAWT aerofoil profiles were traditionally taken from 2D wind tunnel data, though use of CFD-derived values is increasingly common; between these sources there is a vast catalogue of aerofoil performance data available to the blade designer. Profiles are ideally optimised for low speed (below 100 m s^{-1}) and high lift/drag ratio, and many successful wind turbines used aerofoils originally developed for the aircraft industry in the second half of the twentieth century. A typical example is the NASA LS1(Mod), whose characteristics are shown in Figure 3.11, and an

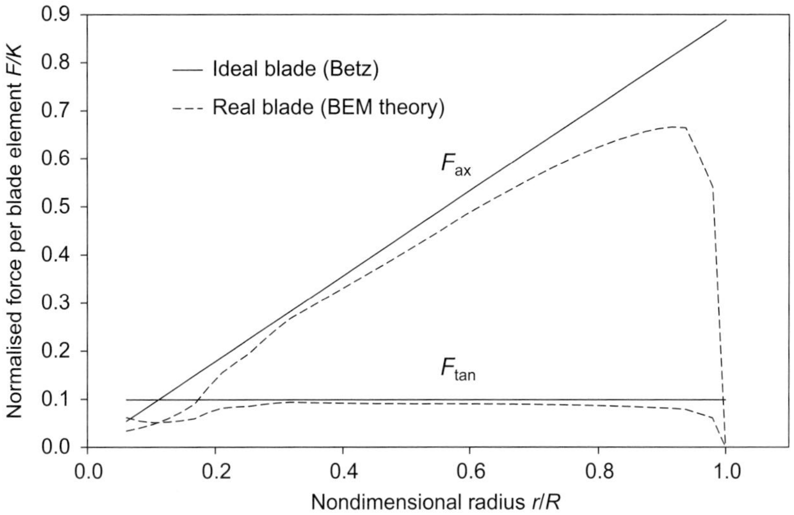

Figure 3.10 Blade element load distributions, comparing ideal (Betz limit) values with a typical BEM design optimised for tip speed ratio $\lambda = 6$. Axial force is proportional to radius, while tangential force is constant.

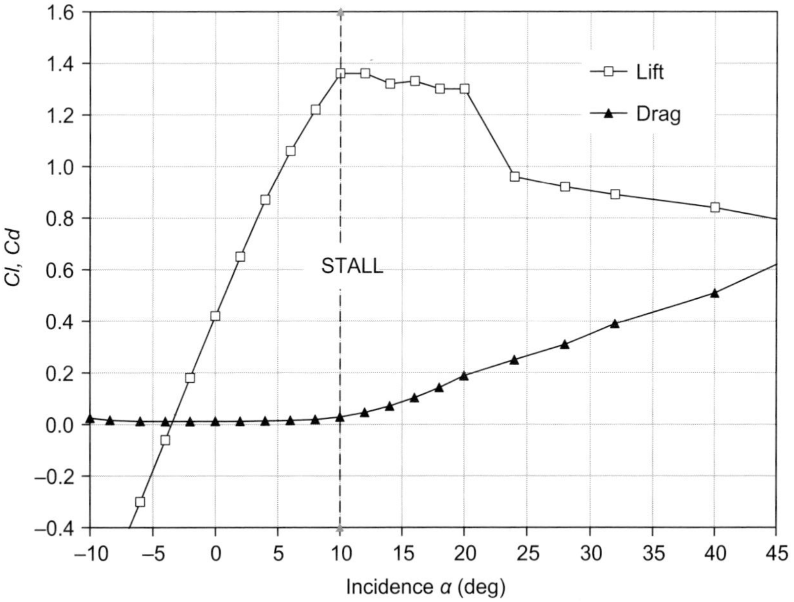

Figure 3.11 Lift and drag coefficients for a 2D aerofoil: wind tunnel measurements for the NASA LS1(Mod) 21% profile. The onset of stall is marked by decreasing lift and increasing drag.

example of a wind turbine blade section using this profile can be seen in Figure 8.22. With the growth of the wind industry, however, aerofoils began to be specifically developed to suit the requirements of large HAWT blades (Björck, 1990). These included restricted peak lift (optimum aerofoil performance is achieved well below $C_l^{\max}$ so reducing the latter lowers extreme loading), thicker sections for increased structural rigidity, and shapes tailored to accommodate certain construction materials.

3.6 Modifications to BEM Theory

The aerodynamic theory described so far includes a number of simplifying assumptions: a BEM code based on it might be sufficiently accurate for some engineering or control studies, but a higher degree of accuracy is called for when calculating the power curve of a commercial wind turbine, or analysing the loading on a rotor operating away from optimal design conditions. The following sections describe some necessary modifications to simple theory.

3.6.1 Inclusion of Drag

While the operation of a HAWT rotor is lift-dominated, real blades are also subject to drag. A revised version of the blade aerodynamic geometry including drag is shown in Figure 3.12: as previously lift force L acts at right angles to resultant velocity V_R, but drag D is now present acting parallel to V_R. The resultant force F_{res} resolves into axial and tangential components according to

$$F_{\text{ax}} = \frac{1}{2}\rho V_R{}^2 cdr(C_l\cos\phi + C_d\sin\phi) \tag{3.30}$$

$$F_{\text{tan}} = \frac{1}{2}\rho V_R{}^2 cdr(C_l\sin\phi - C_d\cos\phi) \tag{3.31}$$

Comparing these equations with Equations (3.14) and (3.15) we see that the inclusion of drag increases the axial force, but decreases the tangential force. Neglecting drag thus leads to an

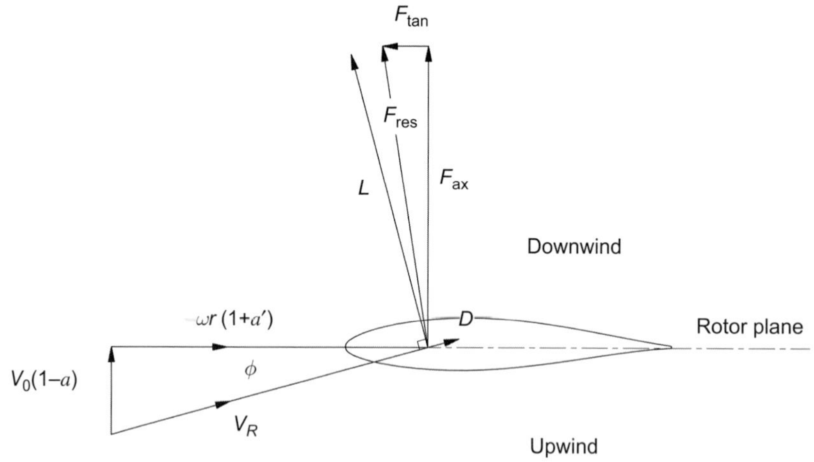

Figure 3.12 Blade element aerodynamics including drag D and tangential induction factor a'. The resultant of lift and drag is F_{res}. Drag increases the axial force component but reduces the tangential.

underestimate of thrust (and hence blade loading) but an overestimate of power (hence energy yield), for which reasons an accurate BEM model should incorporate drag. Some authors argue that because drag does not contribute to axial momentum change it may be neglected when calculating interference factor a; others contend that under some conditions, e.g. stalled flow, drag does influence axial momentum and that Equation (3.19) should be modified to incorporate drag. Although this introduces a little more complexity the equations remain soluble by a similar iterative procedure as before. For a fuller discussion on these topics see (Burton, 2011).

3.6.2 Tangential Induction Factor

As the air exerts a torque on the HAWT rotor, so by the principle of action and reaction the rotor must exert an equal but opposed torque on the air. This causes the wake to rotate in the opposite sense to the rotor. Referring to Figure 3.12, we see that an additional variable a' is present compared to Figure 3.7: this is the tangential induction factor, introduced to take account of the angular velocity (swirl) imparted by the rotor on the incident flow. The tangential velocity at the rotor plane is then increased by an amount $\omega r a'$ (the swirl velocity) where a' is non-dimensional. The correction is generally small, and on an ideal rotor the swirl velocity is inversely proportional to radius, with greatest magnitude on the inboard blade where power production is small. Its effect may, however, be important in respect of the stall characteristics of the inboard blade. The inclusion of the tangential induction factor leads to a modified set of governing equations for BEM theory and in addition to Equation (3.19) the following (for simplicity still excluding drag) now applies

$$\frac{a'}{(1+a')} = \frac{\sigma C_l}{4\cos\phi} \tag{3.32}$$

while Equation (3.21) is modified to

$$\phi = \tan^{-1}\left[\frac{(1-a)}{\lambda'(1+a')}\right] \tag{3.33}$$

The iterative procedure is similar to that previously described, but now the recursive value of ϕ is found using Equation (3.33); the initial estimate can be made assuming $a = 0.33$ and $a' = 0$, and Equation (3.32) is included within the iterative loop to recalculate a'.

3.6.3 Momentum Theory Correction

One significant limitation of BEM theory is the breakdown of basic momentum assumptions at high values of induction factor a, corresponding to conditions of high flow blockage. While the multiple streamtube, and even the simple actuator disc, models yield credible thrust and power predictions for small a, Equation (3.5) is strictly not applicable above $a = 0.5$, as it predicts a progressive decrease in thrust coefficient to zero at $a = 1$. This is inconsistent, corresponding to stationary flow at the rotor plane and a simultaneously unloaded rotor. On a real rotor the opposite is true: the more the flow velocity is retarded, the greater the thrust loading. To account for this anomaly an empirical modification is made to the thrust curve for $a > 0.4$, as shown in Figure 3.3. A maximum C_t value in the range 1.6–1.8 applies (literature sources vary on the appropriate figure) when $a = 1$.

The modified C_t curve is based on experimental measurements, with the extreme value representing a higher thrust loading than that of a solid circular disc. Such a condition is unusual, but not impossible: extreme thrust is associated with very high tip speed operation, for instance if a rotor overspeeds. Severe flow blockage then occurs and the downwind wake is characterised by extreme turbulence and/or recirculating flow, in a 'vortex ring' or 'propeller brake' state (Eggleston, 1987). High blockage operation has been explored in small-scale flow visualisation experiments (see below); at full scale it may lead to structural damage or catastrophic blade failure, though these conditions are usually associated only with control malfunction, and rarely occur in normal operation.

3.6.4 Radial Flow and Stall Delay

The assumption of radial independence between concentric streamtubes holds up quite well for blades in attached flow, and BEM theory gives reliable predictions under these conditions. When the blades are partially or fully stalled, however, there may be a significant radial component of flow at the rotor plane due to centrifugal pressure gradients. This gives rise to stall delay, where the affected blade sections attain higher lift coefficient than under 2D conditions, and the rotor power exceeds that predicted from standard BEM assumptions. The phenomenon was first noted on propellers (Himmelskamp, 1945) and became important to the wind industry with the rise of stall-regulated wind turbines, with commercial operators reporting power levels well in excess of prediction in high winds (Milborrow, 1985). Many studies ensued, including flow visualisation experiments on small rotors in wind tunnels (Ronsten, 1991) and on operational wind turbines (Pedersen, 1988). Figure 3.13 shows a full scale blade tufted to indicate when stall has occurred, in an experiment designed to compare 2D and actual post-stall behaviour (Anderson, 1987).

Stall delay is treated by empirical correction in BEM codes, via adjustments made to the 2D aerofoil lift and drag data in the post-stall region. In the early days when stall-regulated machines were relatively small this was largely a matter of measuring the power curve of a new machine and retrospectively adjusting the design C_l, C_d data to get the theoretical power curve to match; the modified aerofoil properties could then be applied in the design of further wind turbines. This empirical approach was largely successful when scaling up a known design, but less so when a change of profile was involved. More recently CFD analysis has been used to provide more rigorous post-stall correction to aerofoil properties; see Burton (2011).

Figure 3.13 Tufted blade used for full-scale flow visualisation measurements on a 26 m rotor. The tests were intended to explore post-stall correction of the lift curve due to radial flow (Anderson, 1987).

3.6.5 Tip Loss Correction

Simple BEM theory predicts high aerodynamic efficiency at the tip of a wind turbine blade, but flow in this region is highly three-dimensional, and 2D streamtube assumptions are no longer valid. At the blade tip air 'leaks' round from the pressure to the suction side, and in doing so creates a trailing vortex: the effect is commonly seen on aircraft, where vortices trail rearwards from the wing tips; see Figure 3.14. Such flow is highly three-dimensional, and on the outer stations of a HAWT blade there is a progressive loss of performance, with the lift force falling to zero at the extreme tip. Tip loss has been extensively researched, and BEM theory incorporates mathematically rigorous corrections based on the classic propeller theory of Prandtl, Goldstein, and Glauert (Anderson, 1981). Prediction codes are modified to include a loss factor F at each blade element, relating the bound circulation at that radius to the corresponding value assuming an infinite number of blades (BWEA, 1982). The mathematics are somewhat complex, but lead to inclusion of the factor F in the governing equations described earlier.

In practice the loss is negligible over most of the blade and becomes significant only towards the tip; the overall rotor thrust and power output are reduced by a few per cent. The

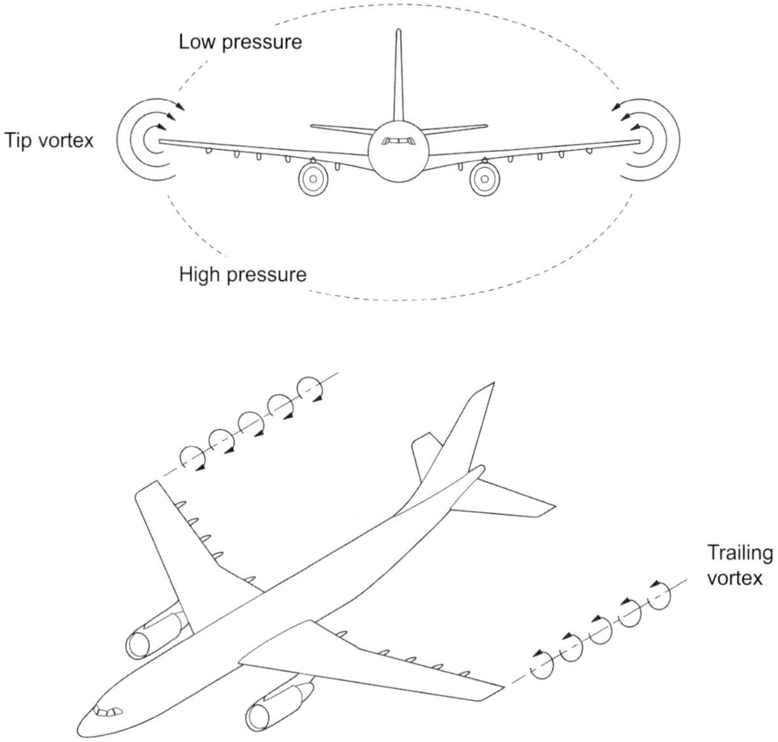

Figure 3.14 Tip vortices trailing from an aircraft wing as air 'leaks' from high to low pressure. On a HAWT rotor blade, the vortices form a helical pattern behind the rotor: this forms the basis of vortex wake theory, an alternative to BEM.

inclusion of tip loss in a BEM calculation can be seen in the predicted axial load distributions along the blade; see Figure 3.10. A similar loss factor should in principle be applied at the blade root but is usually neglected: loading in this region is small (due to the low tangential velocity) and the effect on rotor power is not significant. The tip vortices of a HAWT rotor are shed into the wake and convect downstream at the local axial velocity, forming a helical pattern whose significance is discussed below. Vortex wake analysis in fact offers an alternative to BEM theory, embodying a more physically realistic description of the airflow at and behind the rotor plane, and the topic is discussed in more detail in Section 3.7.2.

3.7 The Rotor Wake

3.7.1 Introduction

Behind a wind turbine in the rotor wake the airflow is characterised by a region of reduced velocity and increased turbulence. These effects extend far downstream, with implications for other turbines

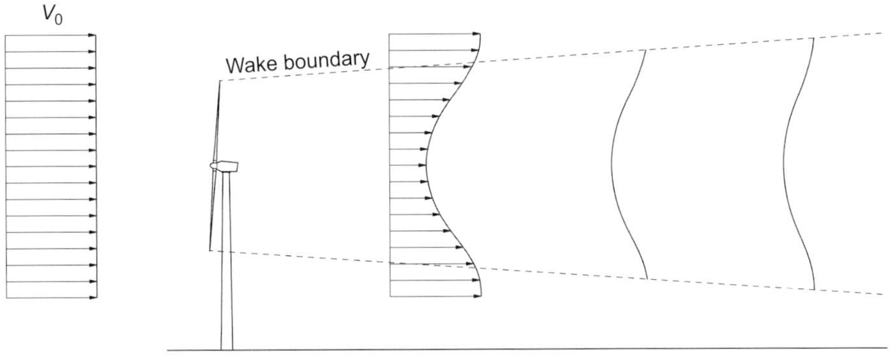

Figure 3.15 Schematic illustration of the rotor wake. The wake boundary expands downwind, and the internal velocity profile is approximately Gaussian, with maximum velocity deficit on the centreline.

sited directly in line: lower wind speeds mean reduced power output, while higher turbulence increases fatigue. The wake is illustrated schematically in Figure 3.15. Its boundary expands downwind in approximately linear manner; within it the velocity profile is quasi-Gaussian with the maximum velocity deficit occurring on the centreline. The wake deficit reduces with increasing distance downstream, and freestream conditions are re-established typically 10–20D behind the rotor (depending on factors including the rotor thrust loading, and ambient turbulence level). Wake recovery seemingly contradicts actuator disc theory, which predicts continuous deceleration of the downstream airflow: in reality, however, the velocity deficit is reduced as air from the undisturbed freestream is entrained by the action of shear forces at the wake boundary, re-energising the flow within the wake. The region immediately behind the rotor is defined as the near-wake, extending to around 2–3D downstream; here the airflow is highly structured due to the vortices shed from the rotor blades (see below). Further downstream, in the far wake, the vorticity breaks down into unstructured turbulence. Analyses of the wake fall broadly into two categories:

- Near-wake studies: using vortex wake theory, these employ a highly mathematical approach to describe the airflow at the blades and in the wake, based on fundamental fluid mechanics. Rotor loading and performance can be calculated using vortex wake theory as with BEM, but the former also allows the near-wake structure and velocity deficit to be modelled. Vortex wake theory is described in Section 3.7.2.
- Far wake studies: the velocity deficit and turbulence content are here estimated using turbulent jet theory, based on fairly simple empirical models. This is typically used for analyses of windfarm arrays, to estimate energy yield and/or wake-induced turbulence levels.

A brief overview of vortex models applicable to the near-wake follows below. Far wake analyses are described in more detail in Chapter 9. For those seeking more detailed information on the properties of the wake there have been a great many theoretical and experimental studies, and several comprehensive literature reviews, including Vermeer et al. (2003) and Stevens et al. (2017).

3.7.2 Vortex Wake Analysis

We have noted that BEM theory is very useful in practice but has several theoretical limitations. For research purposes, more sophisticated aerodynamic models have been developed in which the interaction of the blades with the air is treated using lifting line theory, originally developed by Ludwig Prandtl in the early twentieth century. Prandtl replaced the finite physical wing with a bound vortex whose interaction with the freestream velocity produces the equivalent lift force (Anderson, 1985). On an aircraft wing the bound vorticity is shed in a line from the tip as illustrated in Figure 3.14;[4] with a wind turbine rotor the trailing vortices form a helical pattern in the wake, as shown in Figure 3.16. The helical structure is continuously generated at the rotor plane and migrates downstream at the local axial velocity. The vortex structure plays a fundamental role in the performance of the turbine as the vorticity induces a reverse axial velocity field inside the wake, effectively reducing the incident wind speed at the rotor plane, consistent with power extraction.

The simplest vortex wake codes represent the blade as a line vortex of uniform strength Γ; more sophisticated models assume a spanwise distribution of vortex segments of varying strength to better represent the lift distribution on a real blade. In either case the mathematical treatment involves calculating the induced axial velocity at the rotor plane by integrating the contributions of all vortex filaments in the wake. The Biot–Savart law applies, and the governing equations are essentially the same as those used to calculate the induced magnetic field of a current-carrying wire: a cross section of the helical vortex wake is closely analogous to that of a solenoid, with lines of vorticity replacing current, and induced velocity replacing magnetic field (Figure 3.17). A simple vortex wake model yields the same relationships between axial induction factor, thrust, and power, as blade element momentum theory, leading to Equations (3.5) et seq. Vortex wake codes can thus be used to calculate blade load distributions and net rotor loads, but they require significantly more computation time than BEM due to the large-scale integration of velocity contributions for the entire helical vortex structure.

The real strength of vortex codes, however, is that unlike BEM theory they model the three-dimensional nature of the flow: there is no assumption of radial independence, and the wake structures predicted by vortex wake codes have been verified experimentally (see below). Vortex wake codes are also useful for modelling the time dependency of rotor loading: the inflow velocity at the rotor plane is determined by the entire wake, but this takes a finite time to respond to changes in blade pitch angle or incident wind velocity at the rotor plane. Such behaviour can be simulated with a time-stepping vortex wake model, and this is a powerful capability when developing or analysing power control strategies; see Section 3.9.2.

---

[4] In damp weather, a visible trail is left as the low pressure vortex core condenses moisture from the air.

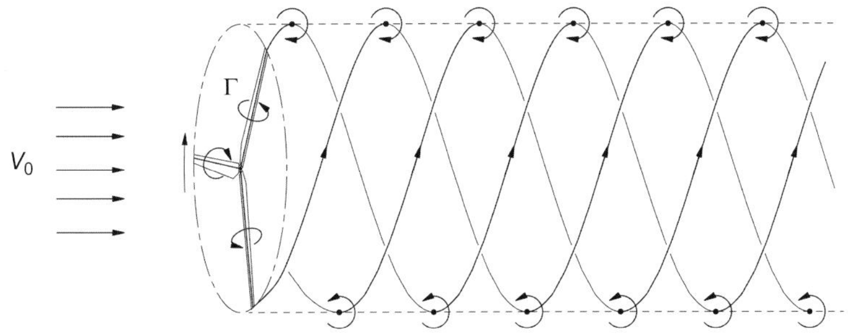

Figure 3.16 Vortex wake model. Each blade is represented as a bound vortex (lifting line) of strength Γ that sheds into the wake at the blade tip, forming a helical pattern behind the rotor. See also Figure 3.17.

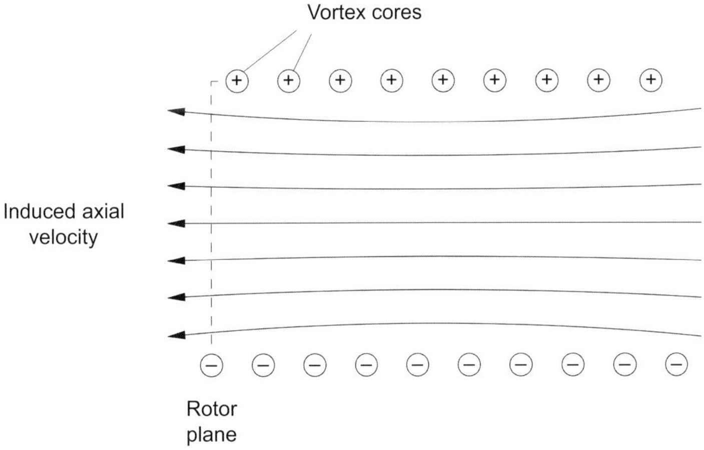

Figure 3.17 Induced velocity due to a helical vortex wake. The axial velocity at the rotor plane is calculated using the Biot–Savart Law: essentially the same mathematical model is used to calculate the magnetic field of a solenoid. In the present case, lines of vorticity replace current-carrying wires.

3.7.3 Near-Wake Measurements

The near-wake structure of HAWT rotors has been investigated experimentally at small scale using particle image velocimetry (PIV), a powerful flow visualisation technique that yields quantitative data in the form of velocity and vorticity maps, giving direct visualisation of the wake structure. Experiments have been carried out in recirculating water tanks (Whale, 1996) and in wind tunnels (Yang et al., 2012). In the former case tests at Edinburgh University employed a 180-mm-diameter rotor operating at tip speed ratios in the range 3–8, with a scanning laser to illuminate the wake cross section. The raw results included 2D velocity plots of the kind shown in Figure 3.18, from which cross-wake velocity profiles were extracted. Figure 3.19 shows wake profiles taken approximately 1D downstream at different tip speed ratios: the profiles exhibit the

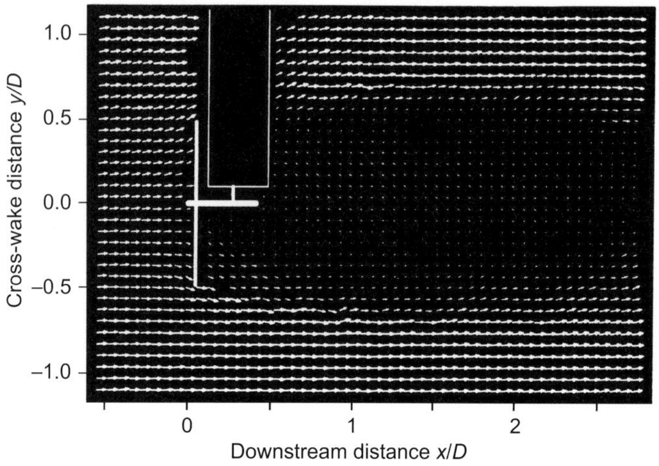

Figure 3.18 Velocity map in the wake of a small-scale model HAWT rotor, measured using particle image velocimetry (Whale, 1996). The example shown corresponds to high tip speed ratio. (Image reproduced with kind permission of Jonathan Whale)

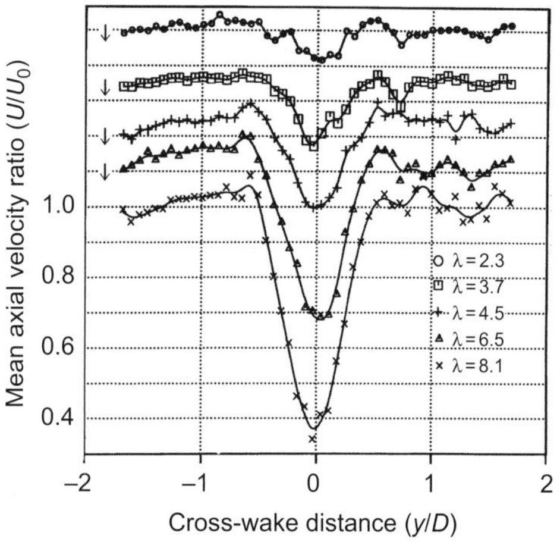

Figure 3.19 Wake deficit behind a model HAWT rotor, from PIV measurements (Whale, 1996). Profiles measured 1.1D downstream of the rotor, for a range of tip speed ratio. The profiles are offset vertically for clarity; $U/U_0 = 1$ outside the wake boundary in all cases. (Image reproduced with kind permission of Jonathan Whale)

classic Gaussian shape, with maximum velocity deficit increasing with tip speed ratio (the profiles shown are vertically separated for clarity). In the extreme case ($\lambda \approx 8$) the centreline velocity deficit is 60%. The patterns observed in these small-scale tests are similar to those observed at full scale: Figure 3.20 shows wake profiles measured 4.4D downwind of a medium-scale commercial wind turbine using hub height SCADA measurements. At this downstream

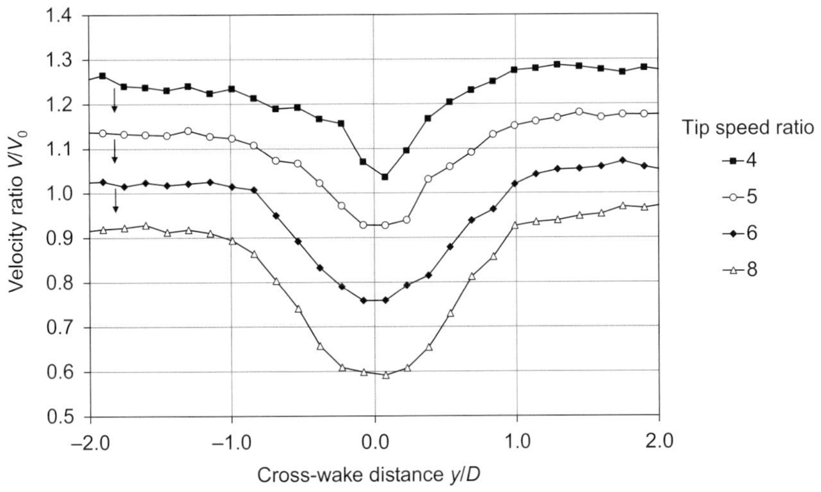

Figure 3.20 Wake deficit behind a real wind turbine. Ratio of wake to freestream velocity measured at hub height, 4.4D downwind from a Vestas V52; profiles are offset vertically for clarity. (From data provided with kind permission of Greenspan Energy Ltd)

separation the wake has expanded more and the centreline velocity deficit is reduced by comparison to the model results at 1D downstream, but the patterns are qualitatively similar. The full scale measurements were taken at hub-height on a Vestas V52 wind turbine which was the middle machine in a linear array of three – see Figure 3.21.

More significantly, PIV measurements enable vorticity to be mapped over the wake cross section, and presented in the form of contour plots (vorticity is derived from the 2D velocity maps via closed-loop integration). Figure 3.22 shows wake vorticity maps corresponding to model rotor operation at (a) low and (b) high tip speed ratio. In the former case the helical vortex spacing is clearly seen, and the image can be compared with the mathematical model illustrated in Figure 3.17. At the higher tip ratio the vorticity is stronger, but the helical pattern has broken down somewhat as the vortices coalesce, with evidence of rapid wake expansion immediately behind the rotor. In both cases the vortex structure starts to dissipate at around 2.5D downstream, marking the end of the near-wake region. Although obtained with small-scale models, vorticity measurements of this kind have been successfully compared with numerical results from an inviscid free-wake model (Whale, 2000), and the fundamental wake behaviour is found to be relatively insensitive to blade chord Reynolds number.

3.8 Operation in Yaw

When a HAWT rotor does not point directly into the wind the aerodynamic conditions at the blade are affected as shown in Figure 3.23. The yaw angle (or yaw error) is γ and the resulting aerodynamic flow conditions are shown for a blade section at the top and bottom of its rotation

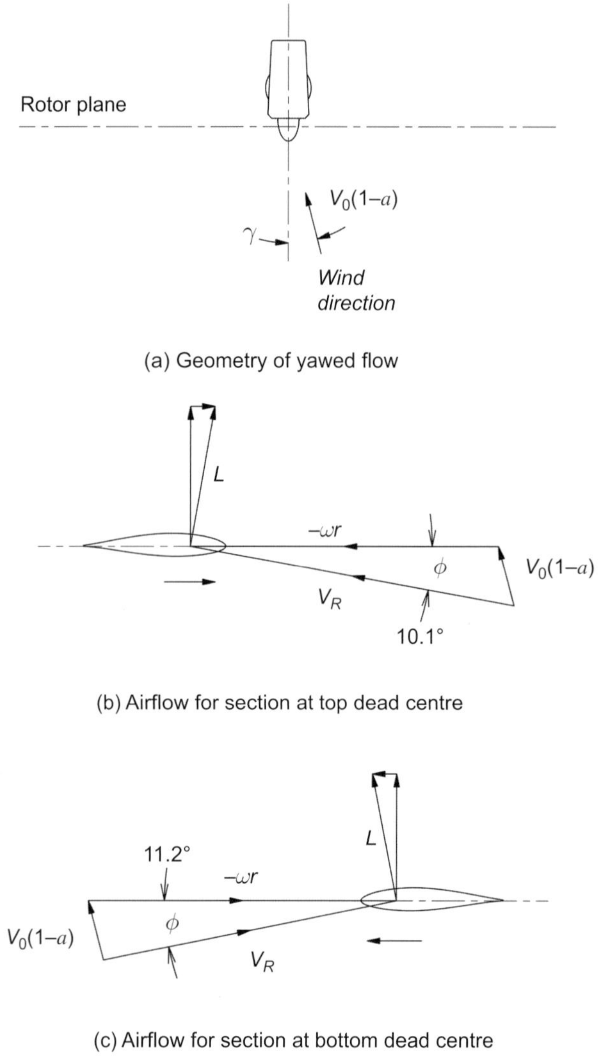

Figure 3.23 Aerodynamics of yawed flow: (a) the rotor is misaligned with the incident wind direction by angle γ; as the blade rotates between (b) top and (c) bottom dead centre the flow angle ϕ and resultant velocity V_R, both change, resulting in cyclic variation in lift.

cycle. Seen in the blade frame of reference the incident wind veers from side to side by $\pm\gamma$ once per revolution and although the axial and tangential air velocities remain unchanged in magnitude, both the flow angle ϕ and resultant velocity V_R are distorted, resulting in a cyclic variation of the lift force. In the example shown the magnitude of lift vector L varies by $\pm$ 10% over the rotation cycle. Yaw error thus introduces cyclic load variation at the blade rotation frequency, and is a potential source of fatigue.

Figure 3.21 All lined up. Measurements from this linear array of Vestas V52s (4.4D spacing) illustrate the properties of rotor wakes. Figure 3.20 shows wake velocity profiles, while Figure 9.6 shows the influence of wind direction on wake power deficits. (Balquhindachy windfarm, Aberdeenshire: photo reproduced with kind permission of Greenspan Energy Ltd)

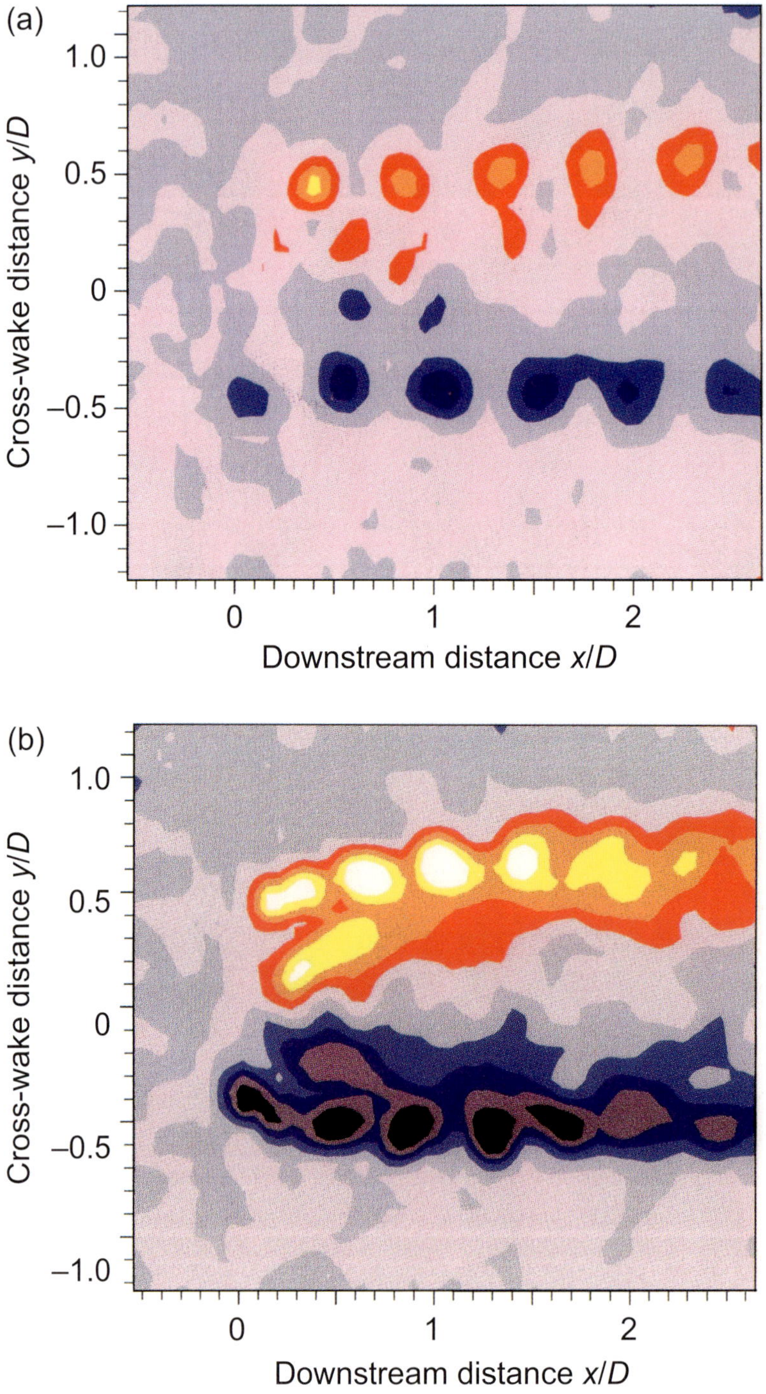

Figure 3.22 Vortex wakes visualised. Vorticity maps of a two-blade model HAWT rotor captured by particle image velocimetry (Whale, 1996). Each plot is a cross section on the wake centreline, with (a) low tip speed ratio and (b) high TSR. Compare the structure seen in the upper plot with Figure 3.17. (Figure reproduced with kind permission of Jonathan Whale)

Power output is also affected by yaw: rotor power is roughly proportional to $\cos^3\gamma$, so while a modest degree of yaw error (5° or so) can be accepted without significant performance loss, higher error will have an economic impact. Restricting yaw error within narrow limits is therefore an important control objective (see Section 6.2). The aerodynamics of yawed flow are in reality more complex than the depiction of Figure 3.23 indicates, although it does represent the analytic geometry assumed for yawed flow calculations in some BEM codes. The results may be somewhat qualitative, however, and more sophisticated aerodynamic models based on vortex wake analysis or CFD are recommended for greater realism.

3.9 Unsteady Aerodynamic Influences

Rotor aerodynamic theory has been thus far described in steady-state terms, but on real rotors unsteady phenomena can have a significant influence, causing transient variation in blade loading and power output. Two types of influence are described below, namely stall-related effects, and dynamic inflow. The former are mainly of concern in the design and operation of stall-regulated wind turbines, whereas dynamic inflow affects all wind turbines, though is of particular interest in relation to rotor pitch control.

3.9.1 Stall Effects

Unsteady stall phenomena include (a) dynamic stall and (b) double or multiple stall. The former can be explained with reference to Figure 3.24, which compares the lift coefficient for a 2D aerofoil under steady-state conditions (solid line) and assuming dynamic variation of incidence α (dashed line). In the first case the lift peaks at the nominal stall angle and then reduces; under dynamic conditions, however, stall is delayed beyond the nominal point, resulting in a significant lift overshoot; similarly, as incidence reduces the section remains stalled at an angle below the nominal stall point. Dynamic changes in incidence thus cause 'stall hysteresis' with greater lift variation than predicted from steady-state theory. Dynamic stall is well understood from research into helicopter rotors, and peak lift coefficients 45% above steady-state maxima have been reported (Bramwell, 1976). Using specially instrumented blades on a 95 kW wind turbine, Madsen measured stall hysteresis loops similar to that shown in Figure 3.24 on a rotor operating at a high yaw angle (Madsen, 1990).

The effect of dynamic stall on energy capture is not necessarily significant, as cyclic variation is averaged out in the nominal power curve. The dynamic variation in blade loading is, however, a source of increased fatigue and must be accounted for in design. Dynamic stall can be included in HAWT simulation codes, and an appropriate mathematical treatment

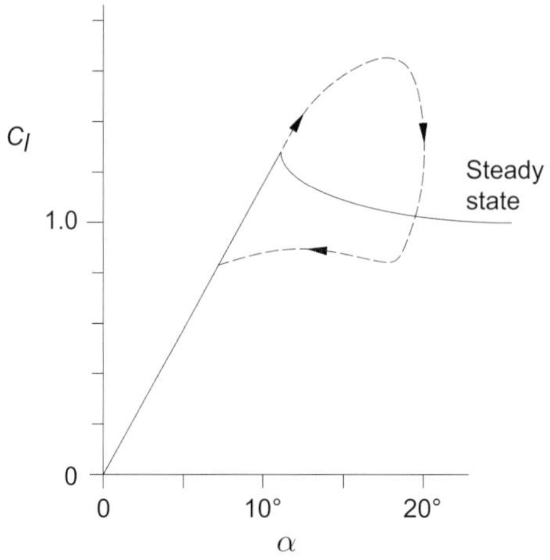

Figure 3.24 Dynamic stall. Lift characteristics of a 2D aerofoil under steady-state conditions (solid line) and with dynamic variation of incidence α (dashed line). Rotors in yaw can experience high cyclic loading due to dynamic stall.

adapted from the Beddoes–Leishman helicopter model is described by Hansen (Hansen et al., 2004).

In the phenomenon of double or multiple stall the power curve of a stall-regulated wind turbine exhibits two or more stable forms, as illustrated in Figure 3.25. Initially this was reported as double stall, but detailed studies revealed that the rotor could occupy one of several discrete power curves, remaining stable for several hours at a time (Snel et al., 1999). An explanation for this behaviour is that each blade has two possible stall states with different power levels: depending on the state of individual blades a three-blade rotor will then exhibit one of four possible power curves. Broadly speaking the two states are determined by whether flow separates from the blade leading or trailing edge: leading-edge stall occurs earlier and results in a lower power curve. Following a novel exercise using stall flags to visualise flow conditions on the blade of a Nedwind 30 wind turbine, Corten proposed that the aerodynamic conditions at the tip dictate the stall state over the rest of the blade (Corten, 1999). A Riso study examining the stall mechanism in detail concluded that double stall should be avoidable with suitable aerofoil design (Bak et al., 1998). The phenomenon of multiple stall is of much less concern with pitch-regulated rotors, which operate in attached flow and well below the stall point in normal operation.

3.9.2 Dynamic Inflow

Dynamic inflow refers to time-dependent changes in rotor loading due to wake influence, and can be explained with reference to the vortex wake model shown in Figure 3.16. Any

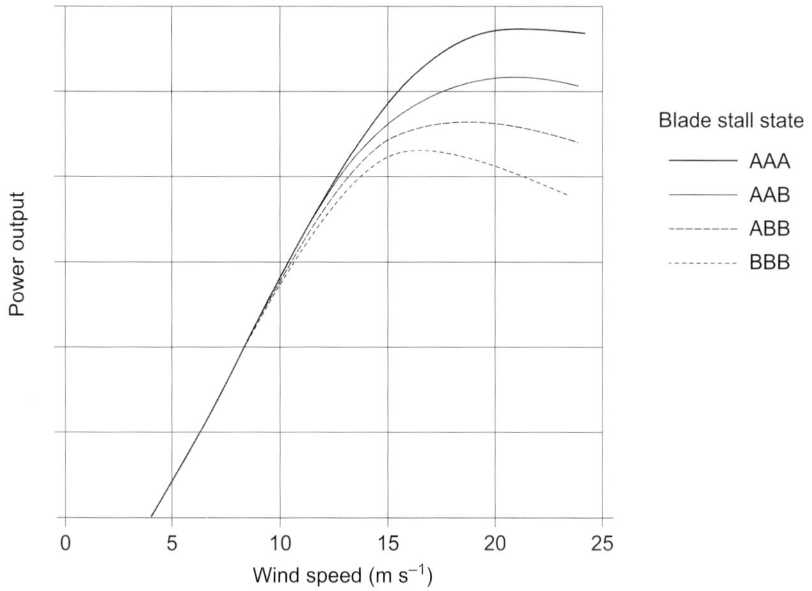

Figure 3.25 Multiple stall. Stall-regulated wind turbines may exhibit several quasi-stable power curves: if each blade has two possible stall states (A and B), a three-blade rotor will have four possible power curves.

sudden change in conditions at the rotor plane due to a sharp gust, change of pitch angle, or variation in rotor speed, will immediately affect the blade lift force. The axial inflow velocity at the rotor plane is, however, induced by the entire helical vortex wake[5] and will take time to register the change, as the vorticity shed from the blades convects downwind at the local axial velocity. The transient change in blade loading is then associated with a sharp increase or decrease in power, but the effect relaxes, and with time the power settles to a less extreme value. Dynamic inflow was first investigated by Øye on the Nibe B wind turbine, in an examination of the effect of pitch change on rotor power output (Øye, 1986). In an elegant experiment Øye drove the blade pitch actuators with a constant-amplitude, variable frequency demand, and measured the simultaneous power output of the wind turbine. The tests were run over a large number of pitch cycles in order to average out wind speed variation, and from the results a time-dependent transfer function was derived relating pitch angle and rotor power. Sinusoidal and square wave pitch inputs were used, with the results in both cases showing that the magnitude of transient power variation increased with pitch frequency.

[5] In theory, the wake extends indefinitely behind the rotor, but in practice, the inflow velocity is determined by the structure within a few rotor diameters.

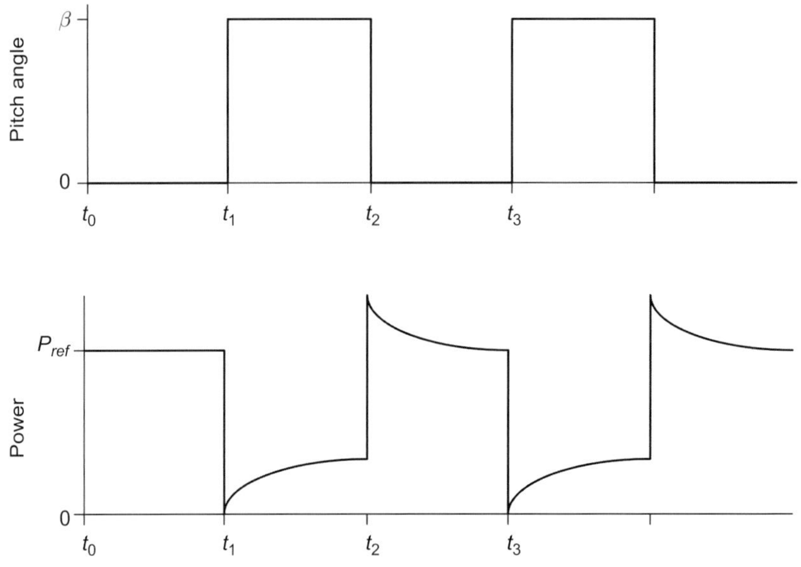

Figure 3.26 Dynamic inflow: effect of pitch change on instantaneous power output. At time t_1, the blade pitch is rapidly changed to reduce power to zero; the inflow velocity takes a finite time to change due to wake relaxation, during which the power recovers to a new steady-state value. At t_2, the pitch is returned to zero, causing the reverse of the above process. See also Figure 3.27.

The basis of Øye's experiment is illustrated in Figure 3.26, which shows simultaneous time series of pitch angle and power; Figure 3.27 shows the corresponding aerodynamic conditions at the blade (constant freestream velocity V is assumed). At time t_0 nominal conditions exist: pitch angle is zero, blade incidence is α, the axial velocity at the blade is $V(1 - a)$ and nominal power P_{ref} is produced. At t_1 the pitch is rapidly increased so as to cancel the incidence, immediately resulting in zero power; the wake takes a finite time to settle to the new conditions, however, so the induction factor a decays slowly. As it does do the axial velocity at the rotor plane recovers towards the freestream value, positive incidence is reintroduced, and the blade again develops lift; by t_2 the power has recovered to a reduced steady-state value. At t_2 the pitch is returned to zero, causing the reverse of the above process: the power now overshoots, before relaxing back to the original steady-state mean. The cycle is then repeated at t_3, and so on. In an attempt to design more accurate power control algorithms for pitch-controlled HAWTs, the dynamic inflow effect (sometimes called wake induction lag) has been incorporated in simulation models as a lead-lag filter in the drivetrain transfer function (Leithead et al., 1989). This was particularly important when wind turbines were predominantly constant-speed, and rotor power variations were carried through directly into the electrical power output; dynamic inflow is arguably less of an issue for the control of variable-speed wind turbines (see Section 6.3.3).

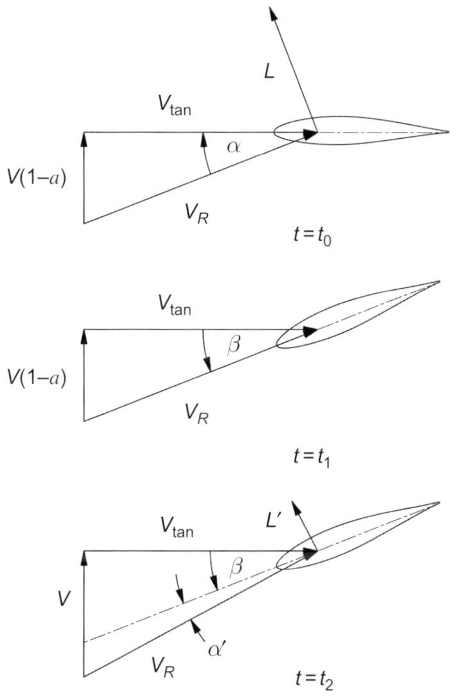

Figure 3.27 Aerodynamic conditions at the blade corresponding to pitch action shown in Figure 3.26. Initially (t_0), nominal power is being produced; at t_1, blade lift is cancelled so power goes to zero, but due to wake time delay, induction factor *a* takes time to decay; at t_2, the freestream velocity has recovered (ignoring a small induction factor) and some lift is again generated.

3.10 Exercises

3.10.1 Actuator Disc

In a simple analysis a HAWT rotor is modelled as an ideal actuator disc operating at maximum aerodynamic efficiency. In this case if the undisturbed upstream wind speed is 15 m s^{-1}, what are the corresponding wind speeds (a) at the rotor plane and (b) far downstream of the rotor?

3.10.2 Blade Aerodynamics

The diameter of a large HAWT rotor is 70 m, its rotational speed 18 rpm, and the undisturbed freestream wind speed 11.0 m s^{-1}. Assuming the rotor is operating at the Betz limit calculate (a) the axial wind velocity at the rotor plane, (b) the tangential air velocity at the blade tip, and (c) the resulting inflow angle ϕ at the blade tip. Ignore tip loss and tangential induction factor.

3.10.3 Elemental Thrust

For a BEM calculation the blade of a 160-m-diameter three-blade wind turbine is divided into 40 elements of equal span. If the entire rotor is operating at optimum (Betz) efficiency in a freestream wind speed of 9.0 m s^{-1}, what is the thrust force on a single blade element (a) at mid span and (b) at the blade tip? Ignore tip loss.

3.10.4 Blade Loading

A two-blade wind turbine has a rotor diameter of 17 m, blades of constant chord length 0.90 m, and fixed rotation speed of 75 rpm. If the axial wind velocity at the rotor plane is 8.0 m s^{-1} and the blade lift and drag coefficients at the tip are respectively 0.90 and 0.02, calculate the axial and tangential forces on the outer 1 m of blade span. Ignore the tangential induction factor.

3.10.5 Power Curve

The output power curve for a wind turbine with 44 m rotor diameter is shown in Figure 3.28. Calculate its power coefficient C_p in hub height wind speeds of (a) 11.0 m s^{-1} and (b) 20.0 m s^{-1}. Assume an air density of 1.225 kg m^{-3}.

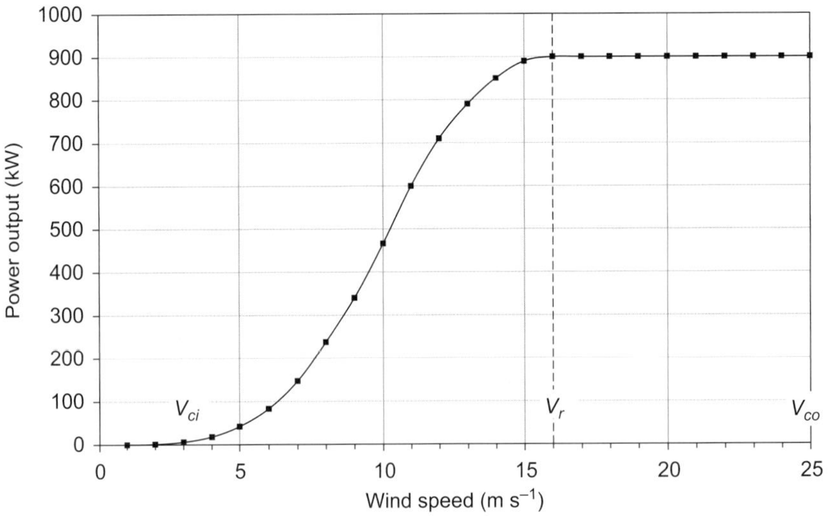

Figure 3.28 See Exercise 3.10.5.

3.10.6 Vortex Wake

Figure 3.22(a) shows a wake vorticity map captured behind a two-blade model HAWT rotor using particle image velocimetry. The image represents a cross section of the helical vortex structure in a vertical plane on the rotor centreline. Taking measurements from the vorticity map, estimate the approximate tip speed ratio λ of the model. Explain your calculation method.

3.10.7 Yaw Error

A wind turbine is rated at 3 MW and its annual capacity factor is 32% when operating with zero yaw error. Estimate its annual output in MWh under the above conditions, then repeat the calculation assuming a constant yaw error of (a) 5° and (b) 15°.

CHAPTER 4 ROTOR DESIGN AND PERFORMANCE

4.1 INTRODUCTION

The strong dependency of aerodynamic power on wind speed ($P \propto V^3$) presents a two-fold challenge to the wind turbine designer. On the one hand, for operation below its rated power level a HAWT rotor should be designed to extract as much power from the air as possible: some wind turbines now achieve $C_p^{max} > 0.5$ and thus capture more than half the available power in light or medium winds. On the other hand, in high winds the rotor must extract only a fraction of the available power – perhaps as little as 5% – in order to avoid the potential for electrical overload or structural damage. The twin objectives of high aerodynamic efficiency and accurate power control are achieved by a combination of rotor blade design and electromechanical control. The present chapter contains an overview of rotor aerodynamic design, while Chapter 6 goes into more detail on the subject of power control.

4.2 POWER, THRUST, AND TORQUE

The net aerodynamic loading on a HAWT rotor is characterised by power, torque, and thrust, for which the governing equations are

$$\text{Power}: P = \frac{1}{2}\rho V^3 \pi R^2 C_p \tag{4.1}$$

$$\text{Thrust}: T = \frac{1}{2}\rho V^2 \pi R^2 C_t \tag{4.2}$$

$$\text{Torque}: Q = \frac{1}{2}\rho V^2 \pi R^3 C_Q \tag{4.3}$$

where ρ is air density, V the freestream wind velocity, and R the rotor radius; the dimensionless power and thrust coefficients C_p and C_t were introduced in Section 3.2, and torque coefficient C_Q is related to C_p according to

$$C_p = \lambda C_Q \tag{4.4}$$

where λ is tip speed ratio.

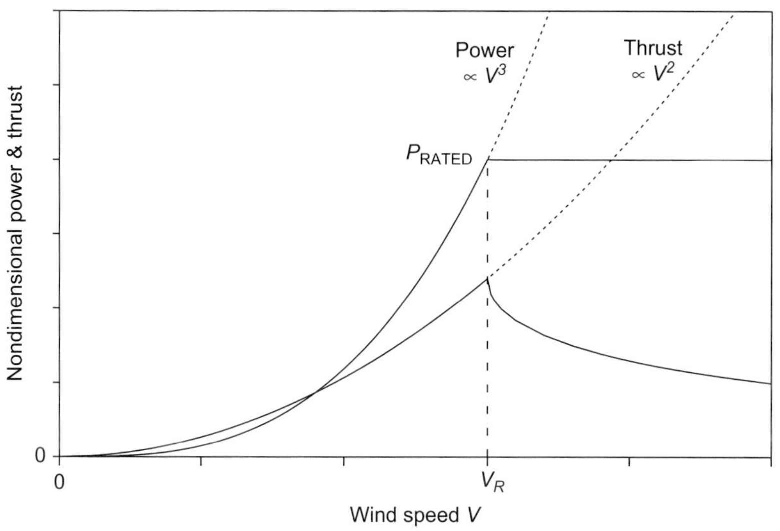

Figure 4.1 Rotor power and thrust as a function of freestream wind speed for an ideal pitch-controlled rotor. Optimum power extraction is assumed below the rated wind speed V_R, and constant power output thereafter. Dashed lines indicate the trends assuming no control is invoked.

For a given rotor geometry and tip speed ratio power output increases as V^3, while thrust and torque are proportional to V^2. These dependencies are observed on real wind turbines, but only across a part of their operational range. Figure 4.1 shows calculated power and thrust curves for a variable-speed pitch-controlled HAWT: in low winds (where constant λ operation is assumed) power and thrust show the expected dependencies on wind speed. Above V_R – the rated wind speed – the power is maintained at a constant level by pitch control, which reduces the blade lift and consequently the thrust. Power control is necessary to prevent excessive rotor loading in high winds, and the dashed lines in Figure 4.1 indicate the power and thrust trends that would apply in the absence of control.

The flat-topped power curve is characteristic of pitch control, while the decreasing thrust force above V_R indicates ideal aerodynamic behaviour: power is the product of thrust and axial wind velocity, so to maintain constant power in a rising wind the thrust must fall. Not all wind turbines demonstrate this characteristic (see Section 4.7.2) but those with positive pitch control do, as evidenced in Figure 4.2: this shows the blade flapwise bending moment (which is approximately proportional to thrust) measured on a pitch-controlled 330 kW wind turbine. The reduction in aerodynamic loading above V_R is evident.

The power curve of a wind turbine is its characteristic 'fingerprint', which in conjunction with the wind frequency distribution for a given site yields the expected annual energy output: this topic is discussed in Chapter 9. For the HAWT designer, however, a more fundamental

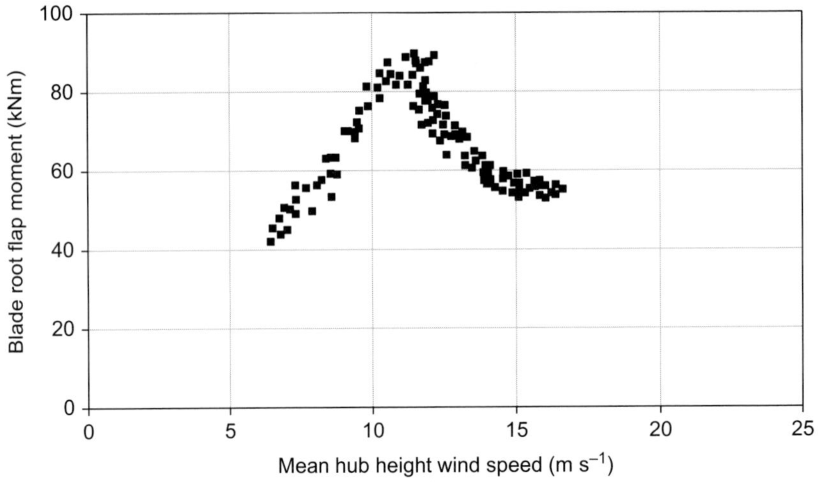

Figure 4.2 Blade root flapwise bending moment on a 330 kW wind turbine with positive pitch control, showing characteristic load reduction above rated wind speed. (From measurements in Anderson and Jamieson, 1988)

indicator of aerodynamic performance is the dimensionless power curve, or C_p, λ characteristic of the rotor.

4.3 The C_p, λ Curve

For a given rotor geometry the C_p, λ curve is unique and from it the power output can be predicted for any combination of freestream wind speed and rotor rotation speed. To illustrate, Figure 4.3 shows the C_p curve for a stall-regulated wind turbine nominally designed to run at 60 rpm; the same rotor can in principle be operated at higher speeds (by suitable choice of generator and gearbox) and the effect of running at 60, 70, and 80 rpm is seen in Figure 4.4. Higher rotation speed gives increased power, and at 80 rpm the peak output has increased from just under 60 to nearly 140 kW.[1] Due to geometric similarity peak power will always occur at the same tip speed ratio and is related to rotor speed Ω by

$$P_{\max} \propto \Omega^3 \tag{4.5}$$

Note that maximum power P_{max} does not correspond to maximum efficiency C_p^{max}. The latter occurs at a lower wind speed, on the steepest part of the power curve. By the time P_{max} is reached the power coefficient has already fallen from its optimal value, and continues to do so with rising wind speed

[1] Dimensioned power curves and C_p, λ curves are essentially reflected images: low wind speed corresponds to high tip speed ratio, and vice versa.

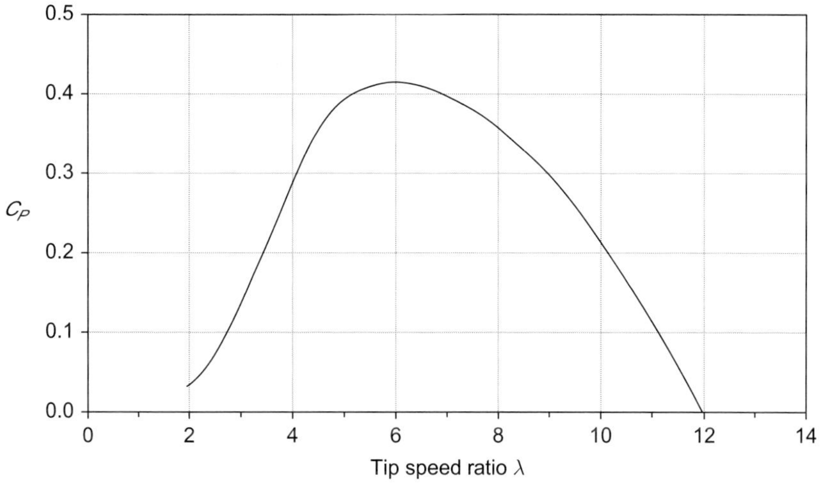

Figure 4.3 Dimensionless C_p, λ curve for a fixed-pitch rotor. The curve is a unique characteristic of the design, and from it the power output can be calculated for any combination of wind and rotor speed.

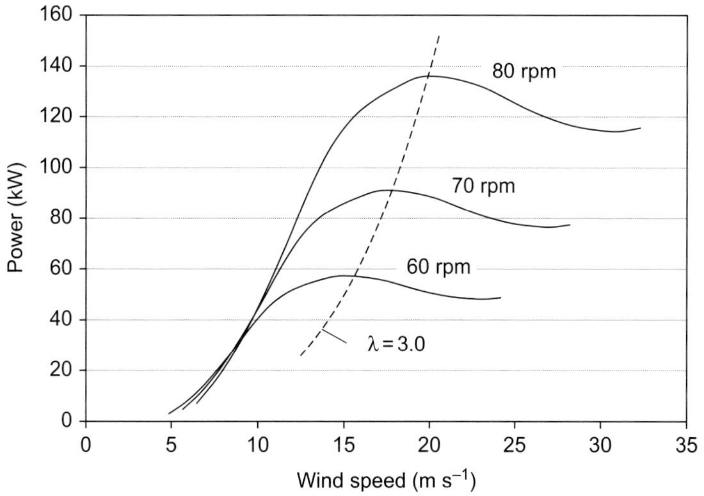

Figure 4.4 Power curves derived from the C_p, λ curve shown in Figure 4.3 by operating at different rotor speeds. Peak power is proportional to Ω^3 and occurs at the same tip speed ratio, but does not correspond to C_p^{max}.

(which for a fixed-speed rotor corresponds to falling tip speed ratio λ). As noted earlier, aerodynamic efficiency becomes less important in high winds.

The C_p characteristic shown in Figure 4.3 was derived from measurements on a stall-regulated wind turbine, and the relatively modest C_p^{max} of 0.42 is not unusual for this type of machine; the design tip speed ratio of 6.0 is also typical. Stall-regulated rotors operate at fixed speed and are designed to lose aerodynamic efficiency in high winds as a means of power limiting (see Section 6.3.1). By comparison, a variable-speed, variable-pitch, wind turbine is designed for high efficiency up to the rated wind speed when power limiting begins; a design tip speed ratio of 8.0 and $C_p^{\max}$ exceeding 0.50 are not uncommon. Once pitch control begins the rotor C_p, λ characteristics are actively changed to manipulate power output and/or rotor speed: this topic is discussed in Section 6.3. The tip speed ratio at which a rotor achieves $C_p^{\max}$ is a key feature of its design and is intimately related to the rotor solidity, as now discussed.

4.4 Tip Speed Ratio and Solidity

Comparing early wind turbines (or historic windmills) with modern high-speed HAWTs we see an inverse relationship between rotor solidity σ and design tip speed ratio λ: faster running rotors have more slender blades. This is illustrated in Figure 4.5, which contrasts the 22 kW design of Christian Riisager from 1975 with the 8 MW Vestas V164 from 2014. The Riisager machine was designed for constant-speed stall-regulated operation, with design tip speed ratio of around 4.0 and rotor solidity of 10%; the comparable figures for the V164, which has variable speed and pitch, are design tip speed ratio of 8.0 and 4% solidity.

The relationship between tip speed ratio and solidity can be explained by considering the optimal blade planform for an N-bladed rotor. Referring to Equation (3.19) and substituting $a = 0.33$ we specify the conditions for maximum power extraction; assuming the same lift coefficient C_l' applies at all points along the blade, the optimum chord length at radius r is found from:

$$c(\lambda, r) = \frac{16\pi R^2}{9NC_l' r\lambda^2} \tag{4.6}$$

The optimal chord is then seen to be inversely proportional to both r and tip speed ratio squared. An ideal blade will thus exhibit highly nonlinear taper from root to tip, and the greater the design tip speed ratio, the more slender the blade. In practice real blades rarely adhere to this formula. Towards the hub Equation (4.6) predicts infinite chord length and the real planform is a compromise dictated primarily by strength considerations (the proportion of power produced at the inner blade is in any case small). Most wind turbine types have used blades with linear or near-linear taper as an approximation to the optimum planform. Notable exceptions, however, are the 1980s designs of Jay Carter in the US, and the more recent Enercon models, which feature highly nonlinear taper and a large root chord. Some old, and some more recent, blade planforms are compared in Figure 4.6.

A further factor influencing the optimal blade shape is the method of power control. Because stall-regulated wind turbines operate at fixed speed the blade planform must be optimised for a range of tip speed ratio, and Equation (4.6) is no longer simply applicable. For such rotors the relationship between optimum solidity and tip speed ratio may be closer to one of simple inverse proportionality (Jamieson and Brown, 1992). Some designs such as the Lagerwey 80 kW two-blade wind turbine had almost untapered blades, apparently far from optimal, yet this early wind turbine

Figure 4.5 High and low solidity. The Windmatic 22 kW (a) designed by Christian Riisager in 1975 had 10% solidity and a design tip speed ratio of 4.0. The Vestas V164-8MW (b) has solidity of 4% and design tip speed ratio of 8.0. (V164 photo courtesy Vestas Wind Systems A/S)

was extremely successful: its cost-effective and robust rotor sacrificed high C_p in favour of increased diameter, illustrating that aerodynamic efficiency is not the only variable to be considered when the ultimate goal is reduced cost of energy.[2]

4.5 Blade Twist and Pitch

Looking along a wind turbine blade from the tip towards the root, the blade is seen to be twisted, with the aerofoil profiles along its length set at increasingly 'nose down' angles (Figure 4.7). The angle of the local chord line relative to a fixed reference at the tip is the *twist*, or structural variation in pitch along the blade; the symbol θ usually applies. Twist compensates for the inherent change of inflow angle ϕ due to the variation in tangential velocity ωr along the blade. At the tip the tangential velocity is at a maximum and under normal running conditions ϕ is typically around 5°: this is an ideal angle for aerodynamic efficiency so no twist is needed and the chord line at the tip lies in the rotor plane. Towards the hub, however, ϕ tends to 90° and without twist the

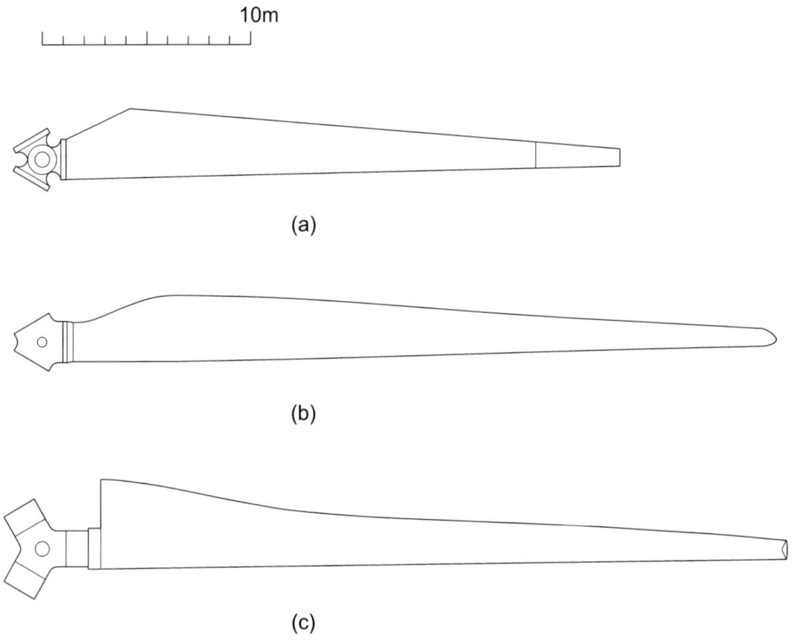

Figure 4.6 Blade planform variations: (a) Howden HWP55 (1988) with linear taper, designed for constant-speed operation and partial-span pitch control; (b) Aerpac APX70 (1997) with non-linear taper, for variable-speed operation at design tip ratio 7.5; (c) Enercon E-70 (2003) with near-optimal taper, design $\lambda = 8$.

[2] The Lagerwey machines were also among the earliest HAWTS with broad-range variable speed.

blade would be heavily stalled over its inboard stations. For optimum performance each blade section is set to achieve an incidence of typically 5°–6°, corresponding to the maximum lift/drag ratio of the aerofoil profile. Referring to Figure 4.7, the general relationship between inflow angle ϕ, incidence α, and twist θ is

$$\theta = \phi - \alpha \tag{4.7}$$

The required twist θ at a given location along the blade is found by subtracting the design incidence α_0 from the local inflow angle, which is determined by design tip speed ratio λ. The following expression can then be derived for optimum twist (Jamieson, 2011):

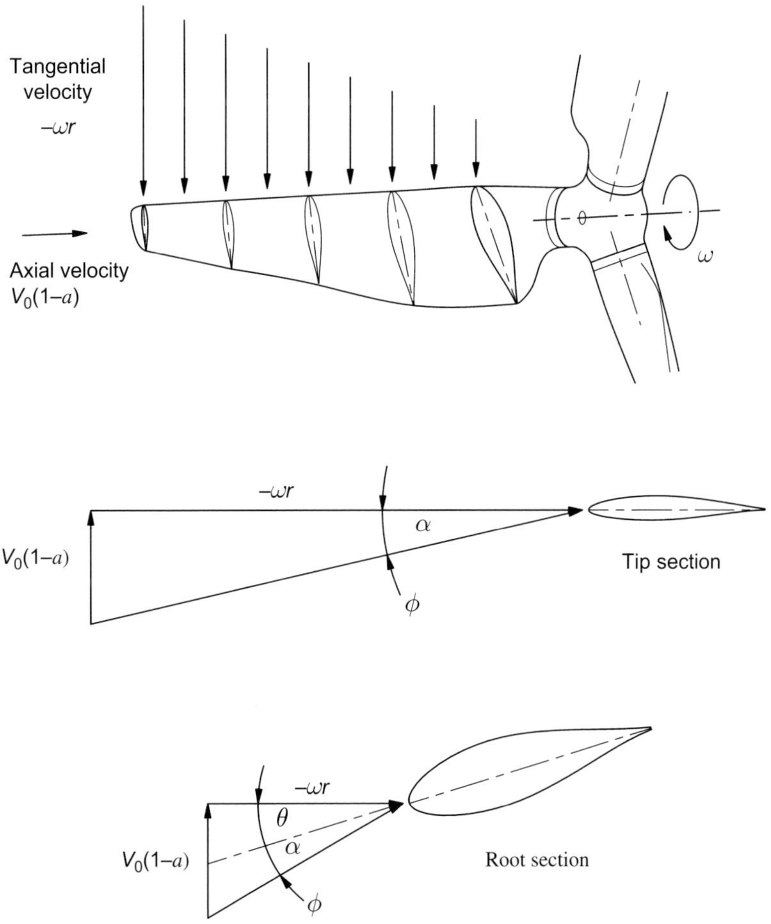

Figure 4.7 Blade twist. At stations towards the root, higher twist θ compensates for the increase in flow angle ϕ, thereby maintaining ideal incidence α. Note that the tangential induction factor is here neglected for simplicity.

$$\theta(x) = \frac{2}{3\lambda x} - \alpha_0 \text{ (radians)} \tag{4.8}$$

where $x = r/R$ and α_0 is the angle of incidence corresponding to maximum lift/drag. An optimal blade design is thus both highly twisted and tapered. As a consequence it may not be practical to manufacture and most real blades compromise at the root section, without excessive chord or twist, on the grounds that energy output at the inner radii is small. Nevertheless some optimised blades have achieved commercial production (see above).

The *pitch* angle is the angle at which the entire blade is set at the hub, on the assumption that it can be varied either actively – most large HAWTs are now equipped with variable-pitch mechanisms – or (in the case of stall-regulated turbines) at the design stage, to finely adjust the rotor peak power level. Pitch angle is usually denoted β, and has the same influence on the blade aerodynamics as twist θ; see Figure 4.7. For simplicity only θ is shown in the figure, but if the blade were to be pitched for power control or air braking the effective pitch angle becomes the sum of θ and β; the relationship governing incidence and hence lift is then

$$\alpha = \phi - (\theta + \beta) \tag{4.9}$$

The topic of blade pitch control is discussed in more detail in Chapter 6.

4.6 Blade Number

Optimum solidity can in theory be achieved with any number of blades (see Equation (4.6)) but practical considerations including structural strength favour a small number of wide-chord blades rather than very many slender ones. Large commercial wind turbines are now almost exclusively three-bladed, and the principal reasons for this dominance are

- tip aerodynamic loss decreases with blade number (see Section 3.6.5)
- three-blade rotors have some advantages in regard to pitch control bandwidth and dynamics
- visual appearance

The difference in $C_p^{\max}$ between good two-blade and three-blade designs is in practice quite small, and is not a significant factor in terms of economic design. Three-blade machines have better pitch control bandwidth for constant-speed power control (see Section 6.3.2) but this advantage may not extend to variable-speed operation. The aesthetic argument, however, should not be underestimated, and a common (though subjective) impression is that three-blade rotors rotate more smoothly than two-blade, making them easier on the eye. Nonetheless a number of two-blade designs have achieved commercial success at small to medium scale, and with the removal of visual impact considerations some advocate the use of very large two-blade HAWTs

Figure 4.8 Monopteros: a single-bladed 640 kW wind turbine. Photographed on a foggy day at the DEWI test site, Wilhelmshaven, 1992.

offshore on grounds of reduced drivetrain torque, lower gearbox weight, and lower overall cost (de Vries, 2013). The weight argument is not, however, universally accepted (see Jamieson, 2011).

One configuration that is no longer commercially pursued is the single-blade wind turbine, although machines up to 640 kW rating have successfully run: Figure 4.8 shows the MBB 'Monopteros' M50 at the DEWI Wilhelmshaven test site in 1992. This wind turbine had an advanced CFRP blade and the rotor weight of 12.5 t was relatively low, albeit the need for a blade counterweight limits the weight advantage over a more conventional two-blade machine. Very low solidity has its drawbacks, however, and the maximum tip speed of the Monopteros was 126 m s^{-1}, causing severe rotor noise that limited the commercial potential of the design. The Italian Riva Calzoni company developed a smaller version of this machine, the M30, rated at 225 kW (Dalpane et al., 1988); its tip speed was significantly lower and the turbine saw some commercial success.

4.7 Rotor Aerodynamic Control

The aerodynamic torque produced by a HAWT rotor must be controlled in some way to (a) limit the power output and structural loading in high winds and (b) bring the rotor to rest quickly in the event of an emergency. There are different ways of achieving these aims, of which full-span blade pitch control is now the most common. Older wind turbine types with fixed-pitch rotors relied on the passive technique of stall regulation for power control, and air brakes for stopping. The control philosophies of pitch control and stall regulation are compared below.

4.7.1 Stall Regulation: Fixed Pitch

Any HAWT rotor, if operated with fixed blade pitch and constant rotation speed, will be inherently stall regulated. The power curve will rise to a maximum at a particular wind speed above which it will level off and ultimately fall, as seen in Figure 4.4. This behaviour is simply a reflection of the rotor C_P, λ characteristic in the context of fixed-speed operation: as wind speed increases so tip speed ratio λ is decreased, with aerodynamic efficiency (C_P) falling sharply as the rotor blades enter the stall regime. The great advantage of stall regulation as a means of power limiting is its mechanical simplicity, with no requirement for blade pitch bearings or actuators, and the technique lent itself very well to early commercial wind turbine designs based on fixed-speed induction generators (see Section 5.4). At one time the majority of wind turbines in production in the US and Europe were of this type, with output ratings of typically 50–100 kW, and in large numbers these formed the basis of the first utility-scale windfarms (see Figure 1.2).

The principle of stall regulation can be explained with reference to Figure 3.12. As the ambient wind speed V_0 increases so too does inflow angle ϕ: this is a consequence of fixed tangential velocity (ωr). Lift on the blade increases until the aerofoil stalls (Figure 3.11) after which a combination of reduced lift and increasing drag acts to limit the rotor power. Despite being a somewhat inexact method of power control stall regulation was highly successful and ultimately used on wind turbines up to 1.3 MW rating. Its disadvantages are a need for relatively heavy rotor blades, and a power curve with modest gradient and low C_p^{max}; these factors stem from the need to operate at relatively low rotation speed to ensure that stall occurs at a manageable power

Figure 4.9 Hub and generator of an E44 wind turbine prior to blade installation. Note the geared electric pitch motors: there is a separate pitch motor for each of the three blades. (Photo reproduced with kind permission of Charlie Robb)

level. At larger scale (above ~450 kW) stall-regulated WECs were found to exhibit unfavourable blade vibration characteristics due to low aerodynamic damping, and this was another factor that ultimately prevented their development to multi-MW size (see Section 7.7.3).

4.7.2 Pitch Control

Nowadays most large HAWTs incorporate full-span pitch control, where the entire blade can be rotated about its long axis by a hydraulic or electric actuator mounted in the hub. Figure 4.9 shows the hub of an Enercon E44 wind turbine, with a separate geared electric pitch drive per blade. Active pitch control confers the ability to start and stop a wind turbine smoothly, and regulate its power output accurately in high winds; the same technique is employed to control aircraft propellers, having first been demonstrated early last century (Flight, 1921). The airflow geometry applicable to a variable-pitch HAWT blade section is illustrated in Figure 4.10 (for simplicity the section shown is untwisted), with the following three operating conditions shown:

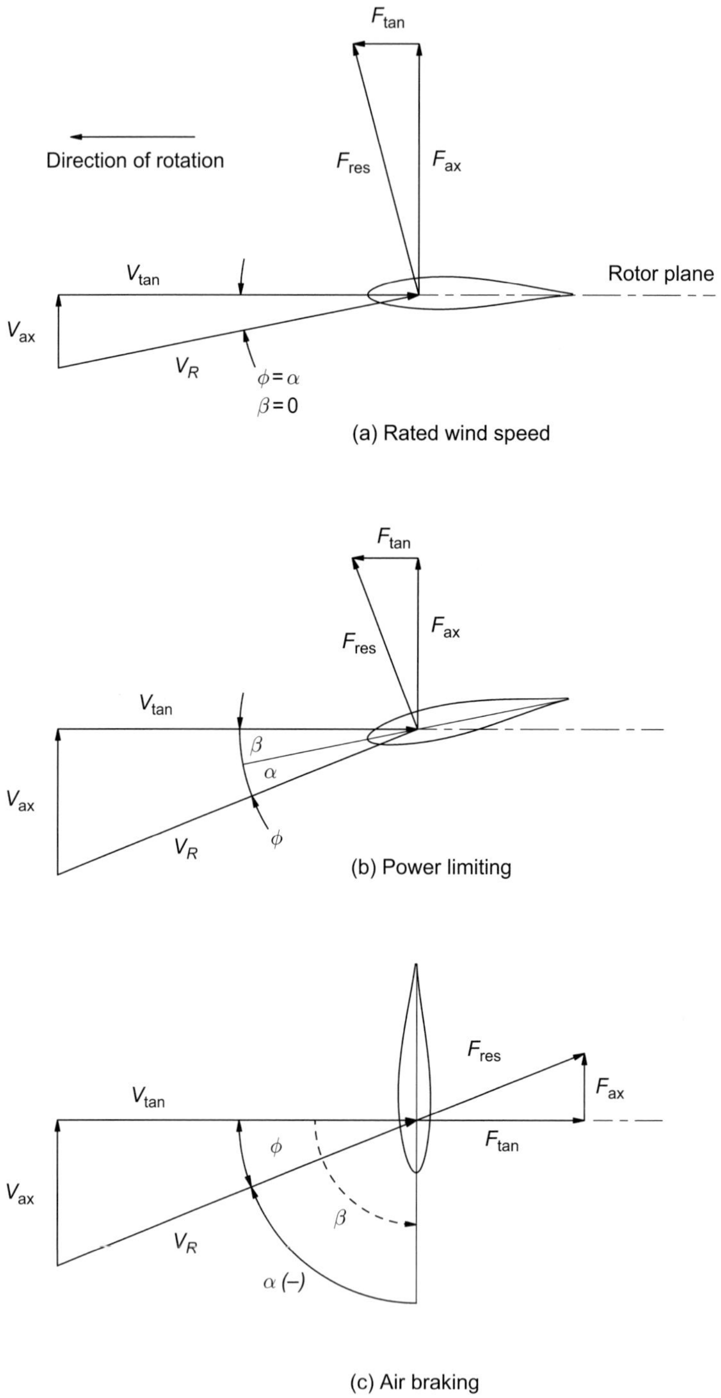

Figure 4.10 Pitch control regimes: (a) at rated power, just before the onset of pitch action; (b) power limiting in high winds; (c) at maximum pitch angle, with the blade acting as an air brake. In cases (a) and (b), the resultant aerodynamic force F_{res} is mainly lift; in (c), it is almost wholly drag.

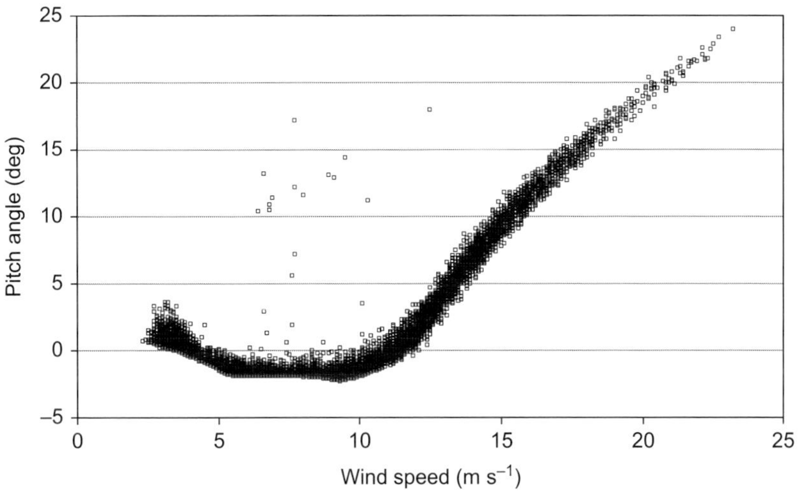

Figure 4.11 Measured pitch control curve for a variable-speed, variable-pitch wind turbine. Power limiting begins at approximately 12 m s^{-1} (hub height wind speed).

(a) at the onset of rated wind speed, just below the threshold for pitch action. The blade tangential force $F_{\tan}$ is at its limiting value, corresponding to rated power.

(b) above rated wind speed. The blade is now pitched nose down, with pitch angle β made between the chord line and rotor plane. As a result the incidence α is reduced, so that $F_{\tan}$ is held at its limiting value (note the corresponding decrease in F_{ax}). As the ambient wind speed increases, pitch angle will be progressively increased to maintain constant power: a typical pitch curve characteristic is shown in Figure 4.11.

(c) air braking, where the blade is pitched by 90° into a high drag configuration, strongly resisting tangential motion. With the generator disconnected this configuration will bring the rotor to a stop within a few revolutions, even in high winds. Wind turbines parked with their blades at full pitch are shown in Figure 4.12.

Power limiting may be accomplished either by positive pitch control, where α is reduced with increasing wind speed, or negative pitch (active stall) where α is increased. Positive pitch control is more common, and more or less standard on very large HAWTs. The effect of positive pitch change on the rotor characteristics is shown in Figure 4.13: increasing the pitch angle reduces $C_P^{\max}$ and shifts the C_P, λ, curve to lower tip speed ratio, so that higher winds are required to achieve a particular power output. Positive pitch control reduces the magnitude of the blade lift force as the wind speed increases.

In the alternative strategy of active stall or negative pitch control, pitch angle β is reduced with wind speed, corresponding to nose-up rotation of the blade section shown in Figure 4.10, so that α

Figure 4.12 Parked rotors. These machines are stopped with their blades fully pitched to 90° in the braking position. (Siemens 2.3 MW turbines at Whitelee windfarm)

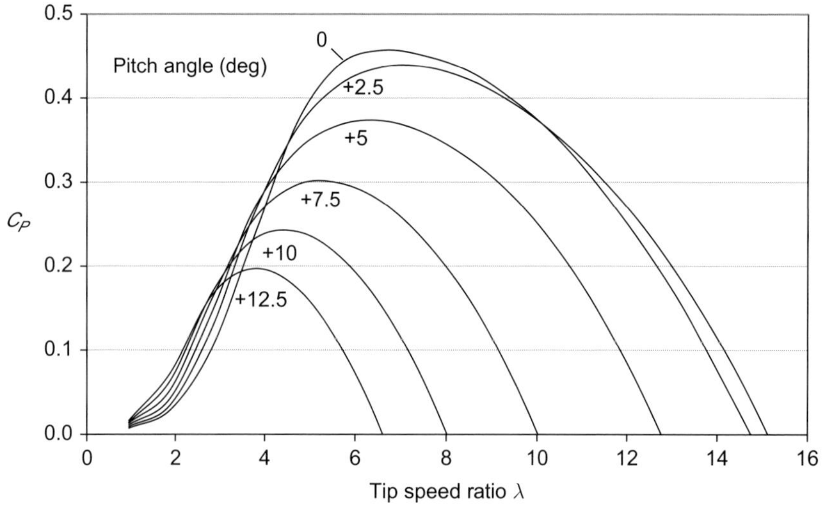

Figure 4.13 Effect of pitch change on C_p, λ curve: positive pitch control.

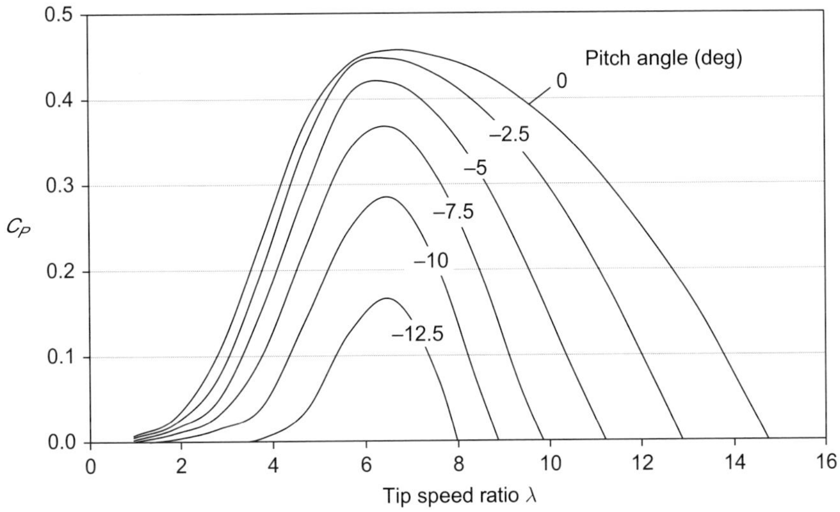

Figure 4.14 Effect of pitch change on C_p, λ curve: negative pitch control (active stall).

increases and the blade is pushed into stall. The change in the rotor characteristics is then as shown in Figure 4.14. There is again a reduction in $C_P^{\max}$, but in contrast to the positive pitch case no sideways shift of the curve in relation to λ: this reflects a rise in drag rather than reduction in lift. Although active stall is used on some large wind turbines (e.g. Siemens 1.3 MW) it is less common than positive pitch, and is implemented on a relatively slow control loop with periodic rather than continuous pitch change. Power control via stall also results in increased aerodynamic noise and reduced blade damping.

A comparison of power curves for two wind turbines of similar diameter and rating, one pitch-controlled and the other stall regulated, is shown in Figure 4.15. The power curve of the pitch-controlled machine is characterised by a steeper gradient below rated wind speed, and a flatter characteristic above it. As wind turbines spend most of their lives in mid range wind speeds pitch control can yield 5%–10% more annual energy than stall regulation. Above the rated wind speed pitch control also limits power more accurately, at least on average (though power fluctuations about the mean may be greater in some cases; see Section 6.3). Some early wind turbines incorporated partial-span pitch control, where only the outer part of the blade was pitched: examples are seen in Figure 7.24 and Figure 8.2. Although this configuration provided the same level of control and air braking as full-span pitch, practical difficulties were associated with locating an actuator in the outboard blade, and the tip attachment spar was subject to high fatigue loading during pitch activity. With partial-span pitch the fixed part of the blade behaves as a stall-regulated rotor and peak power overshoots are in principle more limited than with full-span pitch (see Section 6.3), but in practice this advantage was ultimately outweighed by the factors noted above (Anderson and Jamieson, 1988).

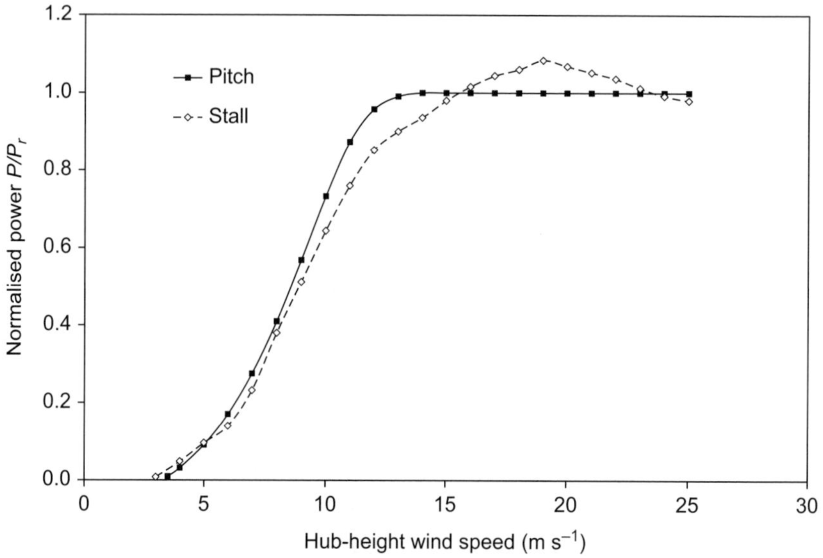

Figure 4.15 Power curves for pitch-regulated and stall-regulated wind turbines of comparable size. Pitch regulation enables greater output in low winds and more accurate power limiting in high winds. (From published performance curves for Vestas V29-225 and Nordex N29-250 wind turbines)

4.7.3 Other Aerodynamic Control Devices

Full-span variable pitch is now standard for large HAWTs, but other aerodynamic control devices have been employed in the past, typically on smaller-scale machines. These include spoilers, which are narrow rectangular control surfaces mounted spanwise in the suction surface of the blade, and raised into the airflow by actuators to ‘spoil’ the lift in a form of active stall. Adapted from sailplane technology, spoilers are relatively crude control devices for WEC power regulation and lack the strong braking capability of pitch control, making them less suitable at large scale. Examples of stall-regulated wind turbines equipped with spoilers included the Danish Windmatic (Figure 4.5) and British Windharvester 60 kW machines.

Hinged vanes at the blade tips have also been used as air brakes on stall-regulated wind turbines such as the Atlantic Orient AOC15/50; see Figure 4.16. Under normal running conditions the vane is held in a low drag configuration (as seen in the photo) by an electromagnet: on loss of the grid the holding force disappears and the brake opens under centrifugal force, generating high drag tangential to the blade motion. When the rotor speed reduces sufficiently the air brakes automatically retract under the action of springs (Hughes et al., 1993). The system is not used for power or speed control. The Dutch ‘Polenko’ 60 kW stall-regulated wind turbine had similar tip brakes, in this case actuated via push rods linked through the hub from an actuator in the nacelle; although their main function was emergency braking, these devices were also used to control rotor speed

Figure 4.16 Vane-type tip brakes on the AOC15/50 wind turbine. The brake plate at the blade tip is held in place by an electromagnetic latch and deploys automatically under the action of centrifugal force.

during the run up to synchronisation. Blade tip vanes are efficient air brakes, but tend to increase rotor aerodynamic noise.

Ailerons or flaps mounted on the blade trailing edge were also trialled on some US wind turbines. These devices are analogous to the control surfaces on an aircraft wing, where they enable lift to be modulated or reversed. Ailerons were tested on the US Boeing MOD-0 experimental wind turbine (shown in Figure 4.17) in the early 1980s (Miller, 1986) and later on the North Wind 250 (Link, 1995), in both cases for power control and air braking. Compared with variable-pitch blades, ailerons have the advantages of allowing a simpler blade root design and requiring smaller actuators. They lack the definitive braking capability of variable pitch, however, and introduce mechanical complexity into the blade structure. Like partial-span pitch surfaces, they are no longer used.

4.8 Downwind Rotors

A downwind HAWT rotor is identical to an upwind one in every respect, save that it is mounted behind the supporting tower rather than in front, so the wind blows over the nacelle before it reaches the rotor. The main advantages of the downwind configuration are as follows:

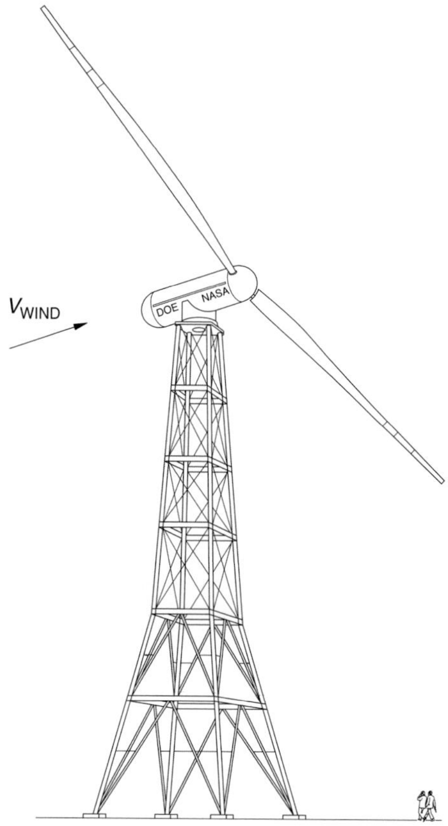

Figure 4.17 Downwind rotor: the 200 kW MOD-0 experimental wind turbine from 1978. Downwind rotors experience greater tower shadow than upwind, though lattice towers helps to alleviate the effect. The MOD-0 was also used to test blade aileron control surfaces.

- The rotor is directionally stable, and automatically turns to face the wind with no need for an active yaw mechanism.
- Aerodynamic thrust loading bends the blades away from the tower rather than towards it, facilitating lighter and more flexible rotor designs.

Several large downwind prototypes were built by commercial engineering companies in the US under a DOE/NASA-sponsored research programme in the late 1970s. These included the Boeing MOD-0 200 kW machine (Figure 4.17) and later MOD-1, which was the first 2 MW wind turbine; an even larger downwind type was the WTS-4, which was built by a consortium of US and Swedish companies and rated at 4.2 MW. These turbines were innovative in many ways, and were used e.g. to explore different blade construction methods and the use of variable-speed generators (Linscott et al., 1984), but they also highlighted some disadvantages of the downwind configuration that limit its appeal at large scale. Despite the free-yaw

Figure 4.18 Proven 600 W free-yaw downwind machine. The streamlined nacelle shroud helps to eliminate tower shadow. (Photo reproduced with kind permission of Charlie Robb)

capability, motors are still required to turn the nacelle to correct for twist in the down-tower electricity cables. More significantly, downwind rotors can be a source of severe noise due to the effect of tower shadow.[3] The wind velocity immediately behind the tower is reduced, and the blades are subject to an impulsive pressure change that can cause an audible 'thump' heard at distance; the problem is particularly acute with tubular towers, and the WTS-4 was somewhat notorious in this respect.

At smaller scale these drawbacks are more easily overcome and there have been a number of successful downwind machines. These generally have lattice towers that present a much lower blockage to the wind than tubular designs, lessening the tower shadow effect and reducing impulsive noise. Alternatively a streamlined shroud is attached under the nacelle, allowing the air to flow smoothly round the tower without creating an area of separated flow behind it; this solution is favoured on some kW-scale wind turbines such as the Proven machine shown in Figure 4.18. Small downwind turbines do not require yaw motors: most have tower-top slip rings so cable twist is avoided; on some larger machines cable twist is corrected by manual rotation of the nacelle on an occasional basis.

[3] 'Tower shadow' is also used to refer to the reduction in wind velocity ahead of the tower; it is more literally correct in the context of downwind rotors, where the effect is much more significant.

4.9 Exercises

4.9.1 Power and Torque

A wind turbine has a rotor diameter of 71 m, rotational speed of 20 rpm, and output of 2.3 MW in a hub height freestream wind speed of 13.0 m s^{-1}. Calculate the resulting (a) power coefficient, (b) tip speed ratio, and (c) rotor shaft torque. Ignore losses, and assume air density of 1.225 kg m^{-3}.

4.9.2 Maximum Power (Stall Regulated)

A fixed-speed stall-regulated wind turbine was designed for peak power output of 60 kW at rotation speed of 55 rpm. When the prototype was run, however, it produced a maximum output of 70 kW in a wind speed of 13 m s^{-1}. To reduce the power, the designers modified the wind turbine to operate at a lower rotation speed: calculate (a) the rotor speed needed to achieve 60 kW peak power and (b) the wind speed at which this would occur. Additionally (c) what physical change to the wind turbine might be made to reduce its rotor speed?

4.9.3 Rated Power Reduction (Pitch-Controlled Rotor)

An 850 kW pitch-controlled wind turbine has its peak rating reduced by 17.7% to 700 kW to comply with local grid limitations, by lowering the power control set-point. The reduction in annual energy output is around 6%. Why is the effect not greater? Explain your answer clearly.

4.9.4 Blade Twist

A wind turbine blade is designed for optimum performance at a tip speed ratio of 7.0. If the blade tip is set at a twist angle of 0.5°, what should the corresponding twist angles be at radial locations of (a) 50% and (b) 25% of tip radius?

4.9.5 Solidity

What is the approximate solidity of the two-blade rotor referred to in Exercise 3.10.4? (Ignore the hub dimensions). If this turbine were instead to have three blades, but retain the same rotation speed and power output, what should be the new blade chord length? Suggest why this might be a challenge for the blade designer.

4.9.6 Optimum Chord

From blade element momentum theory the governing equation for the axial thrust on an arbitrary blade element is

$$\frac{a}{(1-a)} = \frac{\sigma C_l}{4\,\sin\phi\tan\phi}$$

Starting with the above equation, derive the relationship for the optimum chord distribution for a HAWT blade: you will find it in Equation (4.6). Show your working at each stage. Ignore tangential induction factor and make use of the small angle approximation: $\sin\phi \cong \tan\phi$.

4.9.7 Pitch Braking

When executing an emergency stop the blades of a large upwind HAWT are rapidly pitched to a high drag configuration. At one point during the pitch action the blades are seen to bend strongly away from the tower, i.e. in the upwind direction. Explain (in words) why this happens, and at what point during the pitch change the effect is seen.

CHAPTER 5 ELECTRICAL ASPECTS

5.1 INTRODUCTION

Grid-connected wind turbines are essentially power stations, and whether singly or in large arrays they must meet the same electrical specifications and standards as more conventional generating plant. Operating unmanned, they must provide a stable electricity supply and remain safe in the event of a network fault, lightning strike, or mechanical failure. There are, however, several key differences between wind turbines and conventional power stations. In the former case the power output is intermittent depending on the ambient wind speed, and the output power level of a wind turbine cannot always be controlled. In addition a wider range of generator types is used in wind turbines, conferring on them different dynamic and electrical characteristics: not all are equally 'grid-friendly'. The type of generator also influences how a wind turbine operates: for instance the speed at which a HAWT rotor rotates is dictated largely by the generator, not the wind. This chapter reviews these and associated electrical issues at a broad level;[1] some basic electrical principles are first revisited due to their underlying importance.

5.2 FUNDAMENTALS

In any electrical circuit – AC or DC – the instantaneous voltage drop across a resistive element is proportional to the current flowing through it according to Ohm's Law:

$$\Delta V = IR \tag{5.1}$$

where ΔV is voltage drop, I current, and R resistance. The instantaneous power dissipated is P, where

$$P = I\Delta V = I^2R = \Delta V^2/R \tag{5.2}$$

In an AC circuit the voltage varies sinusoidally, with cycle frequency of 50 Hz or 60 Hz used in most electrical power systems. The instantaneous voltage V_i at time t is given by

[1] For more detailed electrical studies, the reader is recommended Freris (2008) and Jenkins (2000), or on the topic of large-scale networks and grid integration (Ackermann, 2012).

$$V_i = V_p \sin(360ft)^\circ \tag{5.3}$$

where V_p is the peak voltage, and f the sinusoidal frequency. When analysing AC power systems, the voltage V generally referred to is the root mean square (RMS) voltage V_{rms}, where

$$V = V_{\mathrm{rms}} = V_p/\sqrt{2} \tag{5.4}$$

Similarly, current I is the RMS value:

$$I = I_{\mathrm{rms}} = I_p/\sqrt{2} \tag{5.5}$$

where I_p is the peak current. The voltage and current waveforms in AC circuits may be subject to a phase difference ϕ expressed in degrees, where one cycle represents 360°. In this case, for instantaneous supply voltage given by Equation (5.3) the corresponding current is

$$I_i = I_p \sin(360ft + \phi)^\circ \tag{5.6}$$

In a purely resistive circuit the current and voltage are in phase and $\phi = 0$. Other circuit elements, however, introduce a phase shift and the terms 'lead' and 'lag' are often used. Across an inductive component the voltage waveform leads the current by 90°; inductive components commonly have some form of windings that store magnetic energy, for instance motors or transformers, although overhead lines also possess inductance. Across a capacitive element the voltage lags 90° behind the current: capacitive elements store charge due to the dielectric properties of insulators, for example in polymer-insulated cables.

All real AC circuits combine resistive, inductive, and capacitive elements, and as a result voltage and current are in general shifted by a phase angle ϕ in the range ±90°. The 'total' or 'apparent' power S flowing from a source to a load in an AC circuit is given by

$$S = VI \tag{5.7}$$

where V is the RMS voltage across the load, and I is the RMS current flowing into it. The 'real' power P is the product of the in-phase components of voltage and current or

$$P = S \cos\phi \tag{5.8}$$

The 'reactive' power Q is the product of the out-of-phase components:

$$Q = S \sin\phi \tag{5.9}$$

These relationships are illustrated by the vector triangle in Figure 5.1. The power factor (often abbreviated pf) is defined as the ratio of real to apparent power:

$$\mathrm{pf} = P/S = \cos\phi \tag{5.10}$$

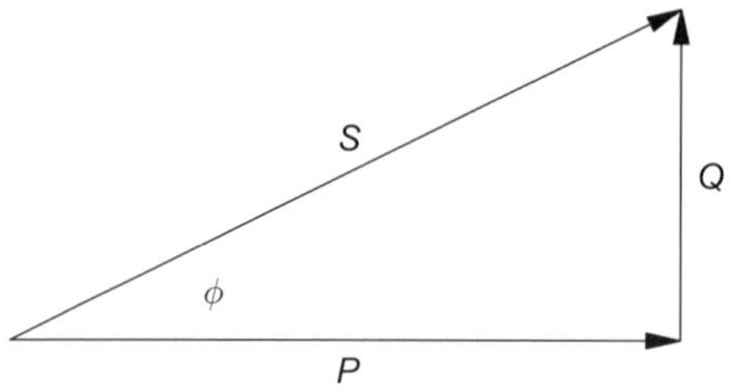

Figure 5.1 Vector relationship of power in an AC circuit: P is real, Q reactive, and S total or apparent power. Power factor is $\cos\phi$, where ϕ is the phase angle between voltage and current.

In practical terms P is useful power delivered and converted, for example, into heat by a resistor or mechanical work done by a motor. Reactive power Q is stored within the circuit as oscillating electromagnetic energy but does no net work, rather like the energy stored in a swinging pendulum. The apparent power S is the resultant of P and Q, with magnitude in general greater than P; apparent power is then an important quantity in power engineering as it dictates the maximum current that will flow in a circuit, hence the thermal rating of its components (cables, circuit breakers, fuses, etc.). The useful power P delivered over time Δt results in the transfer of a quantity of energy E given by

$$E = P\Delta t \tag{5.11}$$

The energy transferred from source to load in an AC power system can therefore be expressed as

$$E = VI\Delta t \cos\phi \tag{5.12}$$

Most distribution networks operate on three-phase AC, utilising three conductors at equal voltage phased 120° apart. In this way the supply can be balanced without the need for a return wire, and a three-phase network can transmit twice as much power per length of conducting wire as a single-phase network (which requires two wires). The apparent power carried by a three-phase circuit is given by

$$S = \sqrt{3}V_L I_L \tag{5.13}$$

where V_L is the RMS voltage measured between any two of the phases (aka the line voltage), and I_L is the RMS current flowing in each conductor (the line current). Real and reactive power are related to S as described above; note that in a balanced three-phase supply the same current magnitude and power factor($\cos\phi$) appear in each of the three conductors. The useful power delivered to a load in a balanced three-phase circuit is

$$P = \sqrt{3}V_L I_L \cos\phi \tag{5.14}$$

and the energy E delivered in time Δt is

$$E = \sqrt{3}V_L I_L \Delta t \cos\phi \tag{5.15}$$

5.3 Measurement and Metering

Energy meters are an essential component of national electricity networks, and are required at both the source and destination of the power in order to record energy flows for commercial reconciliation. At the generator terminals (whether a wind turbine or otherwise) and/or the grid connection point a generation meter accurately measures the quantity of energy delivered onto the network, while individual meters at homes and businesses record consumption. Due to the magnitude of energy transferred in electrical power systems, the smallest unit of measurement is the kilowatt-hour rather than joule (the formal SI unit of energy), where

$$1\ \text{kWh} = 1000\ \text{W} \times 3600\ \text{s} \ = \ 3.6\,\text{MJ} \tag{5.16}$$

In practice the annual output of even a single medium-sized wind turbine will be measured in megawatt-hours (MWh) and windfarms of any size in gigawatt-hours (GWh).

Energy meters integrate power over time based on Equation (5.15) and commonly store the results as half-hourly values indexed against an accurate clock reading. Such meters also record maximum and minimum real, reactive and apparent power flows to inform the commercial contract between generator and grid operator. Accurately calibrated current transformers (CTs) and voltage transformers (VTs) feed representation of the primary current flows and voltages into the meter. Where generators or loads are unbalanced (i.e. the magnitude of current flowing in the three phases differs) CTs and VTs may be fitted on more than one phase to provide accurate results. Energy metering is of less importance in the context of network operation and control, which is dependent on real-time current and voltage measurement rather than integrated values.

5.4 Generators

5.4.1 Introduction

The operating characteristics of a generator depend on its electromechanical design, and a number of different generator types are to be found found on wind turbines, principally

- permanent magnet
- synchronous
- asynchronous (induction)
- doubly fed induction

Wind turbine generators may furthermore be configured for fixed or variable-speed operation, and may be directly driven or geared. The range of possibilities can seem a little daunting, but ultimately all generators obey the same physical laws, including the equations of motion for rotating machinery, and in all cases output power P is given by

$$P = T_G \Omega \tag{5.17}$$

where T_G is the electrical torque developed by the generator,[2] and Ω the angular velocity of the rotating part. It is largely the way in which T_G varies with Ω that differentiates generator types, and the following is a brief overview.

5.4.2 Characteristics of Different Generator Types

Permanent Magnet

The simplest type of generator (at least conceptually) is the permanent magnet or PMG, shown schematically in Figure 5.2. The generator is shown grid-connected, and its stator consists of three wire-wound iron poles connected to a three-phase AC supply, while the rotor comprises a two-pole permanent magnet. Each of the stator poles is connected to a different supply phase, the effect of which is to create a rotating magnetic field whose angular speed n_s is given by

$$n_s = \frac{60f}{(P/2)}\,(\text{rpm}) \tag{5.18}$$

where f is the grid frequency, P the effective number of poles, and n_s the synchronous speed. Assuming a 50 Hz supply the synchronous speed is then 3000 rpm; with double the number of poles (four-pole generator) n_s becomes 1500 rpm. The stator field acts like a virtual rotating magnet, and the tendency of the rotor is to follow it so that when fully aligned there is no lateral force developed between them, hence no torque. If the rotor is externally driven ahead of the stator field a restraining torque is developed between them proportional to the displacement or power angle (symbol δ). Generator power is then the product of this torque and the angular velocity, as per Equation (5.17). Despite the angular difference between them the rotor and external field rotate at exactly the same speed n_s, i.e. the synchronous speed.

Synchronous Generator

A synchronous generator is analogous to a PMG in which the rotating permanent magnet is replaced by an electromagnet, whose excitation field is created by DC current fed via slip rings (or by a small shaft-mounted generator). Synchronous generators can again be multi-pole, and as with the PMG the rotor and stator fields rotate in synchronism at a speed given by Equation (5.18) with torque proportional to power angle δ. A powerful feature of the synchronous machine is the ability to control power factor ($\cos\phi$) by varying the DC magnetising current: this is extremely valuable for supporting the grid voltage, as discussed in Section 5.5, and for this reason synchronous generators are preferred for use in large thermal power stations. They are also used on wind turbines, but nowadays only in the context of variable-speed operation, as the synchronous generator has a 'stiff' torsional characteristic that makes power control difficult at constant speed. Seen in the rotating frame of reference a synchronous generator behaves like a torsional spring, with restoring torque proportional to angular deflection. Synchronous generators are now widely employed on variable-speed wind turbines, which may seem contradictory, but is explained in Section 5.4.3.

---

[2] The usual symbol Q is not used here to avoid confusion with reactive power.

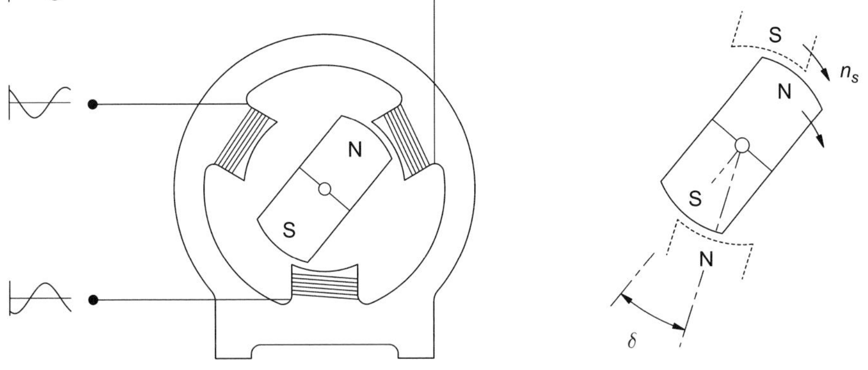

Figure 5.2 Schematic of a two-pole permanent-magnet generator. The three-phase wire-wound stator (left) creates a rotating magnetic field (right) with which the rotor synchronises. Torque is proportional to power angle δ.

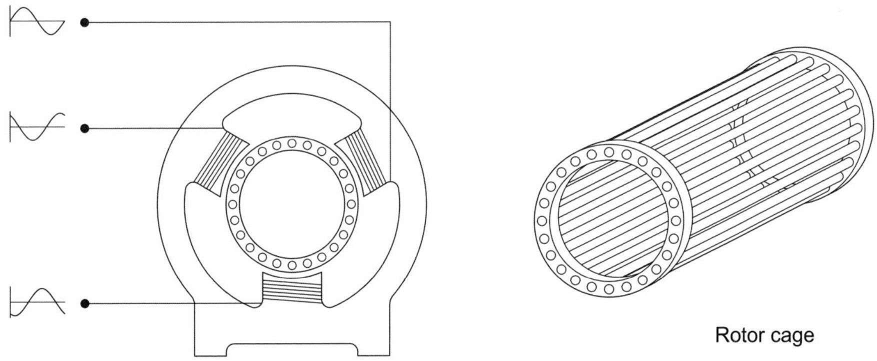

Figure 5.3 Induction (asynchronous) generator. The rotating stator field induces a field in the short-circuited rotor cage. Torque is proportional to the slip speed *s* (Equation (5.19)).

Induction Generator

The induction (or asynchronous*)* generator has a similar stator configuration to the synchronous or PMG machine: grid-connected windings create an external field rotating at n_s and Equation (5.18) again applies. The rotor, however, has no windings or external connection but instead comprises a ring of parallel conducting bars arranged like a circular cage (hence the common name 'squirrel cage') and short-circuited at their ends; see Figure 5.3. The magnetic field on the rotor is created by induction and the resulting torque is proportional to the non-dimensional speed difference, or slip, defined by[3]

[3] For convenience, slip is here defined as positive for $n > n_s$; the normal convention for induction motors (which are physically identical to generators) is the reverse.

$$s = \frac{(n - n_s)}{n_s} \tag{5.19}$$

where n is the speed at which the rotor is being driven. At synchronous speed $s = 0$ and there is zero torque; power is developed by driving the rotor faster than the field speed ($n > n_s$) and full load corresponds to slip typically in the range 0.5%–3%, depending on generator size and design. The slip characteristics of a typical 60 kW induction generator are shown in Figure 5.4. The speed range between zero and rated power is quite small (2.2% in this example), and the generator can be regarded as effectively fixed speed in respect of any change in the wind turbine aerodynamic performance. The power developed by the generator can also be taken as proportional to slip s.

As its torque is proportional to speed, however, the induction generator behaves like a torsional damper, giving it more favourable dynamics than a synchronous machine (which behaves as a spring; see above). This makes the induction machine more 'forgiving' to transient variations in input torque. With absence of rotor windings it is also more physically rugged and for this combination of reasons the induction generator was the preferred choice for the first generation of wind turbines, with many thousands installed. Most wind turbine induction generators are capable of two-speed operation, made possible by a stator design in which the number of active poles can be varied, usually in the ratio 4/6 or 6/8, and for a given grid frequency the result is two possible synchronous speeds, per Equation (5.18). The wind turbine can then be run at a slower speed in light winds, to more closely match the optimum tip speed ratio. When the output power rises above a given threshold pole switching is carried out and the turbine then moves to the higher speed; the generator is briefly disconnected while the speed change is effected.

The disadvantages of the induction generator mostly relate to power quality. The induced rotor current causes a high reactive power demand, and without compensation the power factor may

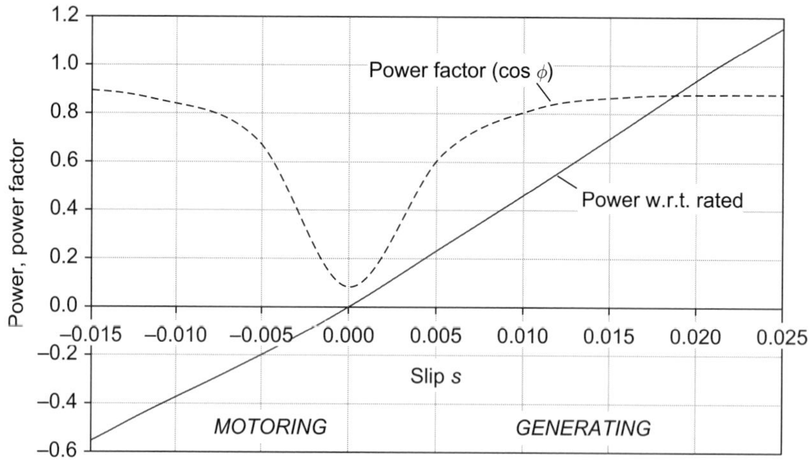

Figure 5.4 Characteristics of a 60 kW induction generator: power (per unit rated) and power factor as a function of slip. On this machine, rated power is achieved at around 2.2% slip.

fall as low as 0.8 at full load (and lower still on part load; see Figure 5.4). The reactive power magnitude may then be 30%–40% of the real power, resulting in high current flows. To combat this tendency power factor compensation is traditionally supplied by fixed capacitor banks. The generator also draws a high transient current (up to 10 times the rated value) when first energised, and some form of current-limiting or 'soft-start' device is required: this may be a simple resistor bank, or a thyristor-switched voltage limiter; in either case the soft start is connected only for the short duration of energisation. Power factor correction and soft-start devices allow simple induction generators to comply with grid regulations, but they lack the flexibility of synchronous machines in regard to power quality.

Doubly Fed Induction Generator

The doubly fed induction generator (DFIG) is a broad-range variable-speed machine. It is a development of the conventional induction generator in which the short-circuited rotor, instead of comprising steel bars, is wire-wound and connected to an external source of variable-frequency AC (a voltage source converter). This solid-state device applies a variable AC voltage to the rotor windings at the slip frequency, allowing the rotor speed range to be extended by ±30% relative to synchronous. The DFIG is shown schematically in Figure 5.5; for a more detailed explanation of the principle of operation see (Fletcher and Yang, 2010). In addition to broad-range variable-speed capability the DFIG enables control of reactive power in a way not possible with a conventional induction machine, and much more like a synchronous generator; DFIGs are therefore 'grid-friendly' generators.

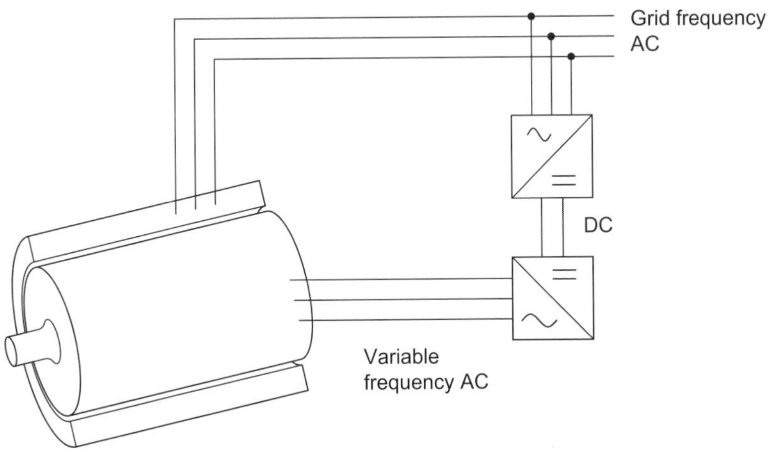

Figure 5.5 Schematic of double-fed induction generator (DFIG). The wound rotor is connected to a variable-frequency AC source, extending the slip range to ±30% of synchronous speed. The stator is directly grid connected.

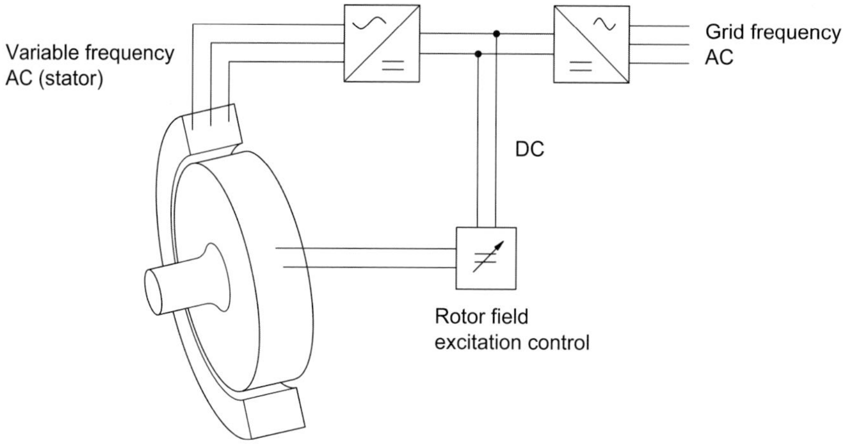

Figure 5.6 Synchronous generator (direct drive) connected via a fully rated converter. The same converter configuration is used with PMG machines, both direct drive and geared.

5.4.3 Variable-Speed Generators

Most large commercial wind turbines, and all those at multi-MW scale, are now designed for broad-range variable-speed operation. Variable speed offers a number of operational benefits, including

- accurate power control in high winds
- improved aerodynamic efficiency and higher energy yield
- lower noise

These advantages are explained in more detail in Chapter 6. The first, accurate power control, can actually be achieved with relatively limited speed variation – typically ±5% about the nominal mean – but for increased aerodynamic efficiency and reduced noise broad-range speed control is required, where the maximum rotor speed may be twice the lowest. In principle any type of generator[4] can be operated at variable speed by connecting it to the grid in series with a fully rated frequency converter, a solid-state device that enables the generator supply frequency to be varied independently of the grid, allowing smooth continuous speed control. This solution is most widely used with synchronous and PMG machines, as shown schematically in Figure 5.6: the figure represents a direct-drive synchronous generator connected via a fully rated converter (as used by manufacturers such as Enercon and EWT).

The frequency converter offers great flexibility not only in the operation of the wind turbine, but also in improved power quality at the point of connection via reactive power control

[4] This includes the conventional induction generator, though this option is uncommon in the wind industry.

(see below). The main disadvantage is cost, as the converter represents an additional expense. The DFIG offers most of the same advantages as the above configuration, but is potentially more cost-effective as its converter is sized for only around 30% of rated power and is therefore proportionally cheaper.

5.4.4 Geared and Gearless (Direct Drive) Generators

A wind turbine rotor may be connected to the generator either directly or via a speed-increasing gearbox. Examples of the two types are shown schematically in Figure 5.7: the Siemens 2 MW platform has a fixed-ratio gearbox, so the generator speed is always proportional to the rotor speed whether the turbine operates at constant or (as in this case) variable speed; the Enercon 2.3 MW machine is gearless, with rotor hub connected directly to a low-speed generator. As can be seen, the two drivetrain philosophies result in visibly different nacelle shapes: the direct-drive generator transmits high torque at low speed, necessitating a large-diameter design; by contrast the geared machine runs at high speed and low torque, so the generator is smaller and lighter. A general rule of rotating electrical machines is that their power output is proportional to the product of physical volume V and rotation speed n:

$$P = KVn \tag{5.20}$$

where K is a constant. As volume roughly equates to weight a direct-drive generator will inevitably be bigger and heavier than an equivalent geared machine, and characterised by a greater diameter and/or length. The lower weight of the high-speed generator may, however, be offset by the gearbox, with the two drivetrain philosophies having similar all-up weights. Direct-drive generators are almost exclusively synchronous or PMG; induction generators and DFIGs are unsuitable as they require a smaller air gap between rotor and stator, which is difficult to achieve at the larger diameters required for low-speed operation (Burton, 2011). Consequently these types are only found on geared wind turbines.

5.4.5 Historic Developments

The small wind turbines that proliferated in the 1980s were predominantly stall regulated, using fixed-speed induction generators of typically 50–100 kW rating. They were simple and robust; the favourable characteristics of this configuration had been demonstrated in the 1950s with the Gedser wind turbine (see Chapter 1) and small induction generators could be mass-produced and purchased off the shelf. At this scale power quality limitations were of less concern to utilities, and generators incorporated fixed capacitor banks for power factor correction; ‘motoring’ a generator up to synchronous speed at start-up was also a common practice.[5] Two-speed operation was an early innovation: initially this was achieved by having two generators, one large and one small, driven off separate

[5] Stall-regulated rotors have low stationary aerodynamic torque, making them difficult to start in low winds. Motoring was a quick and easy solution. An induction motor becomes a generator when driven.

gearbox output shafts with only one generator connected at a time. A more advanced two-speed solution was adopted by the Dutch Polenko company, who by 1980 were using double-wound induction machines (not to be confused with DFIGs) with stator pole switching, a configuration that became standard for fixed-speed wind turbines (both stall and pitch regulated) for the subsequent two decades.

The size of stall-regulated WECs grew progressively, culminating in the Nordex N60 series. Rated at 1.3 MW these widely used machines were the largest commercial stall-regulated HAWTs. Above this size stall regulation becomes uncompetitive due to a combination of lower aerodynamic efficiency (requiring heavier rotors), reduced aerodynamic damping, and inferior electrical power quality. Pitch regulation began to supersede stall in the 1990s, offering higher output and better power control (see Chapter 6). The early pitch-regulated wind turbines continued to use two-speed induction generators, a highly successful example being the Vestas V27 from 1989. On some other types, however, high-wind power control was problematic, with blade pitch systems unable to cope with fast changes in wind speed, and the resulting load excursions causing fatigue damage to blades and gearbox components.

In some cases power control was significantly worse with pitch controlled than stall-regulated wind turbines, and although the inherent damping of the induction generator was beneficial it was not a complete answer to the problem (fixed-speed synchronous generators having been ruled out due to their stiff torsional characteristics). One solution was narrow-range variable speed, initially using a limited-slip induction generator. The Vestas V47-660 was one of the first turbines to incorporate this configuration under the trade name Optislip. In this case the wound rotor is connected in series with a variable resistor, enabling the rotor speed to be varied by around ±10%. This speed flexibility allows the blade pitch control system more time to react to gusts, yielding much smoother power control. Although the resistor generates waste heat it is only connected when power is in the excess; below rated conditions the generator runs at fixed speed as previously.

The potential advantages of broad-range variable-speed generators were understood relatively early[6] but their widespread introduction had to await the advances in solid-state power electronics of the 1990s. This rapidly developing technology not only facilitated speed control, but also improved power quality, which became increasingly important as generators grew beyond megawatt scale. The industry then began to favour more sophisticated WEC designs based on broad-range variable speed in combination with variable pitch (VSVP). The first commercial VSVP wind turbines were geared, among them being the Lagerwey 18/80 kW, and the original Enercon E33.[7] Both these machines were connected via a fully rated AC-DC-AC converter; perhaps significantly the founders of both companies were innovative electrical engineers with a background in power electronics, and both went on to produce direct-drive wind turbines. Other

[6] For a comparison of most of the options that were subsequently used commercially, see Eggleston (1987).
[7] Rated at 300 kW, and not to be confused with the later direct-drive E33-330.

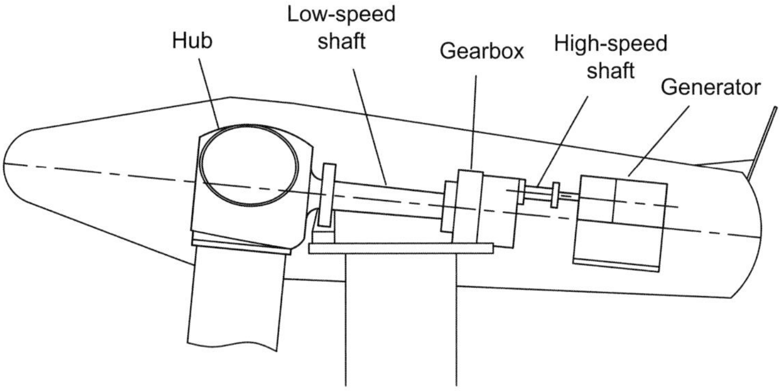

(a) Geared drivetrain (Siemens 2MW)

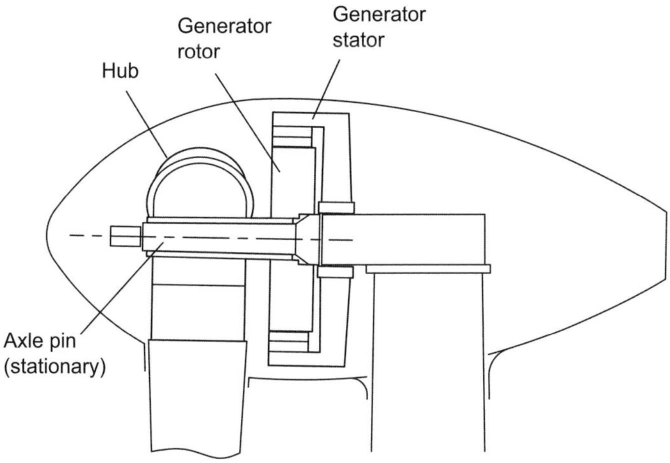

(b) Gearless drivetrain (Enercon 2MW)

Figure 5.7 Geared and gearless. (a) Siemens (Bonus) 2 MW wind turbine with speed-increasing gearbox and DFIG generator; (b) Enercon 2.3 MW with low-speed, direct-drive, synchronous generator (see also Figure 5.8). Both types have broad-range variable-speed capability.

manufacturers opted for the DFIG, which was featured on many wind turbines in the medium power range of approximately 850 kW to 2.5 MW; these included designs from Vestas, Nordex, Siemens, GE, and others. All DFIG machines retain a gearbox.

In 1992 Enercon introduced the gearless E40-500kW, the precursor of a highly successful range of machines based on the concept of a low-speed, multi-pole synchronous generator. The basic configuration, with wire-wound stator and rotor, was hardly new (see Figure 5.8) but the

Figure 5.8 Something old, something new. Multi-pole synchronous generators separated by almost a century: (a) wound rotor for a low-speed US turbo-alternator (Morecroft, 1924); (b) fully assembled 900 kW generator for an Enercon E44, photographed in 2008. Modern power electronics revived the fortunes of the synchronous machine for wind turbine use. (Photo reproduced with kind permission of Charlie Robb)

advent of power electronics allowed it to be employed on a modern wind turbine, combining the benefits of variable-speed operation with superior power quality. Enercon currently produce gearless machines up to 4.2 MW rating based on this configuration. As wind turbine size has increased into the multi-MW range, VSVP control has become the norm, but a variety of generator configurations can still be found. Vestas moved into PMG technology with their 'Grid Streamer' wind turbines, the first developed from the successful V80-2MW machine by replacing its DFIG with a liquid-cooled permanent-magnet generator (de Vries, 2011). The PMG is electrically equivalent to a variable-speed synchronous machine, with fully rated power converter. The Grid Streamer variants, like the DFIG platform from which they were developed, retain a gearbox.

Other manufacturers have dispensed with gearboxes to produce direct-drive permanent-magnet generators that are PMG analogues of the Enercon/Lagerwey synchronous machines, characterised by low speed and large diameter. Examples of large direct-drive PMG wind turbines are the Goldwind PMDD series (currently up to 3 MW) or the Siemens D3 and D7 (3 MW and 7 MW) platforms. As with synchronous machines a fully rated power converter is required, but the direct-drive PMG has a potentially simpler construction, requires no rotor excitation current,[8] and may in time become the most popular choice for large offshore wind turbines. Their main disadvantage may be the cost and availability of the rare earth metals used in the rotor magnets,

[8] This is not always an advantage, as the rotor field is always present: in the event of grid failure, there is a risk of overvoltage with a PMG, unlike a conventional wound-rotor synchronous machine (whose excitation can be cut off).

though rising demand from the wind and electric vehicle sectors is rapidly expanding the markets so long-term cost trends may stabilise. At the time of writing the two largest commercial wind turbines, both designed primarily for offshore use, use permanent-magnet generators: the Vestas V164-9.5MW is geared, and the Siemens SG 8.0 167 (8 MW) is direct-drive. A summary of generator types in current use is included in Table 5.1.

5.5 Power Quality

Operators of electricity networks impose strict regulations on the 'quality' of electrical power produced by grid-connected generators, and wind turbines are no exception. Typically, this means

- maintaining the steady-state voltage at the point of connection within a narrow range
- minimising high-frequency voltage fluctuations (flicker)
- maintaining the power factor within specified limits at the point of connection

Power quality is strongly influenced by the network strength, where a strong network is characterised by low impedance and high fault level (see below). As a general rule, the stronger the network, the more stable the voltage. On weak networks wind turbines may introduce problems due to the inherent variability of their output in turbulent winds, or due to transient voltage changes caused by generators starting and stopping. Power quality is also dependent on the type of generator, and in some cases additional equipment must be installed in order to meet statutory network requirements.

5.5.1 Network Characteristics

The electricity network at the point of connection (PoC) of a wind generator may be represented as a complex impedance $R + jX$ as shown in Figure 5.9. In this simplified diagram the three-phase network is represented by a single line (this is a common convention in electrical power analyses). At the point of connection the generator causes a voltage variation ΔV with magnitude given by

$$\Delta V \approx \frac{(PR + XQ)}{V_{\mathrm{ref}}} \tag{5.21}$$

where

V_{ref}: nominal grid voltage
P: real power
R: network resistance
X: network reactance[9]
Q: reactive power

---

[9] Reactance X is assumed to be inductive, which is reasonable for overhead line networks under high load.

Table 5.1 Summary of Generator Characteristics in Current Use on Commercial Wind Turbines

Type	PMG	Synchronous (fixed speed)	Synchronous (variable speed)	Asynchronous (induction)	DFIG
Drivetrain configuration	Geared or direct	Geared	Direct	Geared	Geared
Electrical connection	Fully rated converter	Direct to grid	Fully rated converter	Direct to grid	Stator direct to grid, rotor via converter
Turbine size range	All	Medium scale	Medium scale to multi-MW	Stall regulated up to 1.3 MW; smaller pitch regulated	Medium scale to multi-MW
Advantages	Robust; grid-friendly	Grid-friendly	Grid-friendly	Robust; superior dynamics to synchronous F/S	Similar to synchronous V/S, but cheaper
Disadvantages	Cost of fully rated converter; rare earth magnets	Poor dynamics; requires variable-slip gearbox	Cost of fully rated converter	Not grid-friendly: high starting current and reactive power	No direct-drive option
Examples	Siemens D7 Vestas V90 GE 6 MW Goldwind	Windflow 500	Enercon E48/70/126 DWT 750	Vestas V27; Nordex N60	Vestas V80; Siemens 2.3 MW; Nordex N90

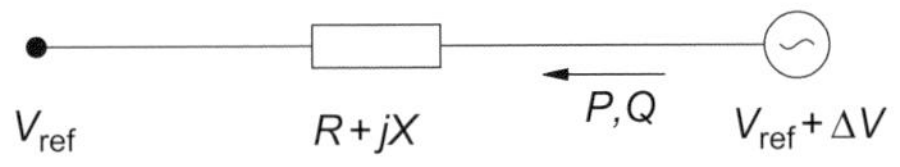

Figure 5.9 Single line diagram for generator connection. The grid is represented by complex impedance $R+jX$; P and Q are real and reactive power flows (Q may be positive or negative).

In the common case ΔV is a steady-state voltage rise due to the injection of real power by a generator into a network with finite resistance. Although Equation (5.21) is an approximation, it is highly useful for exploring the impact of a generator on the network and the effectiveness of reactive power control (Freris, 2008). In this context an important parameter is the network fault level, which is defined as the maximum power that would flow to the point of connection in the event of a short-circuit fault. The symbol S_k is commonly used for fault level, and for a three-phase network:

$$S_k = \sqrt{3}VI_f \tag{5.22}$$

where V is the phase voltage and I_f the fault current; S_k is usually quoted in MVA. The fault level is inversely proportional to network impedance, and a weak network is characterised by low fault level.

5.5.2 Steady-State Voltage

A stable grid voltage is essential to ensure the correct operation of customer's equipment, from large-scale industrial machinery to domestic appliances such as TVs, lights, washing machines, etc. The allowable variation usually depends on the connection voltage: current UK statutory limits are shown in Table 5.2. At the generator point of connection, however, the voltage variation permitted by the network operator may be significantly lower than these. On the SSE North of Scotland network, for instance, generators connecting at 33 kV must not exceed nominal grid voltage by more than 3%, while at 11 kV the applicable limit may be as low as 1.2%. Voltage rise ultimately limits the amount of generating capacity that can be installed on a network, and is often a more significant constraint than thermal capacity in network design.[10] Some network operators also stipulate that generators above a certain size must be capable of real-time voltage control.

With reference to Equation (5.21) voltage control is achieved by manipulating the reactive power Q drawn by the generator in response to changes caused by real power P:

[10] An overhead line with thermal capacity of 4 MW may be restricted to accepting only 3–400 kW of generation due to voltage rise limitations, e.g. if the generator is at the end of a long distribution line.

Table 5.2 Statutory Voltage Limits for UK Networks

Category	Typical voltage	Allowable range (%)
LV	230 V, 400 V	+10/−6
HV < 132 kV	11 kV, 33 kV	±6
HV > 132 kV	132 kV, 275 kV	±10

Note. Data from ESQCR (2002).

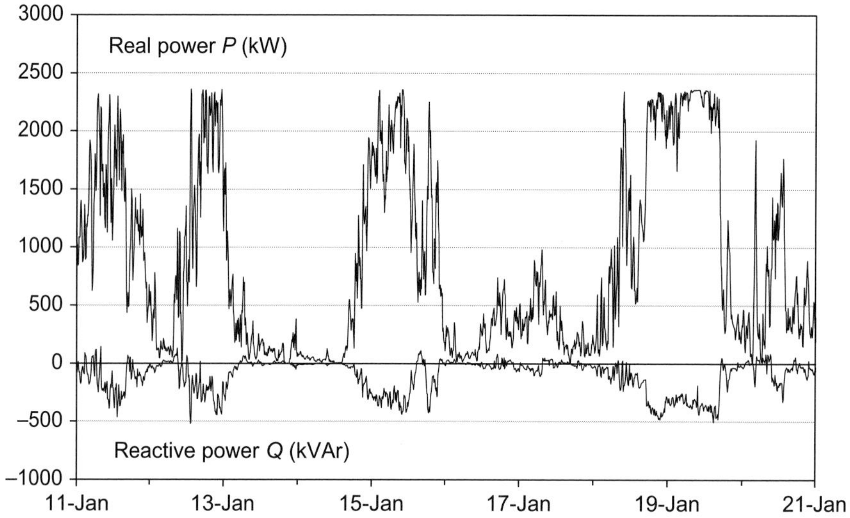

Figure 5.10 Voltage control on a 2.3 MW wind turbine: real and reactive power over a 10 day period.

voltage rise can then be partly offset if the generator operates with an inductive power factor (noting that Q is negative when inductive, positive when capacitive). Most generators now have this capability (traditional synchronous generators always did) and DFIGs and synchronous or permanent-magnet generators with fully rated power converters can all be programmed to carry out real-time voltage control. Conventional induction machines are less versatile, however, and may require the addition of statcoms or SVCs (see below). An illustration of voltage control by a converter-connected generator is given in Figure 5.10, which shows measurements of real and reactive power from a 2.3 MW wind turbine over a period of 10 days (the data are 10 min averages). Increases in real power P are clearly mirrored by corresponding variation in reactive power Q as the WEC controller continuously acts to limit voltage rise. In this case the control algorithm is based on continuous adjustment of the power factor in proportion to the measured grid voltage at the PoC: the governing characteristic can be seen in Figure 5.11.

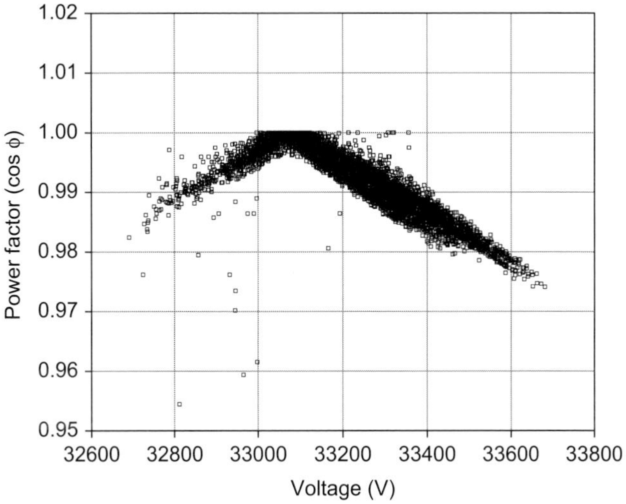

Figure 5.11 Relationship between power factor and grid voltage corresponding to Figure 5.10. Power factor is inductive when the voltage goes high and capacitive when low. Nominal grid voltage is 33 kV.

5.5.3 Flicker

Voltage fluctuations due to sudden changes in network conditions are known as 'flicker'; the term originates in the behaviour of electric lights in the vicinity of large and intermittent electrical loads.[11] Wind turbines can cause flicker due to the variation of their power output in gusty winds, or when a generator energises with a high inrush current. Network operators impose statutory limits on voltage change, with the allowable magnitude related to frequency of occurrence: the larger the voltage dip, the less often it is permitted. In the UK Engineering Recommendation P28 stipulates a maximum transient voltage change of 3% occurring no more than once every 10 min (ENA, 1989); for more frequent events the permissible change is less. Figure 5.12 shows the allowable voltage dip as a function of the time between events. A conservative estimate of the effect of a sudden load change on the network is given by

$$\delta \approx 100 \frac{S}{S_k} \tag{5.23}$$

where δ is percentage voltage dip, S the apparent power related to the event, and S_k the short-circuit fault level at the point of connection. Flicker severity is thus greatest on weak networks with low fault level. Generator starting current can be a concern, particularly that due to the magnetising inrush on an induction machine, which may be up to 10 times the rated current. Figure 5.13 shows

[11] A domestic example is a house with poor wiring: when a high load (e.g. an electric kettle) is switched on, the lights go dim.

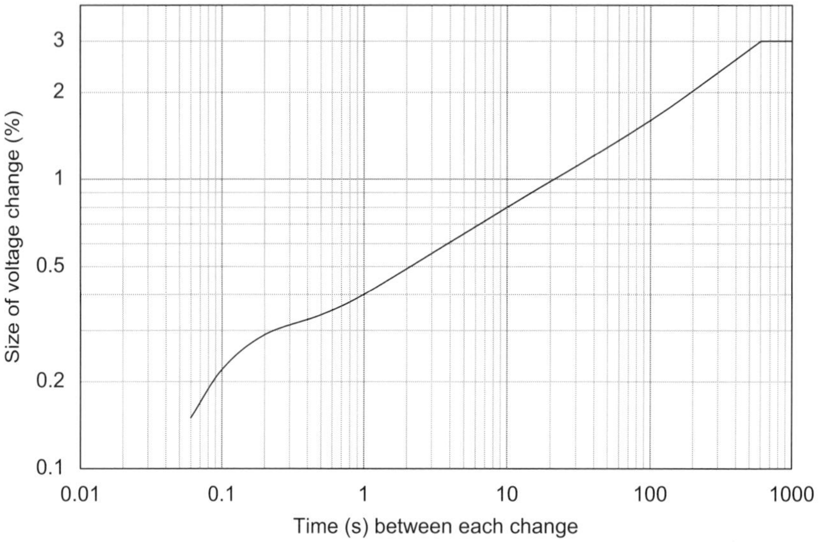

Figure 5.12 Permissible voltage change as a function of event frequency, from P28 (ENA, 1989).

the inrush current measured on a typical 60 kW induction generator. If such a machine were cold-started on a network with fault level S_k = 12 MVA the applicable value of S would be 0.6 MVA[12] and the resulting voltage dip would be 5%, which would exceed permissible limits in the UK.

In practice induction generators are always equipped with current-limiting devices to suppress the peak inrush current. These may be simple resistor banks connected in series for a fraction of a second during the connection sequence, or more sophisticated 'soft-start' modules with thyristor-controlled switching to limit the excitation voltage (in a similar way that a dimmer switch controls the lights in a room); the soft-start module is likewise connected only for the brief period of energisation. Although they do not completely eradicate the inrush current, these measures typically reduce its magnitude to a value between 1 and 2 times the rated current.

Flicker can also be caused by the rapid variation of a wind turbine's power output in turbulent winds, particularly when operating below rated power. Under these conditions the power varies roughly as the cube of wind speed, to an extent dictated by ambient turbulence. The resulting voltage variation is smoother than that due to generator switching, but may nevertheless breach statutory flicker limits. A detailed method for flicker evaluation is recommended in IEC 61400-21, as part of an overall power quality assessment procedure for commercial wind turbines (IEC, 2008). Prototype WECs are subject to a measurement campaign based on the IEC criteria, including measurement of a non-dimensional 'flicker coefficient' that determines flicker severity for a given fault level at the PoC. As with generator starting, flicker severity during normal operation is inversely proportional to the

[12] Ten times the generator rating, with unity power factor assumed.

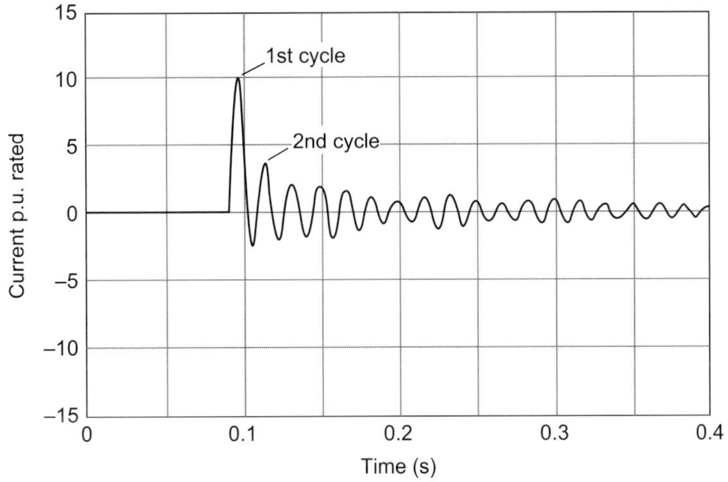

Figure 5.13 Measured inrush current (unsuppressed) on a 60 kW induction generator. This current is due to magnetisation of the rotor and arises even when the generator is unloaded.

local fault level. For a more comprehensive review of the procedures involved in power quality assessment see (Ackermann, 2012).

In addition to nuisance value, flicker can have implications for equipment safety and even for network management. Problems may arise during the energisation of large supply transformers, which experience a high magnetising current (similar to induction generators) with peak inrush current up to 8 times the rated value.[13] In some cases several transformers may be connected via a common circuit breaker, as shown in Figure 5.14. In arrangement (a) closure of the windfarm breaker results in simultaneous inrush to all three transformers and a potentially major dip in the network voltage, or in the extreme case a network blackout. An alternative is to connect the transformers individually as in arrangement (b) with sequential switching to limit the maximum inrush to that of a single unit. In some cases a pre-insertion resistor (PIR) may be installed: this is essentially a scaled-up version of a resistor-based 'soft start' that is briefly connected in series during the current transient, and switched out immediately afterwards. A PIR may need to be rated for an instantaneous power level of several megawatts and connected to the HV network.

5.5.4 Statcoms and SVCs

In some situations additional equipment must be installed on a network in parallel with a windfarm in order to ensure voltage stability, again via reactive power management. This is often the case where older wind turbines are involved, particularly those equipped with conventional induction

[13] This is a rule of thumb figure. Both higher and lower figures can be found in the literature, but a detailed analysis taking account of inrush current harmonic content suggests it is conservative (Bathurst, 2009).

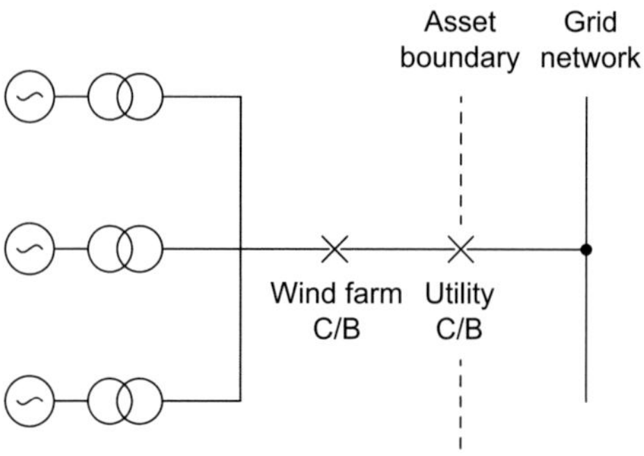

(a) Wind farm connection via common circuit breaker (C/B)

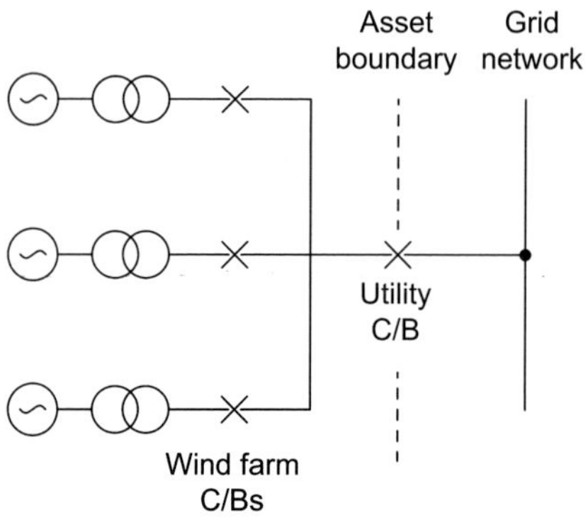

(b) Connection via individual circuit breakers

Figure 5.14 Windfarm connection alternatives. Arrangement (a) results in high inrush current to the three WEC transformers. Arrangement (b) allows for sequential energisation of transformers in order to minimise inrush current (C/B denotes circuit breaker).

generators. Two options for independent control of reactive power are the statcom and the SVC. The statcom or Static Synchronous Compensator is a fully solid-state device: based on a voltage source converter, it is similar in principle to the power converter of a variable-speed wind turbine. Statcoms are fast acting, bi-directional (able to supply capacitive or inductive power), and can respond to sub-cycle voltage changes to smoothly control the voltage at the point of connection. Statcoms may be sized from around 100 kVAr for small applications, up to multi-MVAr capacity to manage the output of large windfarms.[14] Major manufacturers include ABB and National Semiconductor.

[14] The 72MW Braes of Doune windfarm near Stirling in central Scotland is equipped with a 25MVAr ABB statcom.

The Static VAR Compensator or SVC incorporates banks of capacitor and reactors (inductors) to control reactive power: these components are controlled by thyristors (solid-state switches) so that SVCs, like statcoms, can achieve sub-cycle reaction time and real-time voltage control (Saad-Saoud, 1995). The use of traditional power components makes the SVC significantly cheaper than the fully solid-state statcom and SVCs have been widely used to enable older wind turbine types to achieve grid code compliance. An 800 kVAr SVC is installed at the Gigha community windfarm in Argyll (see below). In principle the statcom is a more reliable device, though outages due to component failure are not unknown. The discrete components (especially capacitors) embodied in SVCs have limited lifetime, but may fail progressively, so that the overall device loses capacity gradually while still remaining operational.

5.5.5 Fault Protection

While voltage stability is to some extent an issue of customer convenience, fault protection is a matter of public safety. Supply networks must remain safe under conditions such as transformer or generator short-circuits, or failure or damage to overhead lines and undersea cables. On detection of a grid fault embedded generators must disconnect immediately. In the UK the applicable standard is Engineering Recommendation G59/3, which stipulates the conditions under which automatic disconnection should occur (ENA, 2013): these include over and under voltage, over and under frequency, excessive rate of change of frequency (RoCoF), and phase imbalance. The circuit breaker that connects the windfarm or single turbine to the grid is equipped with a protection relay to sense these conditions, and rapidly disconnect when required. Specific relay settings may vary from one grid operator to another but the requirements given in Table 5.3 are fairly typical.

5.5.6 Harmonics

Networks must also be protected against voltage harmonics introduced by high-frequency switching of solid-state devices such as generator soft starts, frequency converters used by variable-speed generators, or switches embodied in SVCs and statcoms. Typical switching frequencies are 2–6 kHz. The UK standard for compliance is G5/4 (ENA, 2001), and the issue is addressed by incorporating appropriate harmonic filters in generator circuits. This appears to be largely successful, as evidenced by the large number of converter-connected wind turbines and reactive power control devices now operating on networks in the UK and elsewhere.

5.6 Grid Capacity

Introducing large numbers of embedded generators inevitably leads to network capacity issues, with competition to secure connections in areas of limited grid strength. In the UK

Table 5.3 Grid Fault Disconnection Criteria for Embedded Generators

Criterion	*Limit*	*Tripping time (ms)*
Over voltage (%)	110	500
Under voltage (%)	90	500
Under frequency (Hz)	47	500
Over frequency (Hz)	50.5	500
Rate of change of frequency (RoCoF)	Trip setting equivalent to 0.125 Hz s^{-1}	

capacity is awarded on a 'first come, first served' basis. Prospective generators apply for connection at a specific geographic location, and network analysis is then carried out to assess the impact in terms of voltage rise or thermal constraint. Capacity limits are based on a worst-case scenario, which assumes maximum output from all contracted generators occurring simultaneously with minimum network demand. The policy is conservative, ensuring that that stable grid conditions are always maintained, but leads to a significant underutilisation of network capacity. For instance, a distribution network may have available capacity of 10 MW, which is allocated to a single 10MW windfarm. If the capacity factor of the windfarm is 30% the network transmits only 30% of the energy for which it was designed – an average of 3 MW – yet due to the worst-case design criterion no other generator can be connected. This is arguably wasteful, as the critical conditions may occur very infrequently (for instance in the UK maximum wind generation occurs in winter, but minimum network demand in summer).

The conventional solution to increase grid capacity is network reinforcement, but this is often costly and may take years to complete. More innovative solutions are therefore being pursued including active network management (ANM) in which generators are offered 'non-firm' connections whereby the network operator can independently control their output or disconnect them altogether when conditions dictate. The network is continuously monitored at key points and when voltage or thermal limits are reached non-firm generators are curtailed in order of priority (contracted capacity is awarded on a 'last on first off' priority basis). The first example of an ANM network in the UK was the Regional Power Zone (RPZ) developed by SSE for the Orkney Islands (Kane, 2014). Under this pioneering scheme around 21 MW of additional renewable generation (mainly wind) was accommodated on the Orkney 33 kV distribution network, avoiding the need for a new undersea connection to the mainland and an estimated cost of £30M (Cleijne, 2012). By contrast, the cost of the ANM scheme was approximately £500 000. Non-firm connections are now being offered elsewhere in Scotland, and ANM is currently seen as the most economical solution to increasing capacity on limited networks. There are, however, indications that constraint margins in the Orkney RPZ are now at levels that discourage further large-scale generation, at least until further network reinforcement takes place.

5.6.1 Weak Network Example

Some of the foregoing topics can be illustrated by the example of the Gigha community windfarm. This small project was commissioned in 2004[15] in Argyll, on the west coast of Scotland, in an area where limited grid capacity presented a significant challenge. The prevailing network map is shown in Figure 5.15. A 33 kV feeder on the Kintyre peninsula transforms to 11 kV at Ballure substation, from where a network of 11 kV overhead lines and a subsea cable supply the island of Gigha. The proposed windfarm site was at the southern end of the island at the end of a long distribution line.

Figure 5.15 Weak grid. The 11 kV distribution network serving the island of Gigha in 2004. To maintain voltage stability, the windfarm incorporates a static VAr compensator. (Reproduced with permission of Scottish and Southern Electricity Networks; © Crown copyright and database rights 2019, OS licence number 100037385)

[15] The wider background to this project is described in Section 11.5.2.

Network analysis indicated a grid fault level of just 8.3 MVA at the windfarm point of connection, with both flicker and steady-state voltage rise identified as potential issues. The study concluded that without significant network reinforcement the installed generation capacity should be limited to 500 kW. The Gigha project had, however, been planned on the basis of a 675 kW windfarm and the potential cost of network reinforcement threatened its economic viability. Further work, however, confirmed that the desired capacity would be acceptable if real-time voltage control was incorporated to offset steady-state voltage rise and limit flicker due to inrush current. While these conditions would be met fairly comfortably by a modern converter-connected wind turbine, the machines proposed for Gigha were of an older generation, equipped with conventional induction generators and fixed capacitor banks for power factor correction.

The solution was to install a static VAr compensator (SVC) connected in parallel with the windfarm, as shown in the single line diagram in Figure 5.16. The SVC has enough switchable capacitance to offset the inherent inductance of the three generators (whose fixed capacitor banks were disconnected for simplicity) and provide the reactive power control envelope seen in Figure 5.17. This facilitates continuous control of power factor from near unity down to a minimum of 0.85 inductive during online operation. The SVC also has sufficient capacity to suppress flicker due to one generator starting with the other two online. Voltage variation is thus kept within statutory

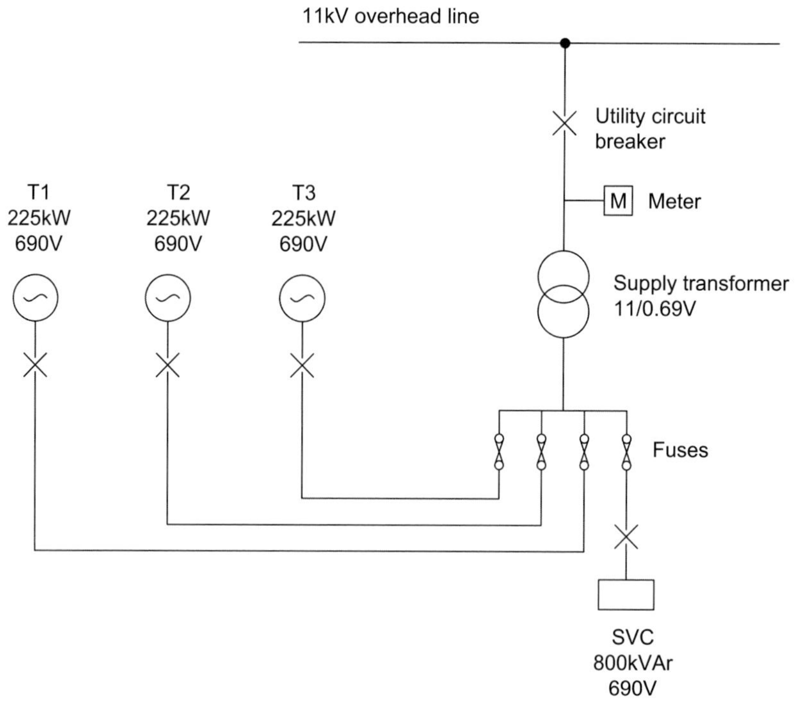

Figure 5.16 Single line diagram of Gigha windfarm, including its static VAr compensator (SVC). The device incorporated fast-switchable capacitance; inductance was inherently provided by the three generators.

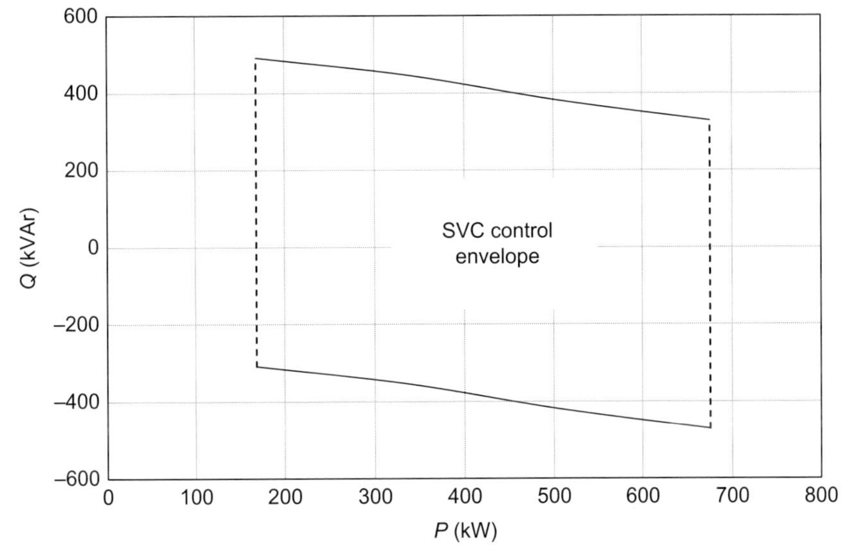

Figure 5.17 Reactive power characteristics of the Gigha SVC; for given real output *P*, reactive power *Q* is variable within the envelope shown. The lower boundary is defined by generator inductance.

limits and the inclusion of the SVC enabled the Gigha project to achieve the desired 675 kW capacity – installing three Vestas V27s rather than two – at an additional cost around one-tenth of that otherwise needed for network reinforcement.

5.7 Lightning Protection

Large wind turbines incorporate comprehensive lightning protection systems, which are now essential given their extreme tip heights and degree of exposure to the elements. When commercial turbines were relatively small lightning protection systems were not generally incorporated; it was, however, recognised quite early that if non-conductive GFRP and wood laminate blades were directly struck by lightning they could be seriously damaged or completely destroyed (Dodd et al., 1983). Figure 5.18 illustrates such a case: the heat generated by the lightning strike has caused an instantaneous pressure build-up inside the blade, effectively blowing the surfaces apart. The risk increases with turbine size, and as the length of commercial blades began to exceed around 20 m lightning protection started to become an integral part of wind turbine design, rather than an optional extra.

Rather like church spires, HAWT blades incorporate one or more metallic components (receptors) that are designed to preferentially attract lightning so that it can be safely conducted to ground. There are several types of receptor, including metallic blade tips, conducting strips along the blade length, or one or more small metallic spots on the blade surfaces. In all cases the receptors are connected to a common internal cable that conducts

Figure 5.18 Lightning damage. Heat and pressure generated by the strike have blown the surfaces of this blade apart, with evidence of charring. All large blades now incorporate effective lightning protection.

the charge safely down inside the blade. An example of a tip-mounted receptor under test at the KEMA high-voltage labs in the Netherlands is shown in Figure 5.19. The conducting path between the blades and the tower structure requires careful design, particularly with variable-pitch blades. On some wind turbine types it is achieved using spring-loaded sliding contacts to maintain the current path between moving parts (e.g. from blade to hub, or hub to nacelle); on others, narrow spark gaps are maintained between specially designed conducting rods and metallic rings to allow the lightning discharge to pass safely between the moving parts without requiring direct contact.

The conducting path between tower sections and ultimately into the ground must also be carefully designed, though this is more straightforward, with no moving parts involved. The tower itself forms part of the conducting path, with short cables used to connect across the flange interfaces. The recommended minimum cross-sectional area of all conducting elements (including the down-blade cable) is 50 mm^2, and a comprehensive earthing mat of copper or aluminium is buried just above the WEC foundation: Figure 5.20 shows a typical example. The size and extent of the mat are dictated by local soil resistivity, and once installed it must demonstrate a resistance in

Figure 5.19 Lightning test on the tip of a 19 m Aerpac blade at the KEMA high-voltage labs, Arnhem, 1995. The discharge has successfully gone to ground via a silver-plated receptor. (Photo reproduced with kind permission of KEMA Laboratories)

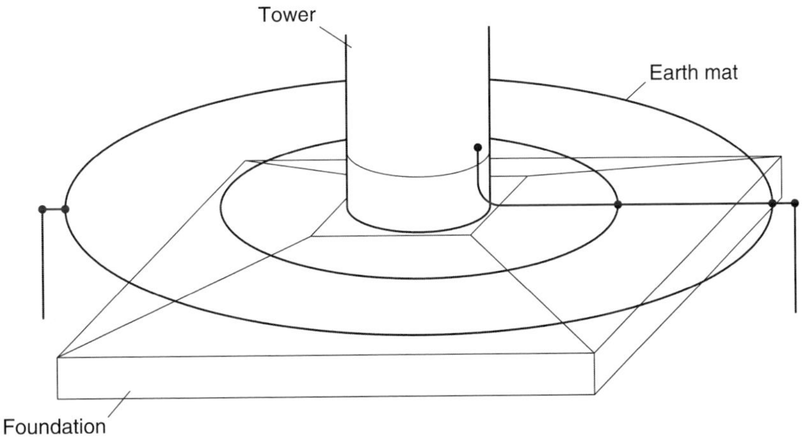

Figure 5.20 Earthing mat round the tower base of a medium-sized WEC. Copper wire is buried above the level of the foundation plinth. Soil resistance testing is required to determine the required extent of the mat, which should have in situ resistance in the range 2–10 ohms.

the range 2–10 ohms (the exact figure depends on regional or national standards, and details of the internal WEC earthing arrangements). Numerous standards apply to WEC lightning protection design, among the most important being IEC 61024-1, which is applicable to general structures (IEC, 1990), and IEC 61400-24, which applies specifically to wind turbines (IEC, 2002).

5.8 Exercises

5.8.1 Induction Generator

An induction generator runs at 1800 rpm at zero load. If full load slip is 2.2% and rated output is 250 kW what is (a) the generator speed at full load and (b) the power output at 1% slip?

5.8.2 Synchronous Speed

What is the synchronous speed (n_s) of a six-pole generator connected to a (a) 50 Hz and (b) 60 Hz grid?

5.8.3 Power Factor Correction

The table below gives the uncompensated power factor characteristics for an induction (asynchronous) generator rated at 300 kW. If a 100 kVAr capacitor bank is connected in parallel with the generator terminals, calculate the resulting corrected power factor at each output level.

Power output % rated	Uncompensated power factor	Corrected power factor
15	0.37	
25	0.53	
50	0.74	
75	0.81	
100	0.84	

5.8.4 Three-Phase Power

A wind turbine is equipped with a three-phase generator with rated output of 3.0 MW and is operated at an inductive power factor of 0.95. The line (phase-to-phase) voltage is 690 V. Calculate (a) the total or apparent power, (b) the current per phase, and (c) the reactive power demand, of the generator.

5.8.5 Direct-Drive Generator

A permanent-magnet direct-drive generator has a diameter of 6.0 m and rated power output of 3 MW. The generator is to be redesigned for the same power rating, but to run at 10% higher speed, and with its axial length increased by a factor of 1.5. Estimate the new diameter of the generator.

5.8.6 Multi-pole Synchronous Generator

The synchronous generator of a 2.3 MW wind turbine has 72 poles and reaches its rated output when the rotor speed is 21.5 rpm. The generator is connected to the grid via a fully rated frequency converter. What frequency of AC supply must be produced by the converter in order to transfer rated power at rated rotor speed?

5.8.7 Network Voltage Rise

A 400 kW wind turbine generator is connected to an 11 kV distribution network with resistance $R = 4.5\Omega$ and inductive reactance $X = 2.5\Omega$ at the point of connection. What level of reactive power must the generator draw in order to limit the voltage rise to 1%, and what will be the resulting power factor?

5.8.8 Flicker

Figure 5.21 shows the instantaneous current in one phase of a three-phase 11 kV supply when energising an 800 kVA transformer. The fault level at the point of connection is 100 MVA. Estimate the percentage voltage dip δ on the 11 kV system when energising the transformer.

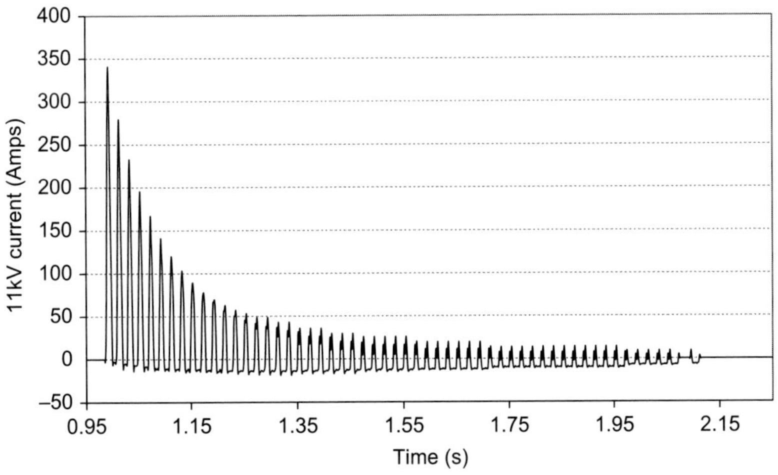

Figure 5.21 See Exercise 5.8.8.

CHAPTER 6 CONTROL

6.1 INTRODUCTION: LEVELS OF CONTROL

A modern wind turbine is a highly autonomous electricity generator, capable of unattended operation for months at a time. Its onboard control systems govern a wide variety of functions: enabling the rotor to steer into the wind, start up and shut down smoothly, and limit its power output in high winds. The turbine will automatically stop generating and shut down rapidly in the event of a fault or emergency, and in addition the controller may regulate electrical power quality, selectively limit rotor aerodynamic noise, or detect ice forming on the blades. Associated with these tasks is an array of sensors and transducers that feed data to the controller, and a network of actuators, valves, and switches that respond to its output commands. The level of sophistication varies with wind turbine size, but the following categories of control are common to the majority of grid-connected wind turbines:

- supervisory control
- power limiting
- starting and stopping
- electrical power quality management
- sector management

Supervisory control covers functions where the controller samples inputs such as wind speed and direction, and takes decisions relating to the operating state of the machine. Data sampling is continuous but the control response may be infrequent, for instance starting up or shutting down when operational wind speed limits are detected, or yawing the rotor into wind when a change in direction is sensed. The importance of yaw alignment has been understood since the days of traditional windmills (see Chapter 1) and its modern implementation is described in Section 6.2.

Power limiting is a continuous control function, but only invoked when the wind speed exceeds the rated power threshold. It is perhaps the most challenging aspect of wind turbine control and the one that has developed the furthest (and arguably absorbed the greatest R&D effort) in the modern era; power limiting is described in Section 6.3. Starting and stopping are requirements common to all WECs, but the way in which they are executed depends greatly on whether the rotor has variable pitch, and to some extent on generator type; see Section 6.4. Power quality management is a more recent development made possible by semiconductor technology (previously discussed in Section 5.5), while sector management is a control option in which the WEC operating characteristics are altered automatically on the basis of wind direction and/or other ambient conditions, for instance to reduce rotor noise. Sector management is discussed in Section 6.5.

6.2 Yaw Control

Operation of a HAWT rotor in yawed flow results in increased blade fatigue loading and reduced energy output (see Section 3.8) so minimising yaw error is an important control objective. The yaw angle is continuously measured by a wind vane or ultrasonic anemometer mounted on top of the nacelle, and when a pre-set limit is exceeded the controller causes the nacelle to rotate to face the wind; most large HAWTs incorporate either geared hydraulic or electric motors for this purpose: an example of the latter is shown in Figure 6.1. The yaw motors are attached to the floor of the nacelle and drive against a toothed ring on the fixed tower top. The nacelle rotates on a large bearing, either of rolling-element type, or a plain ring on which it rests via friction pads that also help to damp the yaw motion.

Yaw correction is carried out periodically and the rate at which the nacelle turns is relatively low, just a few degrees per second: fast yaw response is undesirable as it gives rise to gyroscopic loading on the rotor blades and main shaft. The raw signal from the wind vane is filtered to remove turbulent response and corrected for the directional offset imparted by wake rotation (see Section 3.6.2) to ensure that the signal fed to the controller is an accurate estimate of

Figure 6.1 Yaw control. Two vertically mounted electric motors can be seen rising from the floor of this E44 nacelle (two more are hidden behind). Each motor acts through a gearbox to drive against a ring gear on the tower top. (Photo reproduced with kind permission of Charlie Robb)

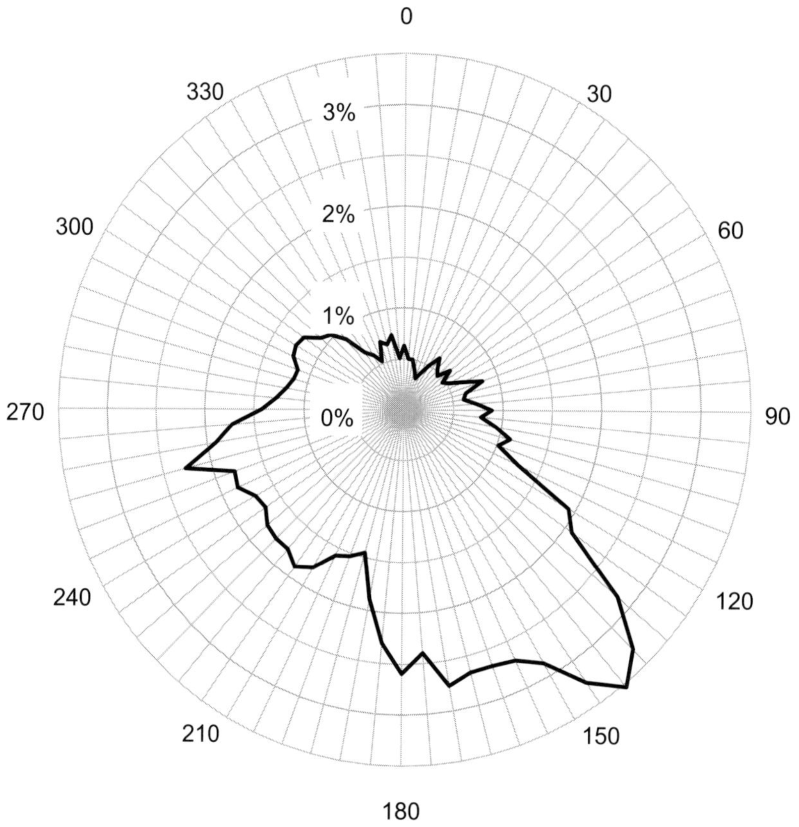

Figure 6.2 Polar frequency plot showing the recorded nacelle orientation of a 2 MW wind turbine over a 12 month period. The prevailing wind direction was south-easterly.

true yaw error. The wake offset is a function of wind speed and rotor loading, and an empirical yaw correction based on measurement may be stored in the controller as a lookup table (Anderson et al., 1993). One recent development is the use of nacelle-mounted LIDAR (laser Doppler anemometry) to measure the wind direction upstream of the rotor and thereby estimate yaw error to a very high accuracy, with no need for wake correction.

During the course of a year a wind turbine rotor will undergo many thousands of yaw corrections, giving rise to the kind of cumulative directional pattern seen in Figure 6.2. One consequence of tracking the wind direction is that the nacelle may execute several complete yaw rotations in the same direction over a period of days or weeks, and the down-tower electricity cables will become twisted. To allow for this the controller is equipped with a twist sensor that detects the number of turns, and during a suitable period of low wind when the turbine is offline the nacelle is automatically rotated several times to correct the cable twist.

6.3 Power Limiting

Above a certain wind speed the aerodynamic power of a wind turbine rotor must be capped in order to prevent excessive structural loading. The need can be illustrated by considering a 1 MW wind turbine that achieves rated power at a wind speed of 12.5 m s^{-1} but continues to operate up to 25 m s^{-1}; at the lower wind speed the rotor will extract about half the available power, but at the upper speed it extracts only around 6% (because power is proportional to V^3) and while 1 MW of fluid power is extracted another 15 MW must pass through the rotor disc. Failure to control the rotor power output accurately in high winds can thus lead to very high aerodynamic loading and potentially catastrophic damage to blades, gearboxes, or in the worst case, the entire structure. Power limiting is a real-time control function and three principal methods have been developed, namely

- stall regulation
- constant-speed, variable-pitch control (CSVP)
- variable-speed, variable-pitch control (VSVP)

Each can be explained with the aid of the dynamic model shown in Figure 6.3. The rotating parts of the wind turbine are represented as a single lumped inertia J, T_R is the aerodynamic torque developed by the rotor, T_G the electromechanical torque applied by the generator, and Ω the rotor rotation speed (hence $d\Omega/dt$ is angular acceleration). The governing equation of motion is

$$T_R + T_G = J\frac{d\Omega}{dt} \tag{6.1}$$

Equation (6.1) is Newton's second law of motion formulated for a rotational system; it is valid for any WEC drivetrain configuration, though probably easiest to envisage for a direct-drive machine where the rotor and generator literally behave as a single lumped inertia.[1] The instantaneous electrical power output is given by

$$P = \Omega T_G \tag{6.2}$$

Between them Equations (6.1) and (6.2) fully describe the response of the wind turbine. If the aerodynamic and generator torques are in equilibrium the rotor will remain at rest or rotate with constant speed, while any imbalance between T_R and T_G will cause it to accelerate or decelerate. Power output P may vary (a) with generator torque at constant rotor speed, (b) with rotor speed at constant torque, or (c) with both torque and speed varying. All these possibilities are seen in WEC operation, depending on the type of control employed. Method (a) is at the heart of the oldest and simplest power control method, namely stall regulation, as now described.

[1] For a geared wind turbine the inertia contribution of the generator is referred to the rotor by multiplying by the gearbox ratio squared; a single lumped inertia can then be assumed.

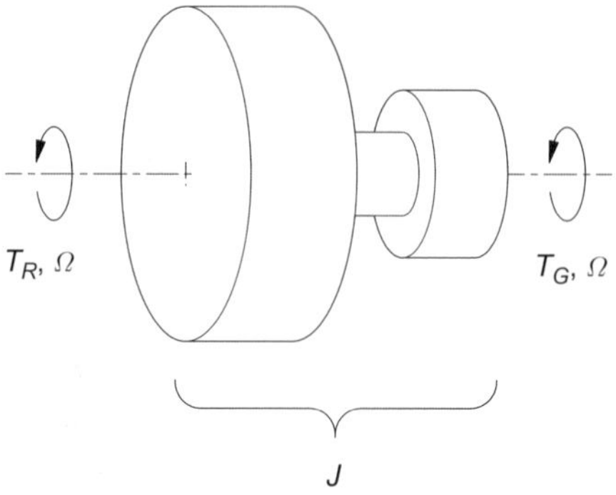

Figure 6.3 Simple dynamic model of rotating wind turbine; see Equation (6.1).

6.3.1 Stall Regulation

Stall regulation is a passive method of aerodynamic power control. The principle is described in Section 4.7.1 but, to recap, the rotor has fixed-pitch blades and rotates at constant speed: when the wind speed exceeds a certain level the blades stall, with progressive decrease in lift and increase in drag. The power curve flattens off or falls as seen in Figure 4.15. The rotor speed is (almost) invariant because of the inherent characteristics of the grid-connected induction generator; $d\Omega/dt$ is then zero, so from Equation (6.1) the rotor aerodynamic torque T_R and generator torque T_G must be equal and opposed at all times and any variation in rotor power in gusty winds will be directly reflected in the electrical power output. As stall is not always a smooth phenomenon this may have consequences for power quality, but in deep stall the power may remain relatively constant. For this reason stall regulation was the preferred power control method for many wind turbines up to about 300 kW rating, and was ultimately employed on turbines up to 1.3 MW. Some incorporated two-speed generators with pole-switching capability, in a basic but effective form of speed control to enable better matching of rotor speed to wind speed.

6.3.2 Constant Speed, Variable Pitch (CSVP)

Variable blade pitch enables rotor aerodynamic torque T_R to be actively controlled. Its introduction effectively allowed rotors to be designed for higher tip speed and reduced solidity, making them lighter and with higher aerodynamic efficiency in light winds. The main advantage of pitch control, however, is the ability to control the power in high winds to achieve a level power curve: the aerodynamic principles are explained in Section 4.7.2. The first generation of pitch-controlled wind turbines retained fixed-speed generators so were subject to the same limitation as stall-regulated machines, i.e. the electrical output power is at all times (neglecting losses) equal to the rotor aerodynamic power. In high winds, power quality is then only as good as the pitch system allows, and some early pitch-regulated WECs suffered from poor control. Although their mean power

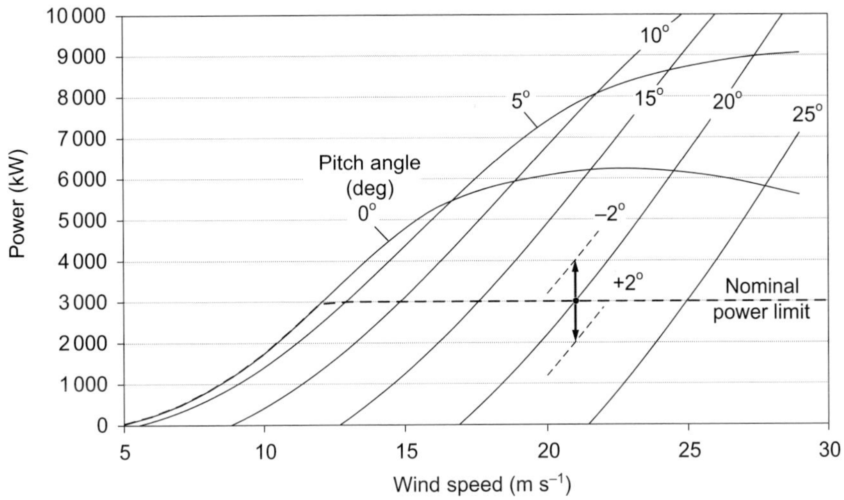

Figure 6.4 Sensitivity of CSVP control. Power curves at increasing pitch angle for a generic 3 MW constant-speed, variable-pitch wind turbine: in a wind speed of 21 m s^{-1}, a pitch error of 2° gives rise to a power overshoot of 1 MW.

could be accurately maintained, in some cases the transient power variation in gusty winds was exacerbated by pitch action and could be greater than that occurring with stall-regulated machines (Hoskin, 1988).

The difficulty is explained with reference to Figure 6.4, which shows the power curve characteristics for a generic 3 MW constant-speed variable pitch (CSVP) wind turbine. Each curve corresponds to a different pitch angle in the range 0°–25° and the nominal power curve (dashed line) is achieved by operating at the appropriate angle in a given wind speed; pitch control begins at around 12 m s^{-1} when rated power is first reached. In high winds the power output is very sensitive to both wind speed and pitch angle, and the penalty for inaccurate control is high: for example at 21 m s^{-1} a pitch error of ±2° results in a deviation from rated power of ±1 MW, or 33%. There is a pro rata variation in the blade aerodynamic loads (flapwise and edgewise bending moments), with significant implications for fatigue. The control sensitivity is high because, unlike a stall-regulated WEC, the blades remain in attached flow in high winds. Large power excursions were common on many CSVP wind turbines, whose pitch mechanisms were unable to respond quickly enough to gusts.

An example was the Howden HWP45, whose measured power characteristics are shown in Figure 6.5 as 10 min averages. In the highest wind speeds, power excursions of ±50% above the rated power level are observed. The HWP45 was rated at 750 kW with partial-span pitch control and a conventional synchronous generator, a combination which was relatively unusual (induction generators became standard for constant-speed WECs). The partial-span pitch philosophy was also not widely used, though in principle it offered some advantages in regard to off-design operation

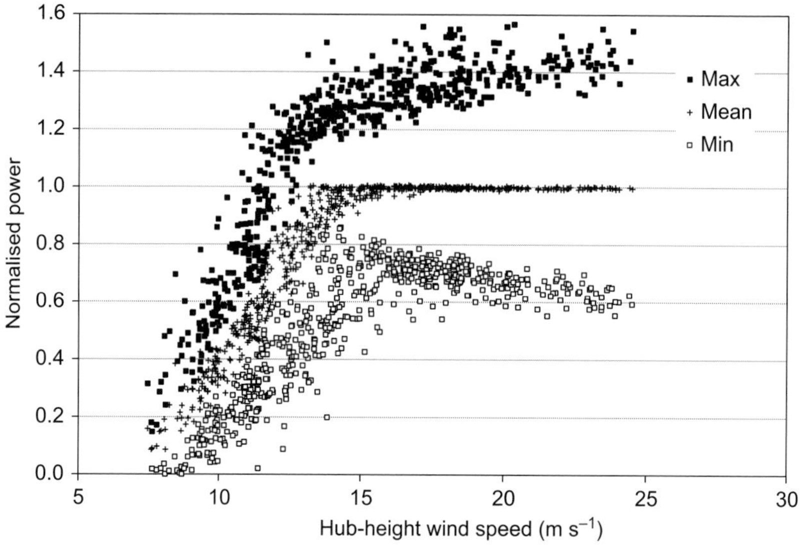

Figure 6.5 Instantaneous power overshoots of 50% in high winds were not unusual for CSVP wind turbines: historic measurements from the HWP45 wind turbine (power normalised with respect to rated).

(see Section 4.7.2).[2] The level of control shown in Figure 6.5 was, however, not unusual for CSVP machines and a number of commercially successful designs were known for their susceptibility to turbulence; this often led to a lowering of the cut-out wind speed, with a modest loss of yield seen as an acceptable penalty. On high-wind sites, however, some turbines suffered blade fatigue damage or broken gearboxes; in some cases the drivetrain dynamic response compounded the problem by amplifying the input torque loading and causing the pitch system to over-respond. A great deal of R&D went into improvements to CSVP controller algorithms and pitch control mechanisms, but without ever completely eliminating the problem. Despite this some CSVP wind turbines (e.g. the Vestas V27) were notably less affected, possibly due to their generally more robust design.

The power control system for a CSVP wind turbine is shown schematically in Figure 6.6. It is a classical single-input single-output feedback loop, in which measured power output is compared with the reference (rated) value, and a corrective demand sent to the pitch actuator. The system is active only in wind speeds above rated. The controller algorithm is typically PI (proportional plus integral) with the option of PID control less favoured as its differential (D) term amplifies transducer noise, causing instability. More sophisticated 'classical' control algorithms have, however, been explored: these include higher-order terms to filter out harmonic perturbations in the measured power (see Section 7.3.4) and prevent undesired pitch activity (Leithead, 1990). One finding of this research was that control of two-blade CSVP wind turbines

[2] This led to a research programme in the early 1990s, including wind tunnel tests of an actively controlled 'flying tip', an artificially stabilised blade section with automatic gust response and fail-safe air braking (Anderson et al., 1998).

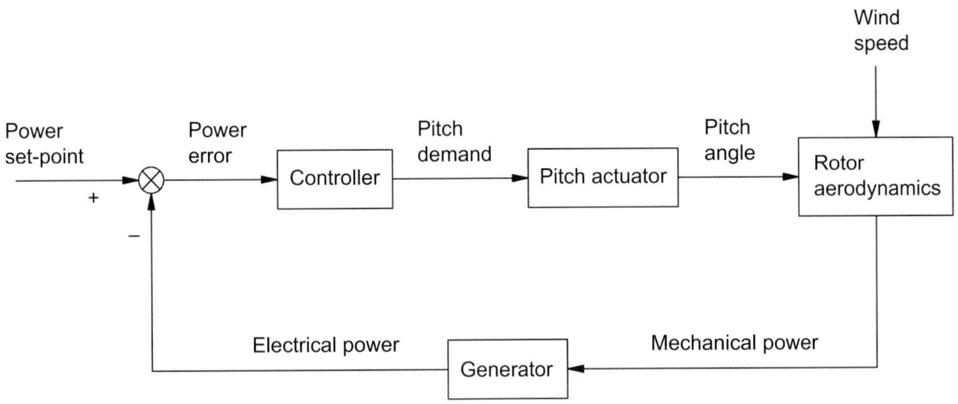

Figure 6.6 Feedback loop for a constant-speed, pitch-regulated wind turbine.

is harder than three-blade, as in the former case the harmonic perturbations in the power signal occur at a frequency closer to the pitch actuator working range, reducing its effective bandwidth.

6.3.3 Variable Speed, Variable Pitch (VSVP)

The definitive solution for accurate power control came with the introduction of variable-speed operation. Referring again to Equation (6.1) if aerodynamic and electrical torque T_R and T_G can be independently controlled they no longer have to be equal, and acceleration $d\Omega/dt$ is in general non-zero. The rotor speed can then be actively varied, and the aerodynamic power it absorbs can differ from the electrical power delivered by the generator, with any mismatch translated into a change of rotor speed. Variable-speed operation brings several significant advantages:

- In low winds, the rotor can be operated at constant tip speed ratio, tracking C_p^{max} in order to maximise energy yield.
- In high winds, the output power can be accurately capped, and the drivetrain components (gearbox and generator) protected against high transient loading.
- Blade pitch can be regulated on a relatively slow speed control loop without incurring high power overshoots.

The advantages of variable speed were identified early (Goodfellow, 1986) and arguably before the technology was sufficiently advanced to fully exploit them. A fundamental requirement is dynamic control of generator torque, which is nowadays achieved using power semiconductors (see Section 5.4.3). Before this, however, limited-range variable speed was demonstrated on conventional grid-connected wind turbines such as the WEG LS-1, which had a ‘reaction drive’ gearbox that decoupled the rotor speed from the fixed-speed synchronous generator (Bedford, 1985). Despite a relatively limited speed range the power output of the LS-1 was very smooth in high winds; the

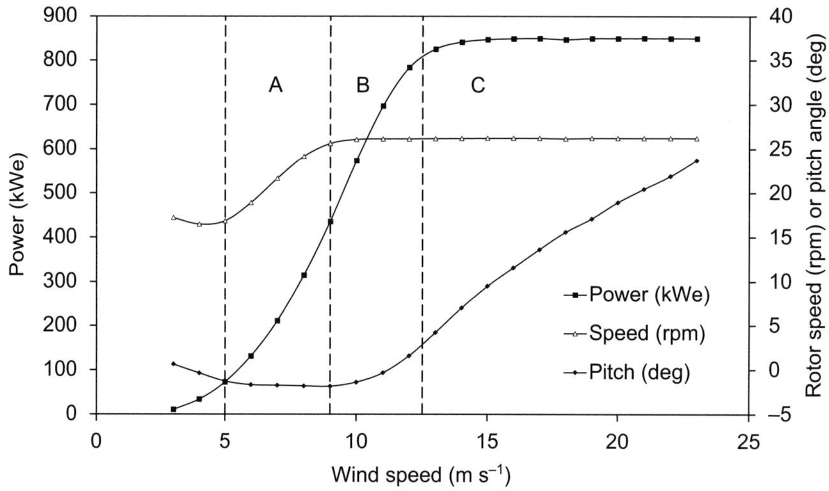

Figure 6.7 Operational characteristics of a broad-range, variable-speed HAWT. Three control regimes are seen: (A) variable-speed, constant tip speed ratio; (B) constant-speed, constant pitch; (C) constant speed, power limiting with blade pitch control.

later WEG MS-3 also achieved limited-range variable speed via a hydraulic torque-limiting mechanism to effectively vary the gearbox ratio, again using a standard synchronous generator (Henderson, 1990).[3]

Variable ratio gearboxes were largely superseded due to developments in solid-state power electronics, which facilitated broad-range variable-speed generators. Among the first wind turbines to benefit were the Lagerwey 80 kW in the Netherlands and the (original) Enercon E33-300 in Germany, both dating from the early 1990s; these machines had conventional generators and a geared drivetrain, but in both cases a fully rated AC-DC-AC converter was inserted between generator and grid. There are a number of potential generator configurations for variable-speed operation, including direct-drive and DFIG: for a review see Section 5.4. In almost all cases variable speed is combined with variable blade pitch (VSVP).

Power control on VSVP wind turbines is more complex than with constant-speed machines as there are multiple control variables, and different control strategies depending on the external conditions. This can be explained with reference to Figure 6.7, which shows the measured characteristics of a medium-scale VSVP wind turbine in terms of its steady-state power, rotor speed, and blade pitch angle, all shown as functions of mean wind speed. Disregarding winds below 5 m s^{-1} (where output is negligible) three control regimes can be identified in Figure 6.7:

A Light winds: the rotor speed is varied in proportion to wind speed in order to maintain constant tip speed ratio and optimum C_p; blade pitch angle is held near the nominal zero position.

[3] This concept is still in use on the Windflow 500 kW wind turbine.

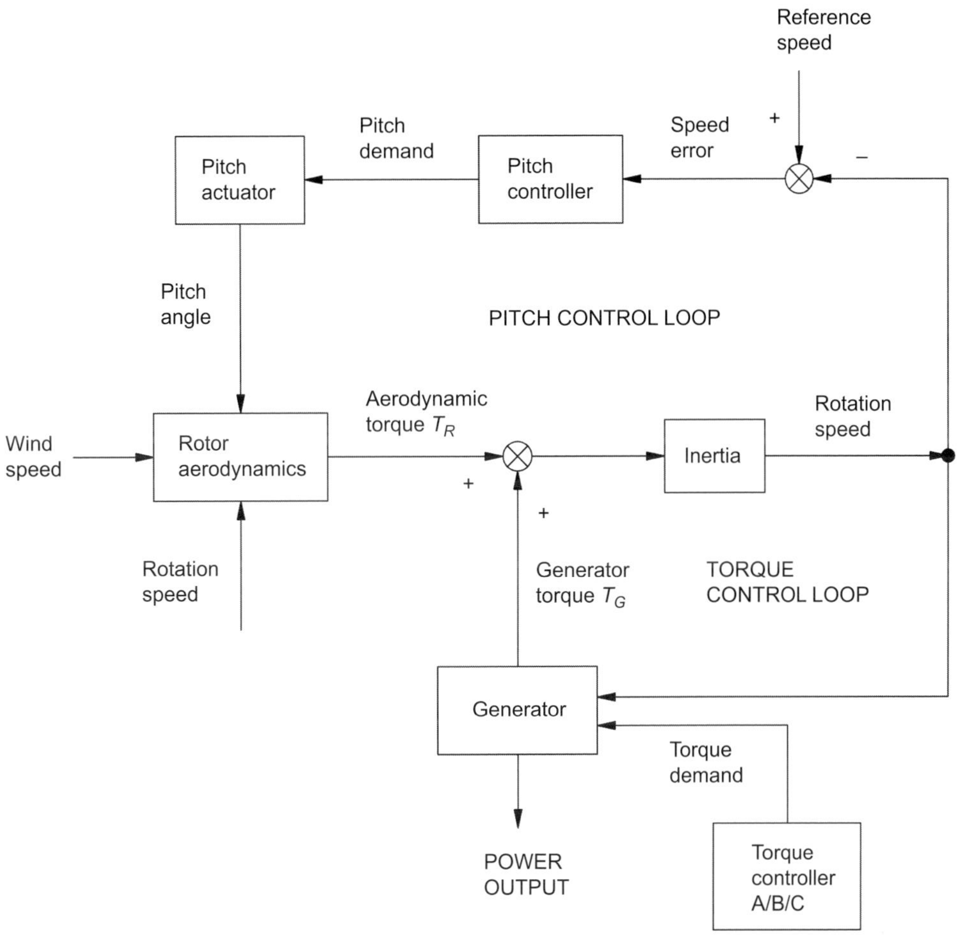

Figure 6.8 Control schematic for a variable-speed, variable-pitch wind turbine.

B Medium winds: rotor speed and blade pitch angle are both held constant; output power continues to rise with wind speed: torque increases but speed does not.

C High winds: power limiting; rotor speed is held constant and blade pitch is progressively increased with wind speed.

The block diagram for a generic VSVP control system is shown in Figure 6.8. With reference to this, the way in which control strategies (A–C) above are achieved is as follows:

A Light Winds: Constant Tip Speed Ratio

The pitch control loop is not used and blade pitch remains constant in the full run position; rotor speed is managed by the generator torque controller, whose objective is to maintain constant tip speed ratio λ corresponding to $C_p^{\max}$. This can be done by controlling torque T_G as a function of speed Ω, according to

$$T_G = \frac{1}{2}\rho\pi R^5\Omega^2 C_p^{\max}/\lambda^3 \tag{6.3}$$

where ρ is air density, R rotor radius, and C_p^{max} and λ are the design optima for the given rotor geometry. T_G is then controlled on a closed loop as a function of Ω^2; all the other parameters in Equation (6.3) are constants.[4] Note that the torque relationship in Equation (6.3) represents a steady-state solution and the time required to accelerate and decelerate the rotor can lead to off-design operation; additional terms may then be included in the control feedback algorithm to account for rotor inertia and it also helps if the C_p, λ curve has a relatively broad peak. For a more detailed review of these topics see (Burton, 2011).

B Medium Winds: Constant Speed and Pitch

Regime B is a transition between constant tip speed ratio and constant power operation. Blade pitch is still held close to the full run position, but rotor speed reaches its nominal maximum value and is held constant, again via generator torque control (with feedback loop now closed on generator speed). In this regime the turbine operates like a fixed-pitch, fixed-speed, WEC on the rising part of its power curve.

C High Winds: Power Limiting

Once rated power is reached the control objective is to maintain both power output and rotor speed at a constant level. This is achieved by invoking both pitch and torque control loops, which operate quasi-independently. Constant power is maintained by controlling the generator torque in inverse proportion to speed (hence maintaining the product $T_G\Omega$ at a constant level). At the same time the rotor speed is managed via blade pitch control, with the pitch actuator enclosed in a PI loop based on speed. Because the pitch response is not sufficiently fast to match turbulent wind variations the rotor speed will vary to some extent about the reference value, but the rotor acts as a flywheel whose inertia helps limit the variation. As the wind speed increases a higher mean pitch angle is automatically demanded to maintain constant rotor speed.

The foregoing characteristics are further illustrated in Figure 6.9, which shows the measured steady-state torque-speed relationship for the same wind turbine as in Figure 6.7; the three control regimes are again marked. With constant tip speed operation (A) a quadratic curve is seen, roughly in accordance with Equation (6.3); in transition regime (B) the speed becomes constant while torque continues to rise; at (C) power limiting is invoked: this regime is now represented as a single point as both torque and speed are constant in the steady state. The effectiveness of VSVP control for power limiting can be seen in Figure 6.10, which shows 10 min power measurements for

[4] The equation is representative of a direct-drive machine: for a geared wind turbine the target value of T_G is reduced by the gearbox ratio. Note also that drivetrain losses are ignored here.

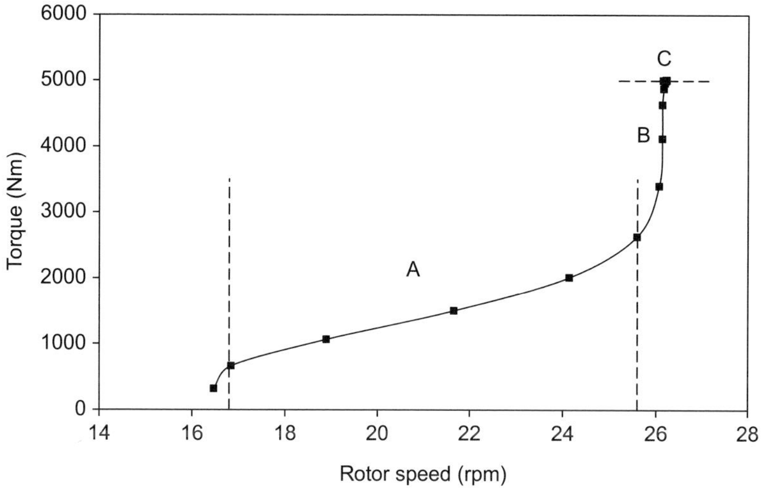

Figure 6.9 Generator torque as a function of rotor speed for a VSVP wind turbine. Control regimes are shown as in Figure 6.7 (note that (C) is represented by a single point on the torque-speed curve).

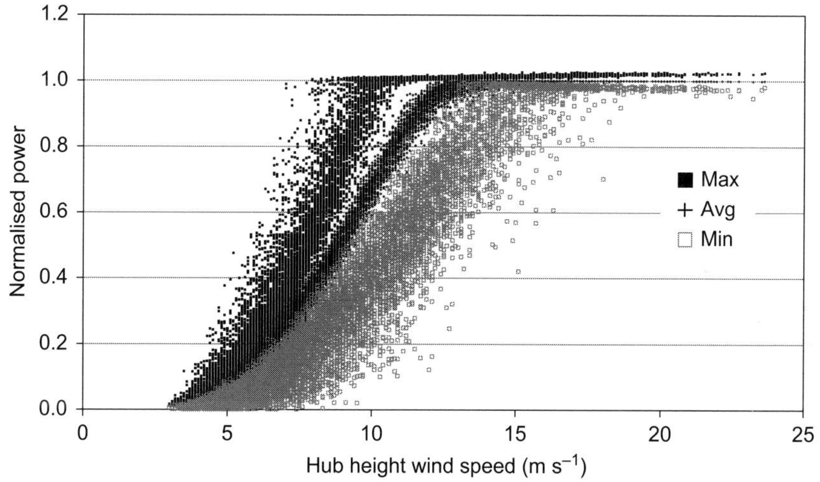

Figure 6.10 Close control. Normalised 10 min power measurements from a medium-scale VSVP wind turbine; power excursions in high winds are within ±2% of the rated level. For a comparison with CSVP operation, see Figure 6.5.

a medium-scale wind turbine. In high winds the maximum power variation remains within ±2% of the rated value; this can be contrasted with the performance of the CSVP machine shown in Figure 6.5.[5]

[5] In its defence, the Howden 750kW was designed in the mid 1980s, with a synchronous generator and 'deadband' pitch control: further development would certainly have included an induction generator and PI control loop.

6.4 Starting and Stopping

In a typical wind regime a wind turbine spends 10%–15% of its time offline, when the wind speed is too low to generate. During these periods the controller continues to monitor wind speed and direction but maintains the turbine in a quiescent state to avoid drawing power from the grid, although some consumption is unavoidable for the computer and essential systems (e.g. anti-condensation heaters and monitoring instruments). Yaw correction is still carried out periodically, though in very calm weather this too may be inhibited. Once the wind picks up enough to allow generation the controller must manage a start-up sequence that begins with the rotor at rest and the generator offline and finishes with the rotor at nominal operating speed and the generator connected to the grid. The way in which this is achieved varies significantly with wind turbine type, and different control strategies are necessary for fixed-pitch stall-regulated wind turbines and those with pitch control. The principal difference is that variable pitch allows offline control of the rotor speed while fixed pitch does not.

This can be further explained with reference to Figure 6.11, which shows the calculated aerodynamic torque coefficient C_Q for a 600 kW wind turbine: recall that the expression for aerodynamic torque is given in Equation (4.3). The rotor shown is nominally designed for stall regulation at a fixed pitch angle of 0°,[6] but the effect on C_Q of varying the blade pitch is shown (see below). When the rotor is at rest $\lambda = 0$, and in the nominal 0° case the torque coefficient is quite low: the blade aerofoil profiles are almost flat on to the incident airflow, producing drag but little lift. This can present

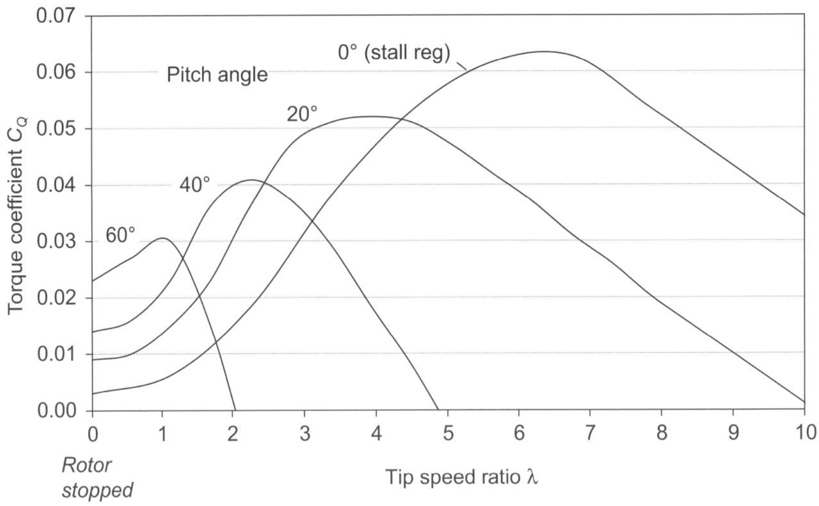

Figure 6.11 Calculated torque coefficient for a 600 kW wind turbine. The rotor is nominally designed for stall regulation at zero pitch angle, but the other curves show the beneficial effect of pitch on starting torque.

[6] For the example shown C_p^{max} is around 0.43 for $\lambda \approx 6.5$. Equation (4.4) applies.

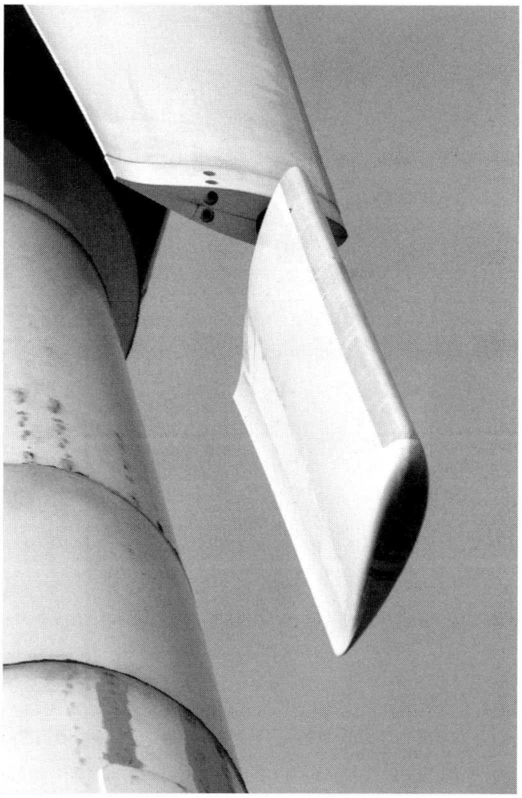

Figure 6.12 Tip air brake. The blade tip is shown pitched to 90° for maximum braking torque: it automatically deploys on loss of grid or on detection of overspeed. On this CSVP machine, the pitch system was also used for power regulation (Howden HWP28, Orkney, 1991).

a challenge for starting up.[7] A small amount of torque is, however, provided by the twisted blade sections near the root and though the effect is modest it is enough to start the rotor turning in a light wind; once turning the torque rises rapidly with rotor speed (C_Q increasing with λ; see Figure 6.11).

The starting sequence for a stall-regulated HAWT then involves shaft brake release, followed by free acceleration of the rotor until it achieves synchronous speed at which point the generator contactor is closed to put the machine online. The timing must be accurate or the rotor will continue to accelerate through synchronous speed to a potential overspeed condition. At the moment of connection there is inevitably some mismatch between the rotor and synchronous speed, but the effect is minimised by an electronic 'soft-start' that momentarily limits the starting current. In the unlikely event that a stall-regulated turbine fails to synchronise, the controller detects an overspeed and deploys blade tip air brakes (see Figure 6.12 for an example) to bring the rotor back to rest or to a safe speed.

[7] 'Motoring' a turbine up to synchronous speed is possible with an induction generator, but not recommended due to the unfavourable impact on the grid.

A similar method is used when disconnecting the turbine from the grid in extreme winds, or under a fault condition: stopping involves simultaneously opening the generator contactor and releasing the blade tip air brakes; the rotor decelerates quickly under the action of the brakes, and when its speed has dropped sufficiently a shaft brake is applied to execute a complete stop. In order to prevent excessive heat build-up or brake wear the shaft brake is rarely applied at full rotation speed, though this may be necessary in an emergency. Some smaller wind turbines incorporate electrodynamic braking, where a resistor bank absorbs the generator output during the braking cycle (capacitive excitation is required). The advantages of this technique are faster shutdown and additional redundancy.

Variable blade pitch brings the capability to control rotor speed with the generator offline, which is a powerful advantage when starting and stopping. It also enables start-up in lower winds: with reference to Figure 6.11, even a modest pitch setting of 20° can be seen to dramatically increase the starting torque. The starting sequence for a CSVP wind turbine is illustrated in Figure 6.13. Initially the rotor is parked with blades feathered at 90°. At time $T = 10$ s the rotor shaft brake is released and the blades are pitched to around 30° to generate strong starting torque. The rotor speeds up and the pitch controller brings it smoothly to synchronous speed and holds it there. At $T = 25$ s the contactor is closed but at the time of synchronisation there is negligible aerodynamic torque so although the generator is connected no power is produced; the blade pitch is then smoothly brought back to the 0° 'run' position to bring up the power. In the case shown the wind speed is below rated, so blade pitch remains in the full run position once online and the power output varies with wind speed; in higher winds, continuous pitch action would be invoked to maintain constant power.

The stopping sequence is essentially the reverse of the above: the blades are pitched to reduce power to zero before the contactor is opened; they are then brought smoothly to 90° to bring the rotor to a standstill. Starting and stopping is far more controlled on a pitch-regulated wind turbine than on a stall-regulated, with fatigue loading consequently reduced. Blade pitch also provides the primary means of braking the rotor and providing a fail-safe emergency stop capability. On loss of the grid, or on receipt of an emergency command signal, the blades are pitched rapidly to 90° to provide maximum braking torque and will bring the rotor to a standstill within a few rotor revolutions. The advantages of pitch-based braking over the use of shaft brakes is that (a) the former remains effective in any wind speed as the aerodynamic torque is effectively reversed and (b) pitch braking removes any torque load from the drivetrain, so avoids brake wear or stress on the gearbox.

Most large HAWTs are nevertheless equipped with mechanical brakes acting either on the high-speed shaft connecting the gearbox to the generator, or on the low-speed shaft immediately behind the hub (see Figure 6.14 for an example of the latter). These are similar to the disc brakes used on trucks and other large commercial vehicles, and are designed to fail safe: the brake pads are actively held off the disc by hydraulic or electromagnetic pressure, but on loss of supply strong compression springs rapidly force them onto the brake disc. Low-speed shaft brakes require a higher torque rating than high speed, and a large disc with

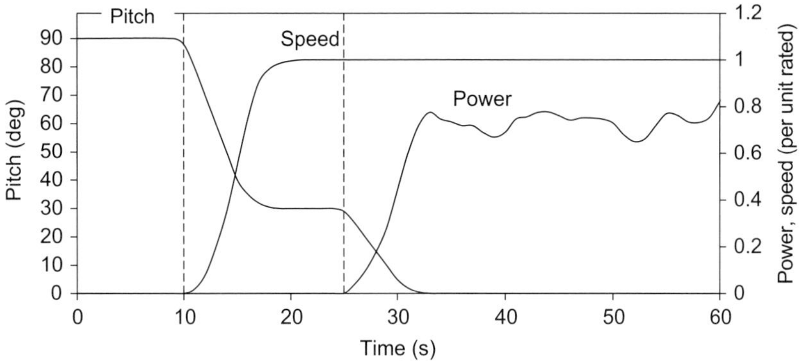

Figure 6.13 Start-up sequence for a constant-speed, variable-pitch wind turbine. At $T = 0$ s, the turbine is parked with blades feathered; at $T = 10$ s, the shaft brake is released, and at $T = 25$ s, the contactor is closed.

Figure 6.14 Shaft brakes on a Tacke 600 kW stall-regulated machine. The disc and callipers are on the low-speed shaft behind the rotor hub. Brakes may alternatively be mounted on the gearbox high-speed shaft.

multiple callipers. The advantages of low-speed over high-speed brakes are (a) better heat dissipation and (b) they act directly on the rotor without imposing any torque on the gearbox. Design codes require a wind turbine to have two fully independent fail-safe braking mechanisms, of which the shaft brakes may comprise one and the blade pitch system the other. In such cases either system must be capable of stopping the turbine from a full-power condition in high winds. Nowadays the pitch system alone may meet the redundancy criteria by incorporating independent actuators for each blade (e.g. Figure 4.9) with a fail-safe backup power supply: batteries for electric actuators and accumulators for hydraulic. The shaft brakes may then be rated purely for parking duty.

6.5 Sector Management

Sector management is a high-level control function in which different operating characteristics are implemented depending on the prevailing wind direction. In principle some form of sector management would be possible on most wind turbines, but it is commonest on VSVP machines, where control of rotor speed and blade pitch offer more flexibility. Sector management strategies include

- noise reduction
- shadow flicker prevention
- fatigue mitigation

The principle of sector management is illustrated in Figure 6.15 in regard to noise reduction. The perceived noise of a wind turbine is greatest downwind of the rotor, and is most noticeable in light winds in the absence of masking background noise (see Section 10.4). Accordingly, properties in the vicinity of a wind turbine can be protected if the machine is operated with reduced noise output for specific combinations of wind speed and direction; on a VSVP wind turbine this is achieved via a combination of rotor speed and blade pitch control. There is some reduction in power output, but the energy yield penalty is usually quite small. The WEC controller can be programmed to execute different strategies for different sectors (Figure 6.15 shows two) and can also take into account the

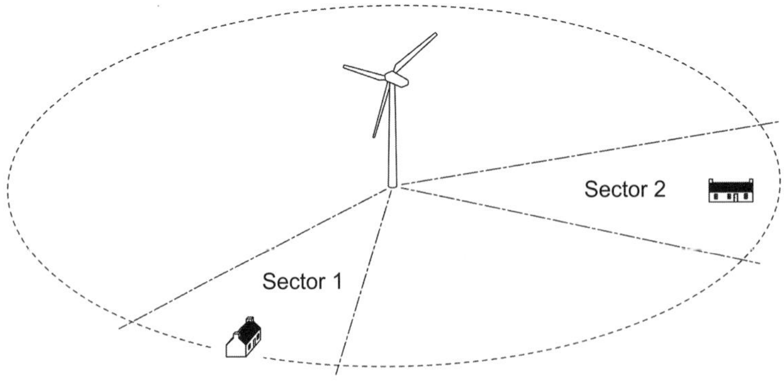

Figure 6.15 Sector management. In the example shown, the wind turbine would be programmed for reduced noise output when the properties shown are downwind of the rotor. Different strategies can be invoked for each sector.

time of day, as different noise limits may apply. The reduction in energy output is then limited to only the times when noise nuisance might occur. For shadow flicker prevention the wind turbine is stopped during periods when the sun is low and the rotor may cause intermittent shadows at nearby properties; this is described further in Section 10.3.3.

Sector management may also be used to curtail operation on sites where ambient turbulence varies with wind direction due to local topographic or array influences. In such cases the supervisory controller is programmed with a lookup table of critical wind directions and wind speed ranges, so that the wind turbine can be operated at reduced power, or stopped altogether, depending on prevailing conditions. This technique may be particularly applicable in complex terrain, or in large windfarms where particular wind directions causes multiple wake interactions; for more on this topic see Section 9.3.

6.6 Exercises

6.6.1 Overspeed

Figure 4.3 shows the dimensionless power coefficient for a fixed-pitch wind turbine. Assuming the rotor diameter is 30 m, if the turbine goes into overspeed on loss of the grid (and following failure of its braking systems) approximately what rotational speed will it achieve in a hub height wind speed of (a) 15 m s^{-1} and (b) 25 m s^{-1}? Neglect drivetrain friction.

6.6.2 Passive Yaw Control

The early wind turbine designed by Christian Riisager (see Figure 1.4) retained a fantail yaw drive mechanism similar to those seen on traditional windmills. Suggest why such a mechanism might be inherently less accurate than an active yaw control system, and why it may suffer from steady-state error.

6.6.3 VSVP Measurements

The scatter plot in Figure 6.16 shows 10 min average measurements (normalised as %) taken from a variable-speed, variable pitch wind turbine; the data are plotted against hub height average wind speed. What is the parameter being measured? Explain your answer.

6.6.4 Torque-Speed Characteristic

The controller of a variable-speed wind turbine is programmed to achieve optimum C_p over a broad range of wind speeds: Equation (6.3) gives the necessary relationship between steady-state generator torque T_G, tip speed ratio λ, and rotor angular velocity Ω. Derive this equation from a suitable starting point, and show your working. For simplicity assume it is a gearless wind turbine.

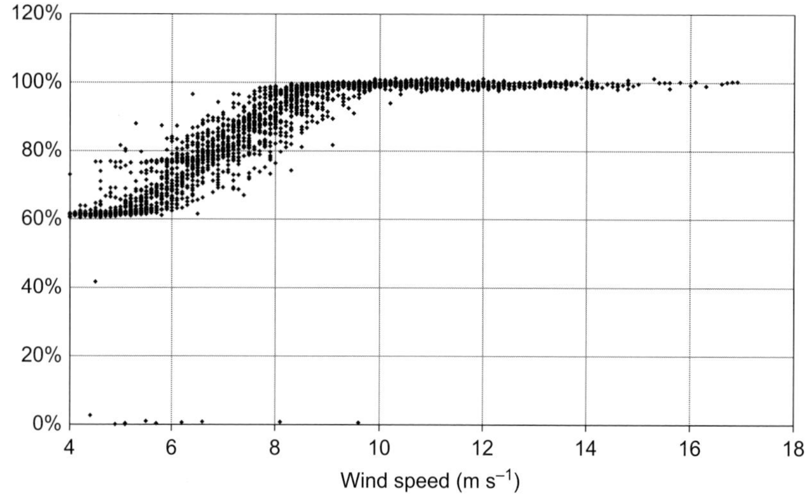

Figure 6.16 See Exercise 6.6.3.

6.6.5 Power Regulation

Variable-speed variable-pitch (VSVP) wind turbines are capable of much more accurate real-time power control than constant-speed variable-pitch (CSVP) machines, as seen by a comparison of the scatter plots in Figure 6.5 and Figure 6.10. Would you expect to see a comparable improvement in the envelope of blade thrust loading (or blade root axial bending moment, which is proportional to thrust)? Explain your answer.

6.6.6 Sector Management

A wind turbine is installed on a site in complex terrain. When the wind comes from certain directions the turbine's operation is curtailed to protect it from extreme turbulence; this is achieved using the technique of sector management. Normally the turbine would operate in wind speeds in the range 4–25 m s^{-1}, but on this site it must be stopped when the following combinations of wind speed and direction arise:

- speed exceeding 15 m s^{-1}, directions 330° to 045°
- speed exceeding 13 m s^{-1}, directions 165° to 185°

The site mean wind speed is 8.8 m s^{-1} with a Rayleigh distribution and an omnidirectional distribution pattern (equal occurrence of all wind directions). Calculate the percentage of online operating time lost due to sector management. Would you expect the percentage of energy lost to be higher or lower than the percentage of downtime? Explain your answer.

6.6.7 Curious Behaviour

An early 60 kW wind turbine was installed on a Scottish farm; the turbine was of three-blade, stall-regulated design, with active yaw drive powered by electric motors. On a calm, very cold, winter day the wind turbine was observed to behave in an unusual way. The rotor was stopped due to lack of wind, but the nacelle would continuously yaw clockwise for several minutes, then reverse and yaw anti-clockwise for a similar length of time, and so on, back and forth. This behaviour continued for over an hour – until the sun came out. The control system was not technically faulty, but what was happening, and why?

CHAPTER 7 STRUCTURAL LOADING AND RESPONSE

7.1 INTRODUCTION

Wind turbines are large and relatively flexible structures, and they are subject to a complex combination of forces of aerodynamic, gravitational, gyroscopic, centrifugal, and electromechanical origin. The loading is also highly time-dependent, with some forces varying periodically, others apparently randomly. A typical example is seen in Figure 7.1, which shows the blade root flapwise moment on a 330 kW wind turbine starting up from rest. Initially the trace contains only a small cyclic component due to gravity,[1] but as the rotor speeds up the aerodynamic loading increases; the magnitude of cyclic load variation, which is due to a combination of aerodynamic and dynamic effects, also grows. At around t = 145 s the generator is connected, and the sudden change in electrodynamic torque when the contactor is closed causes a load transient that is felt at the rotor.

Although complex, traces like this can be analysed to extract the different loads acting on the structure, and allow their magnitude and time dependency to be quantified. With such knowledge load predictions can be made using a combination of blade element momentum theory and traditional mechanical engineering methods. Static load predictions are relatively straightforward, often requiring only the BEM code to predict external aerodynamic loads. Dynamic analyses, however, require a structural model of the wind turbine in which the BEM code provides the external forcing functions. The present chapter describes the basis of such models, drawing on an example that was developed at a time when the forces on WECs were less well understood, and presents field measurements from a mid size wind turbine that was tested at the SCE Palm Springs site in California (Wehrey et al., 1988).

7.2 FUNDAMENTALS

7.2.1 Static and Dynamic Loads

The loads (forces and moments) acting on a wind turbine can be broadly categorised as static or dynamic; in general the net loading is a combination of the two. Static or steady loads are constant or vary with slow time dependency and are independent of inertia: the aerodynamic thrust loading

[1] The gravity bending moment is normally associated with edgewise loading, but on a twisted blade, there is a small gravitational component in the flapwise sense.

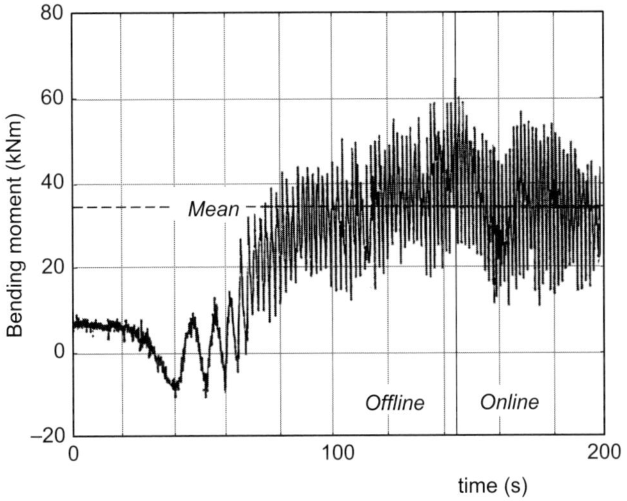

Figure 7.1 Blade root flapwise bending moment on a 330 kW wind turbine during start-up. The trace is a combination of aerodynamic and dynamic loading, with high cyclic component. (Based on analysis of data for the HWP26 wind turbine; Wehrey et al., 1988)

on a rotor in steady winds, for example, depends only on the wind speed and the blade geometry. Dynamic loads vary on a short timescale, measured in seconds, and contain a significant inertial component (forces arising due to acceleration or deceleration): an example would be the braking torque on the rotor shaft during an emergency stop, or the stress in a vibrating blade. To predict the complete forces acting on a WEC requires a combination of aerodynamic and dynamic response theory; the latter is a large topic in itself[2] but we can gain key insights by first understanding the behaviour of a single degree of freedom system.

7.2.2 Dynamic Response of a Simple System

A simple mechanical system is shown in Figure 7.2, defined by a mass, spring, and damper. The mass is constrained to move in one dimension, with governing equation of motion

$$m\ddot{x} + c\dot{x} + kx = F(t) \tag{7.1}$$

where

m	:	mass
x	:	displacement
c	:	damping coefficient
k	:	stiffness coefficient (spring rate)
$F(t)$	:	arbitrary applied force, time-dependent

[2] For a good overview, see Chapter 5 of Freris (1990).

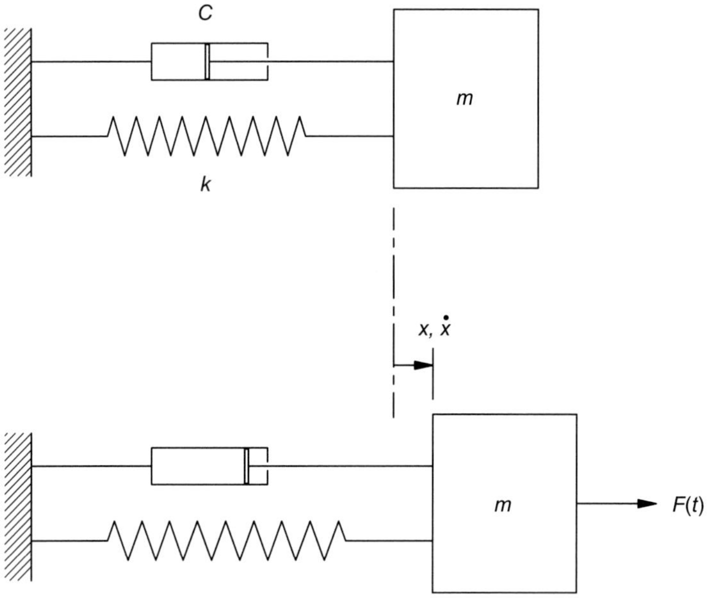

Figure 7.2 Model of a single degree of freedom system, with mass m, stiffness k, and damping c.

In the absence of damping the system has natural frequency ω_n where

$$\omega_n = \frac{1}{2\pi}\sqrt{\frac{k}{m}}\ \text{(Hz)} \tag{7.2}$$

If the undamped mass is displaced from its equilibrium position by an extent X_0 it will experience a spring restoring force $F_0 = -kX_0$, and when released it will vibrate indefinitely with frequency ω_n and amplitude X_0. If there is finite damping in the system, however, energy will be dissipated and the vibration will die away. Typical behaviour of a lightly damped system is seen in Figure 7.17, which shows the edgewise vibration of a HAWT blade on a test stand. The blade has been released from an initial displacement and the slow vibration decay is due to the small amount of material damping present (see Section 7.7.3).

If the mass in Figure 7.2 is subject to a sinusoidal forcing function of the form $F(t) = F_0 \sin\omega t$, where ω is an arbitrary frequency, it will respond at this frequency with amplitude X; the 'dynamic amplification ratio' Q is then the ratio of X to the amplitude X_0 occurring under a static load, or

$$Q = \frac{Xk}{F_0} \tag{7.3}$$

In non-dimensional form the frequency response of the system is then characterised:

$$Q = \frac{1}{\sqrt{\left([1 - r^2]^2 + [2\zeta r]^2\right)}} \tag{7.4}$$

where r is the frequency ratio, and ζ the damping factor:

$$r = \omega/\omega_n \tag{7.5}$$

$$\zeta = \frac{c}{2\pi\omega_n} \tag{7.6}$$

Figure 7.3 shows Q as a function of r and ζ, illustrating how the dynamic response is greatly magnified near resonance ($r = 1$) but diminishes as the applied loading tends towards high frequency, when inertia prevents the system from responding. At very low frequency the response tends to that obtained under static loading. The plot shows the classical response of a system with only a single degree of freedom, and one resonant frequency; a complex structure such as a wind turbine, however, has multiple degrees of freedom, and many natural frequencies. The behaviour of such systems is analysed using the technique of modal analysis.

7.2.3 Modal Analysis

If the mass shown in Figure 7.2 were to be unconstrained in all dimensions it would require six coordinates (three in translation and three rotation) to describe its motion. The system would then have 6 degrees of freedom and consequently six natural frequencies. In general a system of n connected elements has 6 n normal modes of vibration each with its own frequency (aka eigenfrequency), with each mode describing a unique deflected shape of the structure. The normal modes are linearly independent, hence the complete response of the structure can be modelled as a linear superposition of mode shapes. The theory of multi-DOF systems is beyond

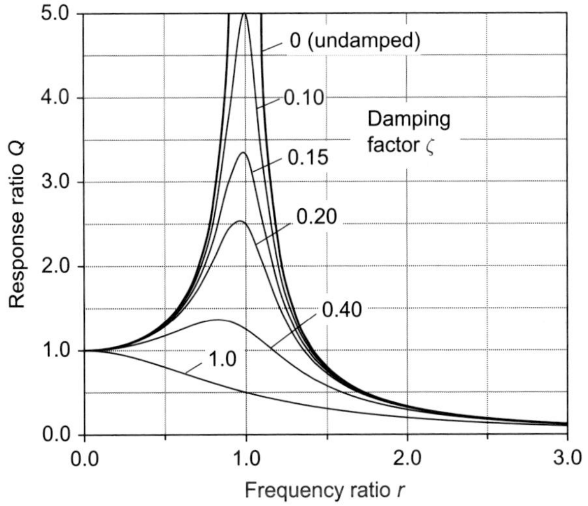

Figure 7.3 Frequency response of a single degree of freedom system.

the remit of the present book,[3] but the following is a condensed summary. The governing equation of motion for a system with N degrees of freedom is similar in form to that of a single-DOF system, but in matrix form:

$$M\ddot{X} + C\dot{X} + KX = F(\mathrm{t}) \tag{7.7}$$

The mass, stiffness, and damping terms (M, K, and C) here represent $N \times N$ matrices, while force F and displacement X and its derivatives are N-dimensional vectors. Equation (7.7) then represents a set of N coupled equations, and if proportional damping is assumed these reduce to N uncoupled equations each equivalent to a single-DOF system. By setting both the damping and forcing function to zero the eigenfrequencies and mode shapes of the system are found as solutions to the reduced equation

$$M\ddot{X} + KX = 0 \tag{7.8}$$

The response of each normal mode is analogous to that of a single-DOF system, and its eigenfrequency to the associated natural frequency. To illustrate, Figure 7.4 shows the finite-element representation of a medium-scale wind turbine used in an early dynamic response study (Wehrey et al., 1988). The blades and tower are modelled as multi-element beams rigidly connected at the hub; the tower–foundation interface is assumed to be rigid. The elemental mass and stiffness values are assessed using traditional engineering design calculations in order to populate matrices M and K; Equation (7.8) is then solved numerically to yield the mode shapes and eigenfrequencies. In principle there are several hundred modes, but the great majority are at frequencies too high to be excited in normal operation and for practical purposes the WEC behaviour can be described by the lowest frequency modes. Figure 7.5 shows the lowest three mode shapes and their corresponding eigenfrequencies.

The first mode contains significant axial (fore–aft) tower motion: for practical purposes it can be considered to represent the first tower vibration mode, and interpreted as a single-DOF system in which the nacelle and rotor are replaced by a lumped mass. Modes 2 and 3 contain blade axial motion while the tower remains almost stationary. These modes have almost identical eigenfrequencies and may be broadly interpreted as the lowest blade flapwise mode, which can alternatively be modelled by treating the blade as a cantilever bean with rigidly fixed root. Higher frequency modes include transverse and torsional motion, but the response of the complete wind turbine can usually be satisfactorily described by a dozen or so mode shapes. The dynamic response of the WEC is then computed as the modal response to a complex forcing function comprising cyclic loads, which may be subdivided into deterministic and stochastic components. These terms are described below.

7.2.4 Deterministic and Stochastic Loads

Deterministic loads are predictable in regard to their frequency and magnitude, and arise from the interaction of the rotating blades with steady but non-uniform airflows or with gravity. Examples of

[3] Recommended reading: *Theory of Vibration with Applications* (Thomson, 1993).

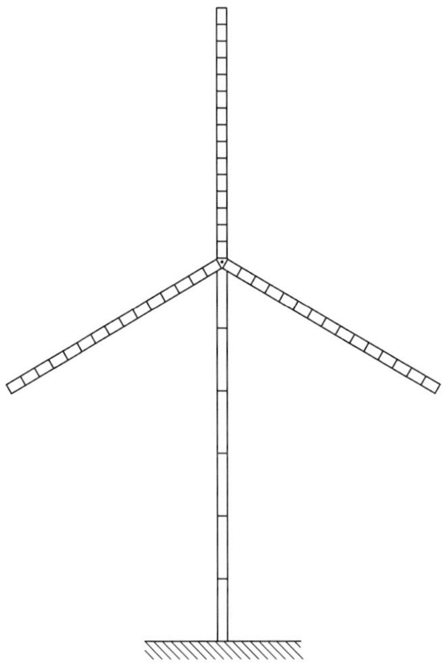

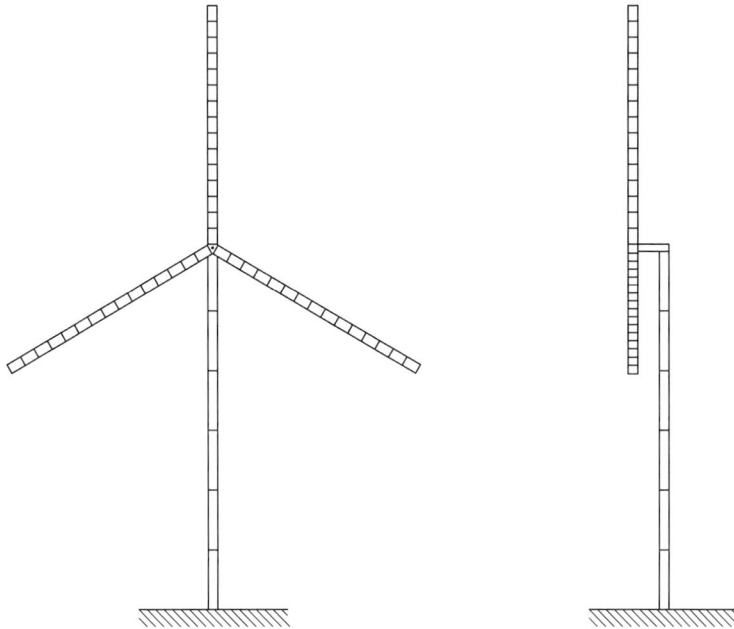

Figure 7.4 Finite-element model of three-blade WEC, based on the analytic model of the HWP26 wind turbine described in Wehrey et al. (1988). Measurements made on this machine are described in Section 7.8.

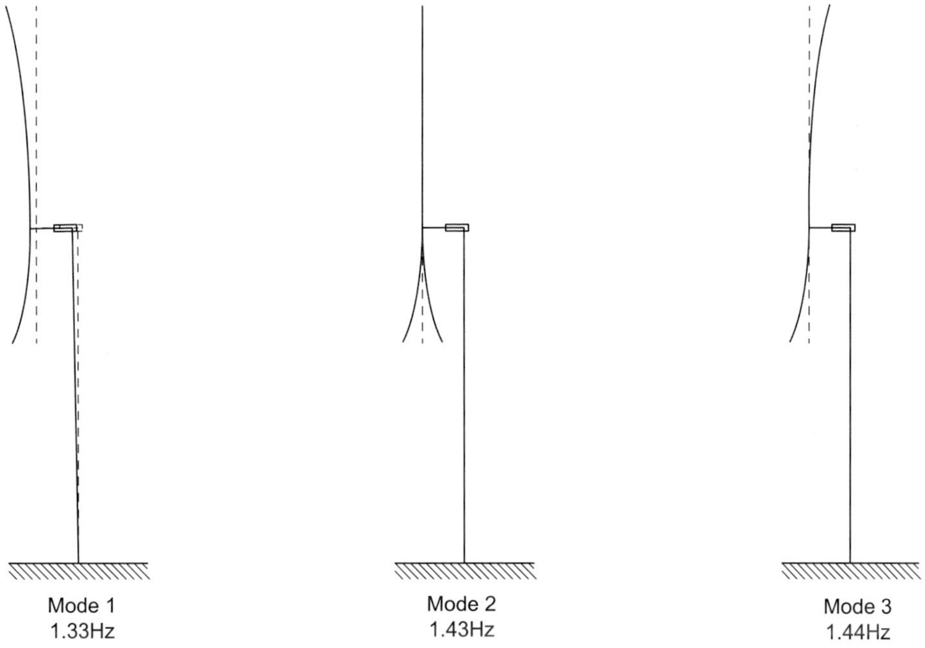

Figure 7.5 Lowest three mode shapes and eigenfrequencies for the WEC model shown in Figure 7.4.

deterministic loads include the blade root gravitational bending moment, or the cyclic aerodynamic loading due to wind shear (see Figure 7.9). Seen in the blade frame of reference these loads vary sinusoidally at the rotation frequency, reversing once per revolution. In non-dimensional notation this corresponds to a forcing frequency of '1P' where

$$\text{Nondimensional frequency}(nP) = \frac{\text{Forcing frequency}}{\text{Rotor rotation frequency}} \tag{7.9}$$

This non-dimensional notation is commonly used to characterise cyclic loads whose frequency is determined by the rotor rotation frequency or its higher multiples (2P, 3P, 4P, etc.). Hence irrespective of turbine size or actual rotation speed, the blade root gravity bending moment will always be at 1P, and the frequency at which blades pass the tower will always be 3P on a three-bladed wind turbine, or 2P on a two-bladed one. The way in which higher harmonic loads are generated by the interaction of the WEC rotor with the airstream is explained in Section 7.3.4.

Stochastic loads are caused by atmospheric turbulence and vary quasi-randomly, depending on the magnitude and spatial distribution of turbulence at the rotor plane.[4] An example of a turbulent wind record with purely stochastic content is shown in Figure 2.4. Although the frequency content of the turbulent wind spectrum is apparently random, however, stochastic blade loads concentrate energy at discrete multiples of the rotor rotation frequency (1P, 2P, 3P) due to the phenomenon of rotational sampling: this is explained in Section 7.3.4.

7.3 Aerodynamic Loads

7.3.1 Reference Frames

Blade aerodynamic theory is described in detail in Chapter 3. Blade loads are commonly referred to coordinate axes in the rotor plane, as shown in Figure 7.6:

- Axial (thrust) force acts parallel to the rotor axis, giving rise to the axial bending moment, which tends to bend the blade out of the rotor plane (as seen in Figure 3.8).
- Tangential force acts in the plane of the rotor, giving rise to torque – hence power – and to the blade tangential bending moment. The tangential moment at the blade root is approximately equal to the blade's contribution to net rotor torque.

For the purposes of blade design the loads are often resolved into the local aerofoil coordinates, taking account of twist or pitch. The local axes are designated flapwise and edgewise (or chordwise) where:

- Flapwise force acts normal to the blade chord line: the flapwise moment causes bending about the chord line, i.e. in the direction in which the blade is naturally flexible.

[4] Although atmospheric turbulence is three-dimensional, the longitudinal (axial) component has the greatest effect on aerodynamic loading due to its influence on the inflow angle ϕ at the rotor plane.

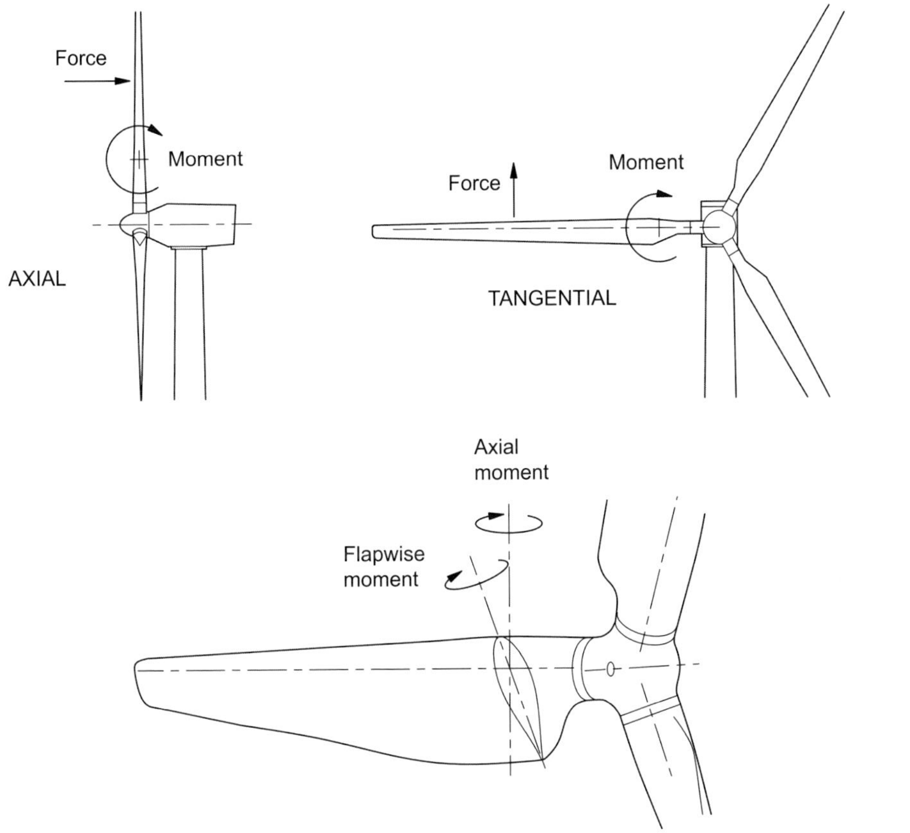

Figure 7.6 Rotor blade coordinate systems. Axial and tangential forces and moments are referred to the rotor plane; flapwise and edgewise are referred to the local aerofoil chord line.

- Edgewise force acts parallel to the chord line; edgewise bending is then normal to the chord line, in which direction the blade is inherently stiff.

On an untwisted blade with zero pitch the two frames of reference are the same, i.e. axial = flapwise, and tangential = edgewise. If the blade is twisted, however, or with non-zero pitch setting, then care is needed when specifying the coordinate system. On a twisted blade, for instance, the flapwise root moment will contain a gravitational component due to coupling of the blade weight into the flapwise axis – this effect is seen in Figure 7.1 – whereas a purely axial moment has no gravity component. Likewise, the high aerodynamic braking load generated by a blade section pitched at 90° (as in Figure 6.12) acts flapwise with respect to the section itself, but tangentially with regard to the rotor. As a general rule, on a rotor in normal power production the blade root flapwise and axial bending moments are usually of similar magnitude, and the corresponding curves of bending moment against wind speed are fairly indistinguishable. The tangential and edgewise root moments may, however, differ significantly in both magnitude and sign (the edgewise moment may be negative) due to the cross-coupling effect of blade twist.

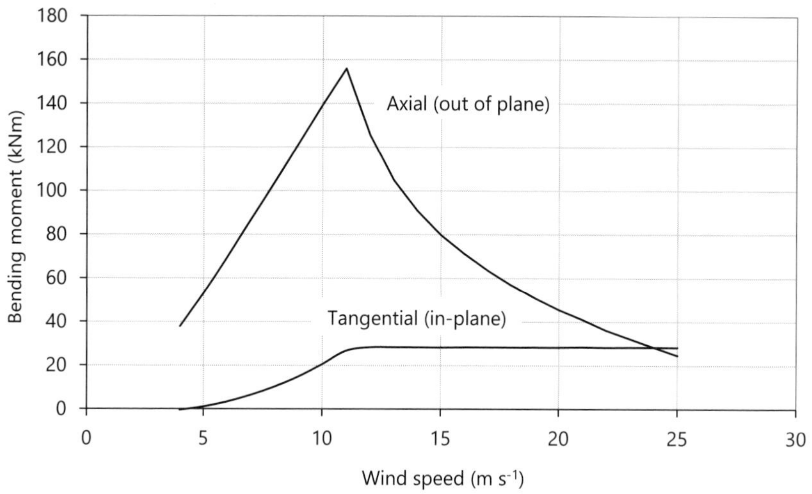

Figure 7.7 Blade root aerodynamic bending moments calculated for a 330 kW pitch-controlled WEC.

7.3.2 Steady Loads

Examples of blade aerodynamic loads calculated from BEM theory are seen in Figure 7.7, which shows axial and tangential root bending moments plotted against mean wind speed. The results are for a constant-speed variable-pitch rotor with power regulation, and they can be compared with the net loads on an ideal rotor seen in Figure 4.1: blade axial bending moment has the same form as thrust, and tangential bending moment the same form as the power curve. The axial moment is generally much larger than the tangential, reflecting the underlying blade element force distributions seen in Figure 3.10. The reduction in mean axial load with wind speed above rated is characteristic of positive pitch control, as explained in Section 4.2.

7.3.3 Deterministic Loads

The principal sources of deterministic aerodynamic loading on a blade are wind shear and tower shadow: these effects are illustrated in Figure 7.8, which shows the non-uniform wind profiles arising in each case. Wind shear is the variation in freestream wind speed with height: during one complete rotation the rotor blade experiences higher air velocity at the top of its sweep than at the bottom, giving rise to a quasi-sinusoidal variation in loading. Tower shadow affects the blade only in the lower half of its sweep: as it passes in front of the tower the blade experiences a region of reduced air velocity, which causes an impulsive change in aerodynamic loading. The effects of wind shear and tower shadow are illustrated in Figure 7.9, which shows the blade root axial moment through one complete revolution; the data are calculated for a 33 m rotor at a mean wind speed of 11 m s^{-1}. The cyclic variations due to shear and shadow are shown separately, and combined in the net blade load.

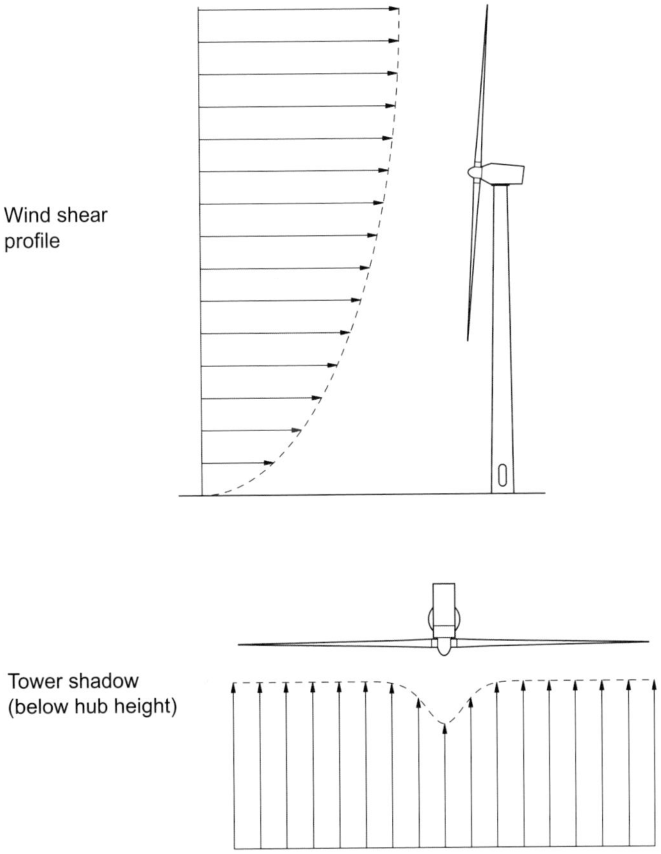

Figure 7.8 Influence of (*top*) wind shear and (*bottom*) tower shadow on the incident wind speed. Wind shear causes a quasi-sinusoidal change in blade load; tower shadow is an impulsive load.

The load variation due to wind shear (dashed line) here corresponds to around 20% difference between the top and bottom blade positions. As this occurs once every rotation (non-dimensional frequency 1P) it has a significant influence on fatigue life: the blade shown would experience more than 10^8 rotation cycles in a 20 year lifetime. The impulsive loading due to tower shadow (solid line) represents a smaller variation than that due to shear but contributes to a higher net cyclic variation: in the combined load waveform (bold line) the cyclic range is increased to 30% of the mean. The magnitude of the tower shadow impulse is a function of the tower diameter and blade clearance, and is normally modelled assuming 2D potential flow upstream of the tower. The axial wind velocity is here given by

$$V_r = U\left[1 - \left(\frac{R}{r}\right)^2\right] \tag{7.10}$$

where V_r is the reduced wind velocity, U is the freestream velocity, R is the tower radius, and r the radial distance of the blade section from the tower centreline.

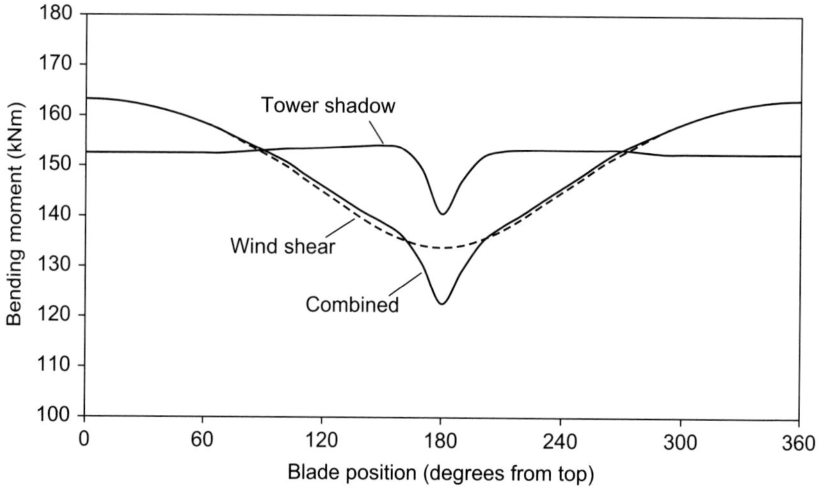

Figure 7.9 Effects of wind shear and tower shadow on blade root axial moment. Loads shown are calculated for a 33 m/330 kW HAWT at a wind speed of 11 m s^{-1}, shear index 0.16, and hub height 25 m.

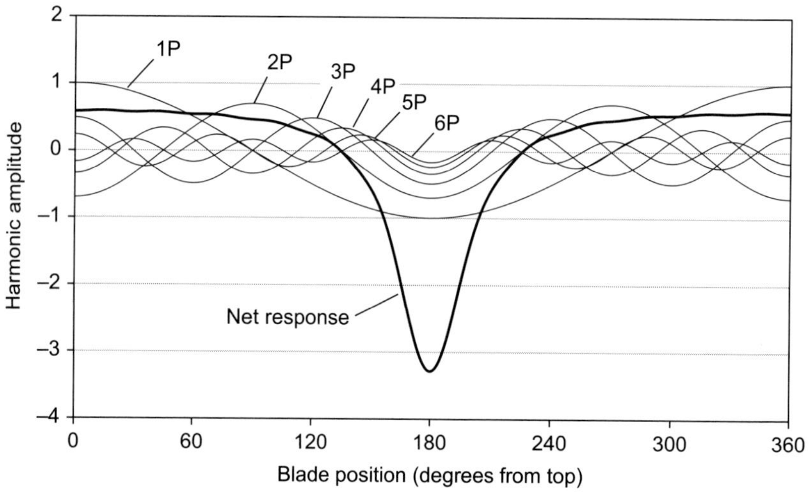

Figure 7.10 Decomposition of tower shadow impulse load (bold line) into constituent *n*P harmonics; the example is illustrative, with harmonic amplitudes normalised by the 1P magnitude.

The impulsive load due to tower shadow is further shown in Figure 7.10, together with a series of harmonics that combine to produce the same observed waveform (alternatively the tower shadow load waveform can be decomposed into the given harmonics). The dynamic response of the blade and the greater WEC structure may then be susceptible to excitation at each of the individual harmonic frequencies (1P, 2P, 3P, etc.). Harmonic load amplitudes decrease with frequency as seen

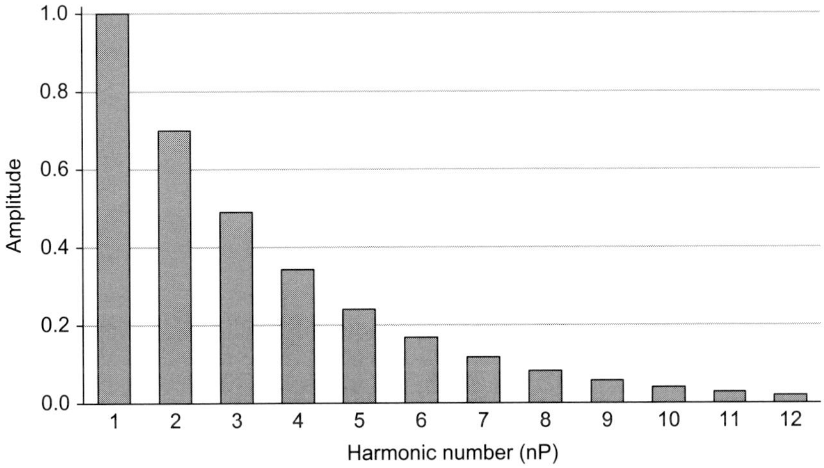

Figure 7.11 Harmonic content of the tower shadow impulse illustrated in Figure 7.10. The blade or other part of the WEC structure may suffer dynamic excitation due to the major harmonics.

in Figure 7.11, and in practice only the first few are significant, but they are important for the following reasons:

- Harmonics may excite a modal resonance: for instance the first blade flapwise frequency may lie well above 1P, but close to the 2P harmonic component of tower shadow.
- Some harmonics appear in the net rotor loads, and in particular the power output: these can adversely impact on the blade pitch control system; see Section 6.3.2.
- A blade load occurring at frequency nP can feed into the stationary tower or nacelle structure at a frequency of $(n \pm 1)$P, giving rise to further resonance possibilities. This is the phenomenon of ground resonance, well known in helicopter dynamics (Bramwell, 1976).

7.3.4 Stochastic Loads

Stochastic aerodynamic loads are caused by atmospheric turbulence, which can be characterised as random eddies superposed on the freestream velocity. The turbulence frequency distribution is commonly described by a Kaimal or von Karman spectrum (Burton, 2011). The spectrum experienced by a rotating blade, however, contains distinct harmonics at multiples of the rotation frequency due to the phenomenon of rotational sampling. This can be visualised as the blade repeatedly ‘chopping’ through turbulent eddies as they cross the rotor plane, as shown in Figure 7.12. Consequently in the blade frame of reference the turbulent energy is concentrated around discrete frequencies, as shown in Figure 7.13. This phenomenon is similar to the effect of tower shadow (see above) with the difference that turbulent eddies occur at random locations in the rotor plane rather than a fixed position. The spectral peaks due to rotational sampling tend to be wider than those associated with structural eigenfrequencies.

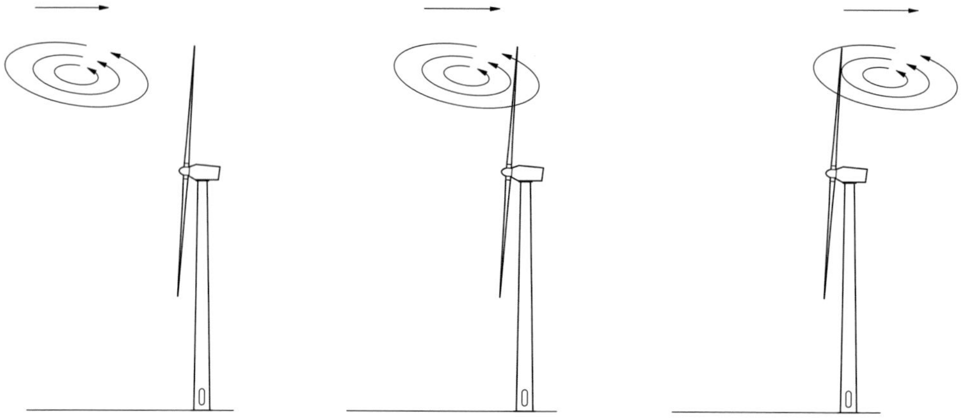

Figure 7.12 Rotational sampling of turbulence. An eddy passing through the rotor plane is repeatedly 'chopped' by each blade at its rotation frequency (1P), giving rise to *n*P harmonics in the blade load spectrum.

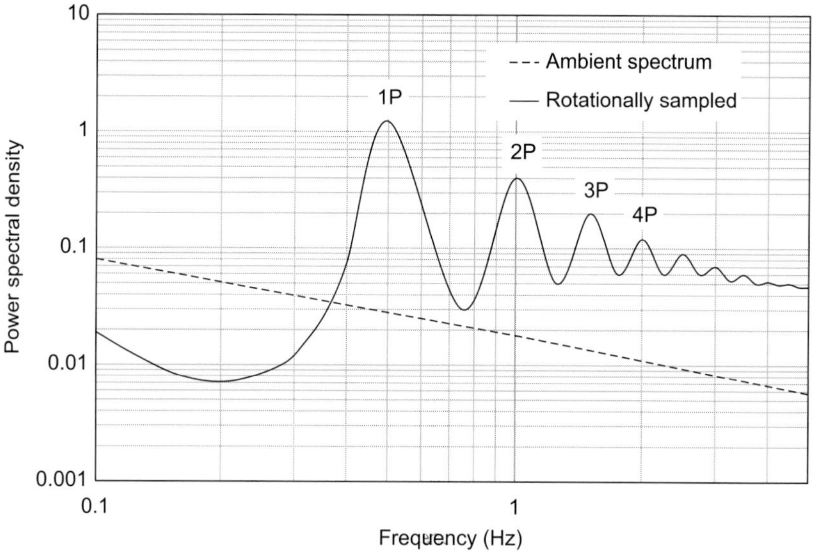

Figure 7.13 Ambient wind turbulence spectrum and the spectrum experienced by a rotating blade due to rotational sampling. The spectral peaks occur at rotor frequency 1P and its harmonics.

Rotational sampling introduces harmonic content in the blade aerodynamic loads at 1P, 2P, 3P, etc., so potential resonance at these frequencies must be avoided in the WEC structural design. Critical rotation speeds may also occur during operation, particularly on variable-speed wind turbines for which 1P and its multiples are variable frequencies and the potential for resonance is therefore increased. Not all the blade harmonic loads appear in the net rotor loads, however, and in general on a balanced rotor with N blades the net rotor loads contain harmonics only at multiples of N. Thus on a three-blade rotor the net thrust and power contain harmonics only at 3P, 6P, 9P, etc., while on a two-bladed machine the harmonic loads are at 2P, 4P, 6P, etc. This is a result of the angular separation of the blades and the phase difference it imparts to their harmonic loads. It explains, for instance, why the 1P cyclic gravity loading on individual blades cancels to zero at the hub and is not seen in the resultant torque or power.

7.4 Gravitational Loads

Gravity gives rise to both steady and cyclic loads. Examples of steady gravitational loads are the dead weight of the nacelle and rotor assembly acting on the tower top, or the weight of the entire WEC structure on the foundation plinth. Cyclic gravity loads are experienced by the blades, most significantly at the root, where the gravity bending moment reverses with 1P frequency. Gravitational loads become relatively more significant with wind turbine size as a consequence of the 'square-cube law' (see Section 1.5) and for very large blades the weight moment is the dominant cyclic load in the edgewise direction, greatly exceeding the in-plane aerodynamic loading. This has important fatigue implications: for instance on a multi-MW wind turbine with an average rotor speed of 10 rpm, each blade will experience more than 10^8 gravity load reversals during a 20 year design lifetime.

7.5 Gyroscopic and Centrifugal Loads

Gyroscopic loading arises principally as a result of yaw motion, when the wind turbine nacelle steers to face into the wind. The resulting gyroscopic torque (or couple) M acting on the rotor is proportional to the yaw rate and rotor angular momentum according to

$$M = J\Omega\dot{\gamma} \tag{7.11}$$

where J is the inertia of the rotor, Ω its angular speed, and $\dot{\gamma}$ the yaw rate. The geometry is illustrated in Figure 7.14. The axis of M is orthogonal to both Ω and $\dot{\gamma}$, hence the gyroscopic couple exerts an overturning moment about a horizontal axis in the rotor plane; in the rotating frame this manifests as a reversing axial bending moment at the blade root with 1P frequency;[5] the rotor shaft experiences a bending moment of twice this magnitude. Due to their inherently high rotation speeds this can be an issue for small wind turbines, particularly those configured for downwind free-yaw operation, which may experience high yaw rates in turbulent winds. Some early machines were prone to main shaft failure for this reason, until redesigned with more substantial components and/or yaw rate dampers (Hughes et al., 1994). In contrast, gyroscopic loading is rarely an issue on very large wind turbines, due to a combination of low rotor speed and yaw rate. The loads will be greater on floating offshore WECs due to their additional degrees of freedom, but rotation rates in

[5] The axis of couple M is fixed in the stationary frame of reference, but in the rotating frame of reference it reverses every revolution, giving rise to cyclic blade bending at 1P.

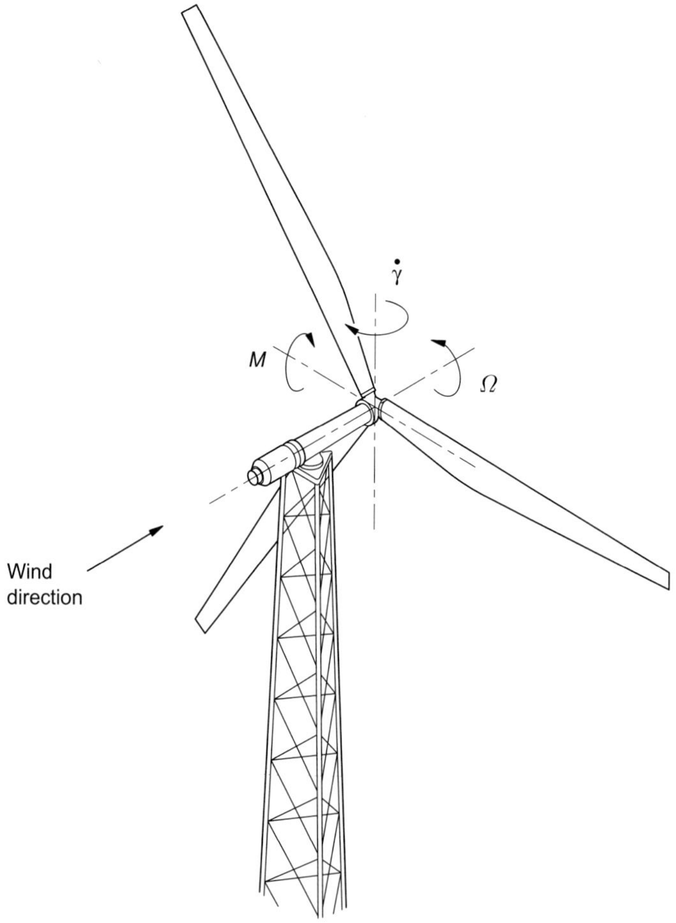

Figure 7.14 Gyroscopic loading due to yaw motion. Gyroscopic overturning moment M causes cyclic loading of the blade root and rotor shaft. Downwind free-yaw machines can be particularly susceptible due to high yaw rates.

pitch (in the nautical sense) are probably still too low for gyroscopic effects to be a significant factor.

Centrifugal loading is – perhaps surprisingly – not a major concern on large wind turbines as it does not give rise to significant blade bending loads. The pull-out force on blade root attachment bolts is dominated by flapwise and edgewise bending, with only a minor contribution due to centrifugal force (which may however be more significant on small machines). Centrifugal force can, however, give rise to cyclic loading on an unbalanced rotor, causing nacelle yaw or tower vibration. These effects are illustrated in Figure 7.15: an offset mass m gives rise to centrifugal force F_c acting radially outwards in the rotor plane; the horizontal component of F_c causes transverse tower vibration at a frequency of 1P, and a cyclic yaw moment with the same frequency and amplitude XF_c where X is the distance between the rotor plane and nacelle rotation

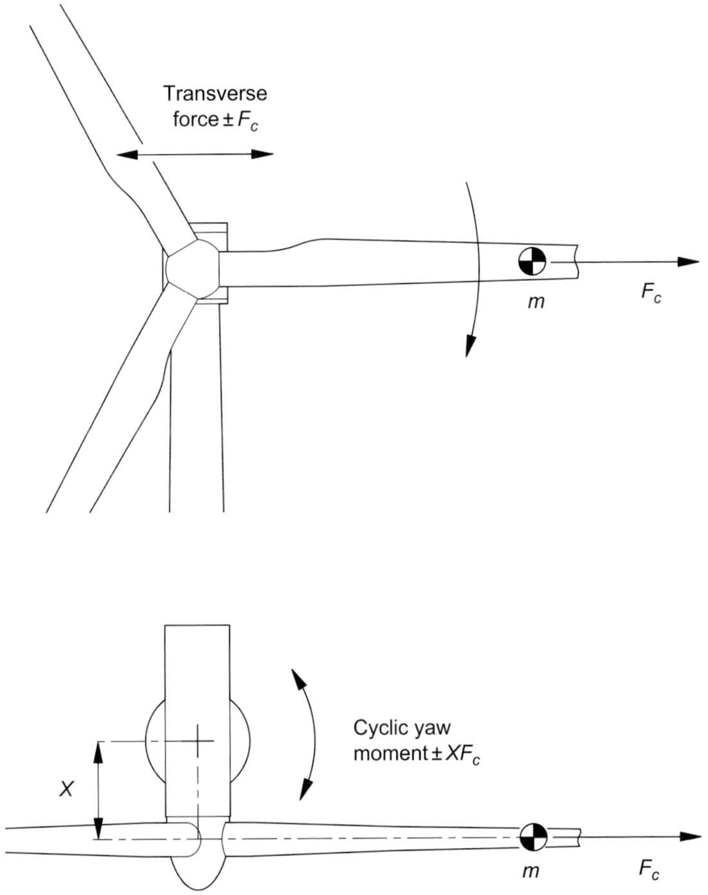

Figure 7.15 Cyclic loads on an unbalanced rotor. Offset mass m is subject to centrifugal force F_c, causing both tower transverse excitation and cyclic yaw moment at 1P frequency.

axis. The same effects may also be caused by aerodynamic imbalance, e.g. if the pitch setting on one blade is slightly different from the others. Both pitch error and rotor mass imbalance can, however, be largely avoided in practice. Blade mass balancing is described in Section 8.5.3.

7.6 Electromechanical Loads

Electromechanical forces originate with the generator, where a torque is mutually developed between its stationary and rotating parts; the electrical characteristics of different generator types and their influence on wind turbine control are discussed in Chapters 5 and 6. Electromechanical torque can be a critical design parameter in terms of extreme or fault load cases, as explained with reference to Equation (6.1). In steady-state operation electromechanical and aerodynamic torque are in equilibrium and the rotor speed is constant; in the event of a short-circuit fault, however, the generator torque may suddenly increase by almost an order of magnitude, giving rise to extreme dynamic loading on the drivetrain and rotor blades. Under these conditions the generator effectively acts as a violent brake and the high inertial loading is a potential source of structural damage. Short-circuit fault is therefore a key design case, and the GL design recommendations at one time specified extreme torques of 8 times the rated value for induction generators, and 10.5 times rated for synchronous machines (Germanischer Lloyd, 1993). A converse case is loss of grid, when the generator torque disappears, leaving the rotor aerodynamic torque unbalanced. This results in a potential runaway condition, and represents an important design case for which fail-safe aerodynamic braking is required (see Section 6.4).

7.7 Vibration and Resonance

As noted earlier, a wind turbine has many normal modes of vibration and is subject to many excitation frequencies, some fixed but others varying with rotor speed. The potential for structural resonance is therefore significant, and presents a challenge in regard to both structural design and operation. The task is compounded by the low level of material damping in a HAWT structure, which can be illustrated by the example below. Figure 7.16 shows a 19 m HAWT blade on a test stand, where its edgewise vibration characteristics are being measured: the tests were part of an exercise to develop a mechanical vibration damper, further described in Section 7.7.3. The blade is seen being given an initial edgewise

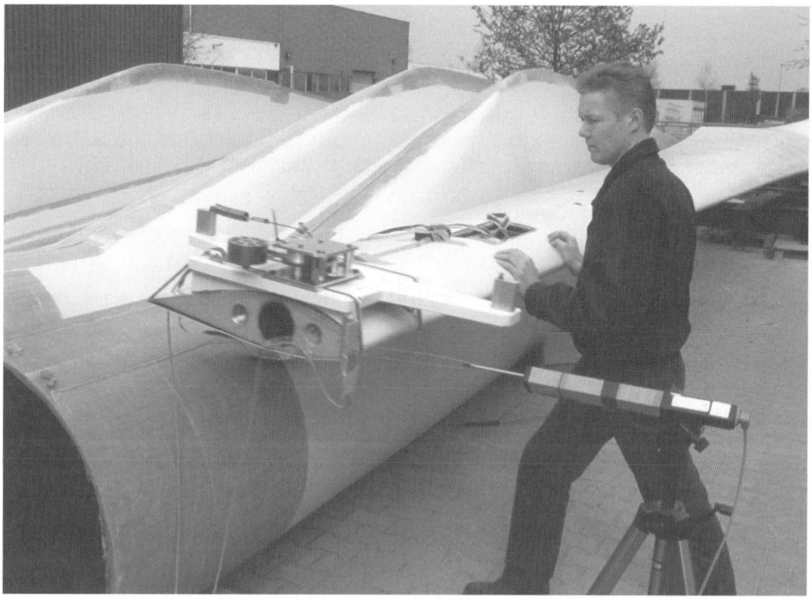

Figure 7.16 Pushing the envelope. Edgewise vibration tests on an APX40 blade at the Aerpac factory, Almelo, in 1997. The apparatus at the blade tip was designed to simulate 16 g centrifugal acceleration on a mechanical vibration damper (see Section 7.7.3).

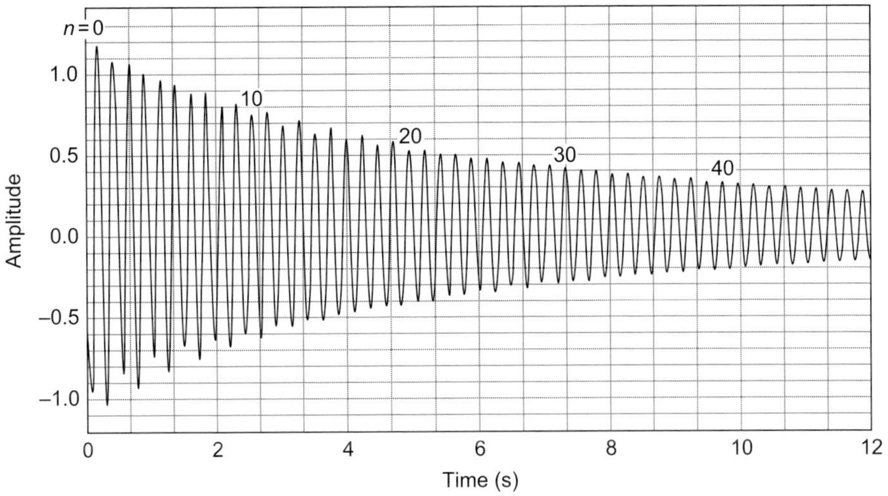

Figure 7.17 Edgewise vibration decay of the blade seen in Figure 7.16. From this trace, a material damping coefficient ζ of 0.006 is calculated, meaning the stationary blade is practically undamped. Equation (7.12) applies.

deflection so that its free vibration amplitude can be measured over time. A resulting decay trace is shown in Figure 7.17, from which the material damping coefficient ζ is found from

$$\zeta \cong \frac{\ln(A_0/A_n)}{2\pi n} \tag{7.12}$$

in which A_0 and A_n are the amplitudes at, respectively, the start of the measurement period and after n cycles (the cycle counts are shown in the figure[6]). In this example a coefficient ζ of 0.006 is obtained, evidence of very low material damping, and for most practical purposes the stationary blade may be treated as undamped. The blade shown was an Aerpac APX40 type made from glass-epoxy, but the above finding extends to other materials (glass-polyester tends to have a slightly higher damping value, and carbon-epoxy a lower).

7.7.1 Aerodynamic Damping

On a rotating blade, however, aerodynamic action can add significant damping. The principle is explained with reference to Figure 7.18, which showing the forces acting on a blade section. Flapping motion in the axial direction modifies the incident wind speed V, such that axial force F_{ax} varies in proportion to flap velocity $\dot{x}$ with $F'_{\text{ax}} = F_{\text{ax}}(1 - \dot{x}/V)$. As the blade flaps forward (into the wind) L increases so as to oppose the motion; conversely, flapping downwind reduces the lift. The resulting restoring force is proportional to flap velocity, and thus has the characteristic of viscous damping. For a blade in attached flow the aerodynamic damping coefficient ζ may be of the order

[6] This calculation has the advantage that the measurements need not be calibrated nor the time base known.

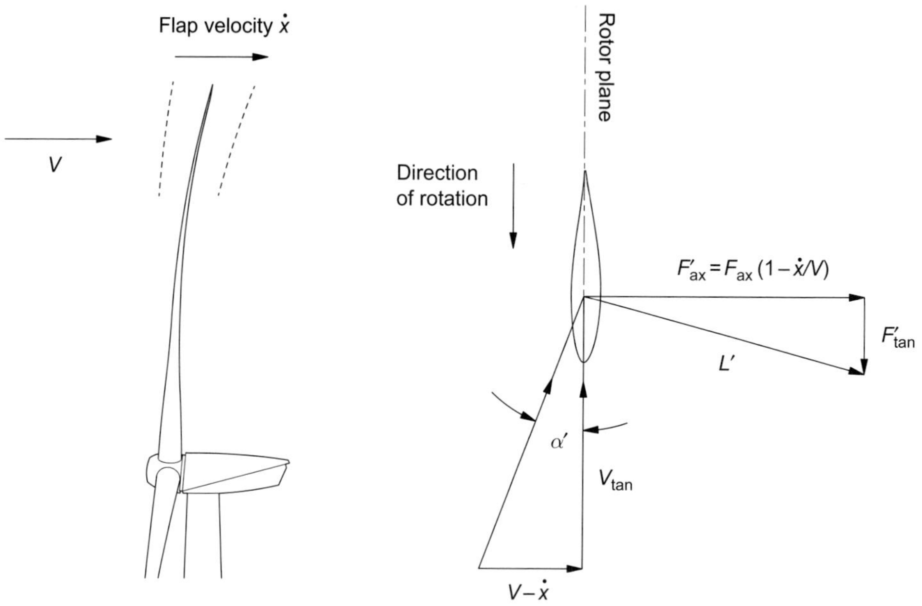

Figure 7.18 Explanation of aerodynamic damping. Positive flap velocity $\dot{x}$ decreases the effective axial wind speed V, causing a proportional decrease in lift L and axial force F_{ax}; negative flap velocity has the converse effect. The result is equivalent to viscous damping, with typical damping coefficient $\zeta \cong 0.15$.

0.15, more than an order of magnitude greater than material damping. Aerodynamic damping has a significant beneficial effect on the dynamic response of the blade, and that of the overall WEC structure.

The linear relationship between lift force and axial wind velocity, with the blade in attached flow, is the normal case for a positive pitch-controlled rotor. On a stalled blade, however, the linear lift characteristic is weakened in high winds as C_l flattens off or reduces with α (see Figure 3.11). Reduced aerodynamic damping and consequently increased flapwise vibration might therefore be expected on stall-regulated rotors. In practice, however, this has not generally been the case and despite the many thousands of stall-regulated wind turbines built, very few suffered from stall-induced flapwise vibration. In a side-by-side comparison of a 250 kW pitch-regulated wind turbine with an otherwise identical stall-regulated machine, flapwise cyclic loads were actually found to be lower on the stall machine, with no evidence of dynamic instability (Hoskin, 1988). This was perhaps an early indication of the high cyclic blade loads induced by pitch control, but also a vindication of the principle of stall regulation (although the mean blade loads were higher on the stall-regulated machine).

The flapwise stability of stall-regulated blades is partly explained by the drag rise that accompanies stall, contributing to positive thrust: flapwise loading then continues to rise with wind speed, which translates into a positive damping characteristic. By contrast, however, a loss of edgewise aerodynamic damping became a feature of large stall-regulated wind turbines and ultimately had major implications for this type of machine. This topic is discussed in Section 7.7.3.

Aerodynamic damping is of particular importance to two-blade rotors whose blades are attached to the hub via a teeter hinge that allows a free 'see-saw' motion of the rotor. The advantage is that unbalanced rotor loads are not transmitted to the drive shaft, but instead result in teeter motion (teeter hinges are also seen on two-blade helicopter rotors). Use of a teetered rotor is estimated to reduce blade root bending moments by as much as 40%, and significantly reduces hub fatigue loading. The teeter motion has the unusual property of being resonant at the rotor rotational frequency so that a stationary observer sees the blades rotating in a fixed plane tilted out of the vertical; aerodynamic damping prevents the resonant motion from reaching high amplitudes. A good description and mathematical analysis of teeter motion is given by Garrad in Freris (1990).

7.7.2 Tower Resonance

The main source of excitation of the lowest tower vibration modes is the *n*P cyclic variation in rotor thrust caused by blade tower shadow and rotationally sampled turbulence. Many large HAWTs are designed with 'soft' towers whose lowest modal frequency lies well below 3P (assuming a three-blade rotor) and in this way resonance due to tower shadow is avoided. Some towers are 'soft-soft' with modal frequency below 1P, so that excitation due to blade mass imbalance is similarly avoided. The picture is, however, more complicated for variable-speed wind turbines as the excitation frequencies vary with rotor speed, so there is greater potential for tower resonance. In this context a useful tool is the Campbell diagram: this is an interference plot on which key eigenfrequencies are shown in relation to harmonic excitation frequencies, where the latter are proportional to rotor speed. A typical example is shown in Figure 7.19, applicable to a variable-speed wind turbine with online speed range of 14.5 to 29 rpm. The horizontal axis represents rotor

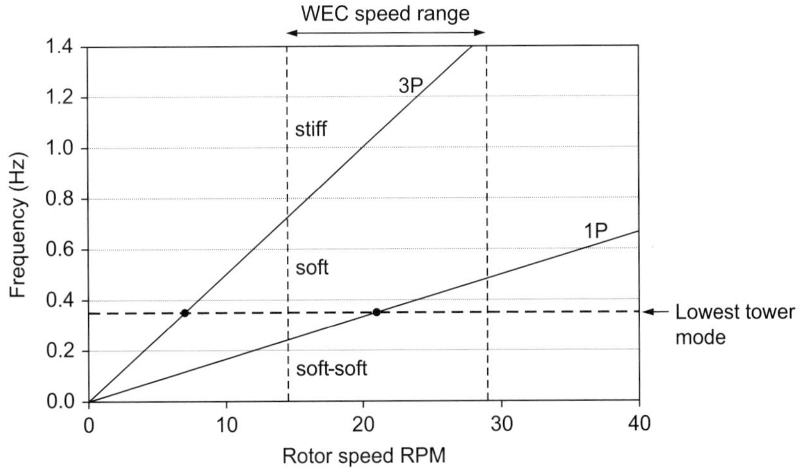

Figure 7.19 Campbell diagram, used to indicate potential resonance conditions. In this case, operation at 21 rpm would cause tower resonance only if the rotor were not well balanced (1P excitation).

speed, and the associated excitation frequencies are plotted as straight lines rising from the origin: in this case for simplicity only the 1P and 3P frequencies are shown.

The lowest tower modal frequency is represented by a horizontal line. Critical rotation speeds are then marked where the 1P and 3P frequencies coincide with this line, indicating potential for resonance. The tower frequency here is 0.35 Hz, and hence susceptible to 1P excitation at 21 rpm or to 3P excitation at 7 rpm. In principle neither should be of concern: the thrust load spectrum of a well-balanced HAWT rotor contains no 1P component, while the 3P excitation only threatens resonance at a very low running speed when the turbine is offline and excitation energy is inherently low. During the start-up sequence the controller is nevertheless programmed to prevent the rotor from 'dwelling' at the critical speed.

7.7.3 Blade Edgewise Vibration

The phenomenon of edgewise stall vibration arose as an issue when rotor diameters grew beyond around 35 m; it also came as something of a surprise as a HAWT blade is inherently much stiffer edgewise than flapwise and stall-regulated wind turbines do not generally experience extreme flapwise vibration (see above). Thousands of stall-regulated turbines rated up to 300 kW had operated for many years with no signs of instability. The first indications came in measurements on a 37-m-diameter machine (Stiesdal, 1994): the blade edgewise loading was seen to be gravity-dominated in light winds, but in high winds increasing vibration occurred at the edgewise eigenfrequency with cyclic load amplitudes increasing by up to 50%. Later, several commercial designs of 500–600 kW rating suffered accelerated fatigue damage, and in some cases catastrophic blade loss (Moller, 1997). An illustration of edgewise stall vibration is seen in Figure 7.20, which shows strain measurements from a medium-scale HAWT blade in high winds: the edgewise eigenfrequency can be seen as a high-frequency waveform superposed on the gravity cycle. In the case shown the vibration amplitude is modest, but under some circumstances it could exceed the gravity loading by a factor of 4 or 5 (Anderson, 1999).

The phenomenon is due to loss of aerodynamic damping. The following is a strictly qualitative explanation based on a more rigorous analysis by Pedersen (1998). A three-blade rotor has several vibration modes containing edgewise motion, and the lowest modes can be represented by A–C in Figure 7.21. In mode A the three blades deflect in phase with equal amplitude: consequently a torque reaction is generated at the rotor shaft and transmitted to the generator, where the vibration energy is absorbed: the generator effectively provides torsional damping. With modes B and C, however, the blades vibrate in anti-phase and no torque reaction is generated – the drivetrain does not 'see' the vibration and the generator offers no damping. In the worst case a blade in stall experiences negative aerodynamic damping and the vibration amplitude in modes B and C will grow exponentially. This is not a response to a harmonic forcing function, but rather divergence or self-excitation. In practice the negative damping coefficient may be small, but left unchecked the vibration amplitude grows progressively until blade damage occurs.

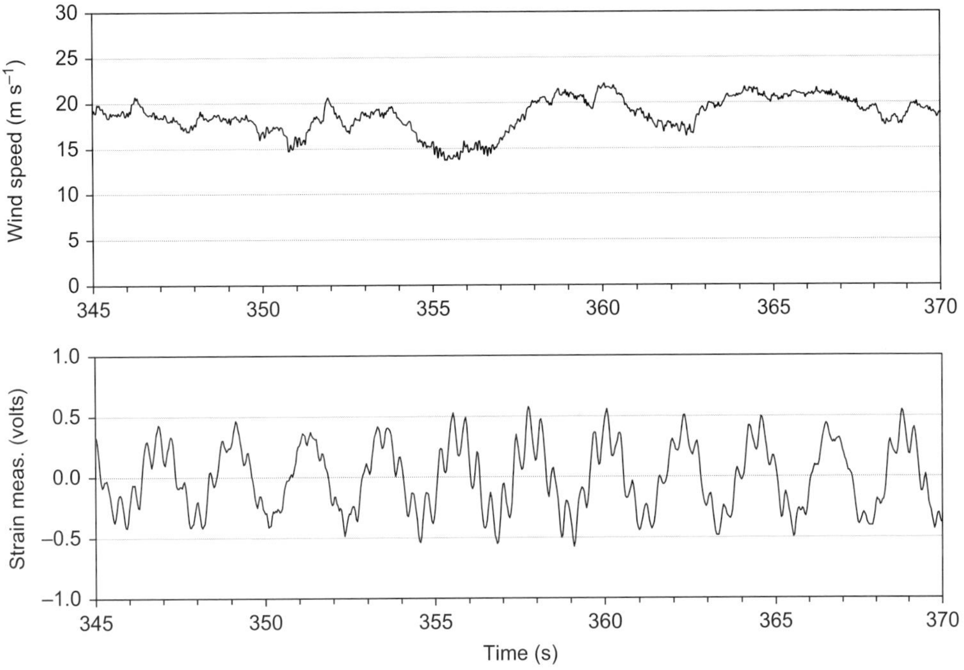

Figure 7.20 Measured edgewise vibration on a 600 kW stall-regulated blade in high winds. The edgewise eigenfrequency is seen as the high-frequency waveform superposed on the normal gravity cycle.

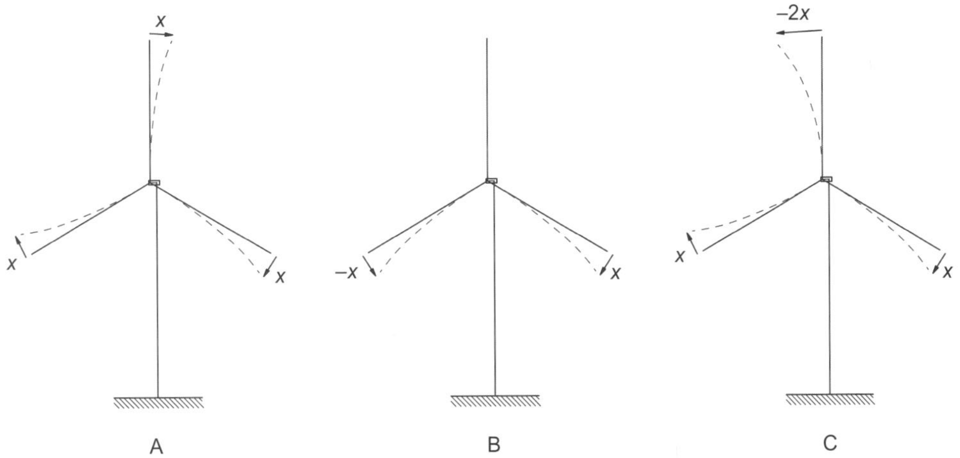

Figure 7.21 Qualitative explanation of edgewise stall vibration. Mode A is reacted by the drivetrain and consequently damped by the generator. Modes B and C cause no net torque reaction at the hub, and the generator then contributes no damping. Under stall conditions, these modes may diverge.

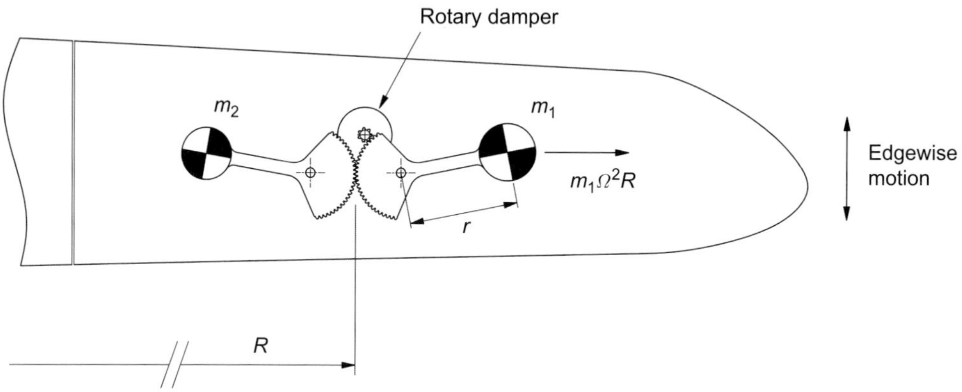

Figure 7.22 Operating principle of the Aerpac vibration damper (schematic, and not to scale). Centrifugal acceleration $\Omega^2 R$ acting on masses m_1 and m_2 makes the coupled pendulum resonant at the blade edgewise eigenfrequency; the rotary damper converts vibration energy to heat.

To combat the problem some large HAWT blades were fitted with internal dampers, tuned to resonate at the edgewise eigenfrequency. One such solution was a mechanical damper employed by Aerpac, which was based on the principle of an opposed pendulum (Anderson, 1999); the damper is illustrated schematically in Figure 7.22. Located in the tip section of the blade, two pendulum masses m_1 and m_2 are mounted in opposition and coupled by gears. In isolation the outer mass m_1 will tune to a high resonant frequency under centrifugal force; the inner mass m_2 is, however, mounted unstably and has the effect of reducing the resonant frequency of the coupled system. For a damper located at mean radius R from the rotor centreline the resulting tuned frequency ω_n is given by

$$\omega_n = \Omega\sqrt{\frac{R(m_1 - m_2)}{r(m_1 + m_2)}} \tag{7.13}$$

where Ω is the rotor angular frequency and r the length of the pendulum arms. By appropriate choice of the masses (for stable operation m_1 must be greater than m_2) the system is tuned to the required eigenfrequency; rotation speed Ω is a constant as the turbine is stall regulated. The mechanism incorporates a small rotary damper driven by one of the pendulum gears, to absorb the vibration energy.

The Aerpac damper was first installed in blades of 19 m length, following ground tests using springs to simulate centrifugal acceleration, equivalent to 16 g in the operating blade (see Figure 7.16). The trials indicated that a damping coefficient of approximately 0.01 could be achieved with a few kilos of moving mass; although quite modest, this effect is enough to ensure net positive damping and prevent divergence of the edgewise vibration. The principle was confirmed in field trials, with damper-equipped wind turbines able to operate in wind speeds up to their normal cut-out limit without excessive vibration. The difference in blade edgewise vibration activity with and without dampers is seen in Figure 7.23.

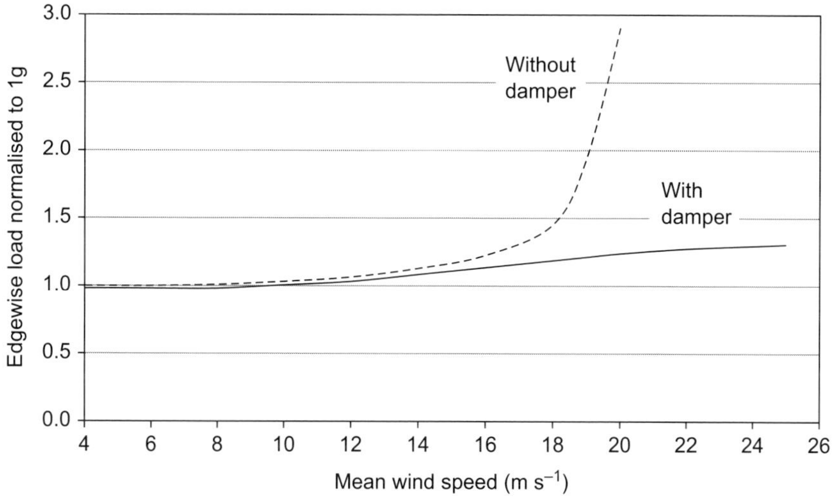

Figure 7.23 Measured edgewise vibration envelope for 19 m stall-regulated blade with, and without, a mechanical damper fitted (Anderson, 1999).

Other solutions to the edgewise vibration problem included liquid dampers based on the principle of the resonant water column, again tuned by centrifugal acceleration (Veldkamp, 1998). Achieving long-term reliability of blade-mounted dampers of any type was, however, problematic and stall regulation fell out of fashion as wind turbine size continued to grow; this configuration is now uncommon for wind turbines larger than a few hundred kW. The reason that edgewise vibration did not affect smaller stall-regulated wind turbines was probably one of scaling. The small blades were relatively heavy and inherently strong, but as wind turbines grew the need to reduce weight in the face of the scaling laws (see Section 1.5) meant that structural over-design would be uneconomic at MW scale. Even without the edgewise vibration problem stall-regulated rotors tend to be less cost-effective, however, and as a result of these factors (and others related to electrical power quality; see Section 5.5) large wind turbines now all employ full-span blade pitch control.

7.8 Dynamic Response Measurements

7.8.1 Introduction

The analytic models used to predict wind turbine loading and performance have been developed over many years, and have benefited from a large body of measurements made on full-scale machines, including both experimental and production types. This process can be illustrated with reference to tests carried out on an early machine built by the James Howden Company of Glasgow. Howden produced several medium-sized WECs, including the HWP26 and HWP31, both rated at 330 kW. In its day the HWP31 was one of the largest series-produced wind turbines and formed the basis of a 26 MW windfarm constructed in 1985 in Altamont Pass, California (Shearer et al., 1986).

The 31-m-diameter rotors suffered early structural problems, however, forcing an extensive and rapid redesign of both blades and hub. A completely new 33 m rotor was installed on all the turbines and the windfarm was recommissioned in 1987, following which the re-bladed machines (now renamed HWP33) remained in operation for 23 years.

The HWP26 had the same rating but a smaller rotor, and was designed for higher wind sites. Only two were built, one of which underwent extensive field trials under a US Department of Energy test programme at the Southern California Edison (SCE) test site near Palm Springs. The HWP26 is shown in Figure 7.24, and its chief characteristics are listed in Table 7.1. The Howden machines were

Figure 7.24 The Howden HWP26 wind turbine at the SCE test site in Palm Springs, 1986.

Table 7.1 Details of the Howden HWP26 330 kW Wind Turbine

Type	Three-blade, fixed speed, upwind
Rotor diameter	26 m
Output rating	330 kW
Rotor speed	42 rpm
Aerodynamic control	Partial-span blade pitch (active stall)
Blade construction	Wood-epoxy laminate
Blade planform	Linear taper, linear twist (16° at root)
Aerofoil profile	NASA GA(W)-1, 17% thickness
Generator	Synchronous, 1800 rpm
Operating wind speeds	6–28 m s^{-1}

relatively modern in concept and their wood-epoxy blades were light and stiff, but they differed from today's wind turbines in key aspects such as the use of partial-span blade pitch control, and power limiting based on a simple dead-band control algorithm (see Section 6.3.2). They were designed, however, at a time when knowledge of HAWT aerodynamics and structural response were less well developed than today, and the aim of the HWP26 test programme was to improve the understanding of these topics.

Over a period of several months a variety of tests were carried out to investigate both mean and turbulence-induced cyclic loading. The wind turbine was heavily instrumented, with strain gauges attached at a number of locations on the blades and tower; a multi-mast anemometer array was erected in the prevailing wind direction to record wind speeds across the full rotor disc area with 3D propeller anemometers used to enable detailed turbulence analysis. Testing on the HWP26 led to a number of published reports, including an assessment of the dynamic response of the structure (Hock et al., 1987), blade loading in relation to turbulent inflow (Madsen et al., 1987; Wright et al., 1988), and a comparison of measured blade load histories with results from a dynamic model incorporating BEM aerodynamic theory (Wehrey et al., 1988). Some of the key results of the HWP26 test programme are described below.

7.8.2 Mean Loads

The first measurements of interest with a new wind turbine design are usually the mean power curve and the blade root bending loads, as these indicate its fundamental aerodynamic performance. The results for the HWP26 are shown in Figure 7.25 and Figure 7.26. The measured power curve (Figure 7.25) exhibits the flat-topped appearance characteristic of a pitch-controlled machine, with the power limit evident at 330 kW; the solid line is an estimate based on BEM theory. The rounding at the 'knee' of the experimental curve indicates an apparent shortfall in power, but is largely an artefact of the averaging period in conjunction with wind speed variance: the higher the turbulence, and/or the longer the averaging period, the more pronounced the rounding effect (Jamieson et al., 1988) and in this case

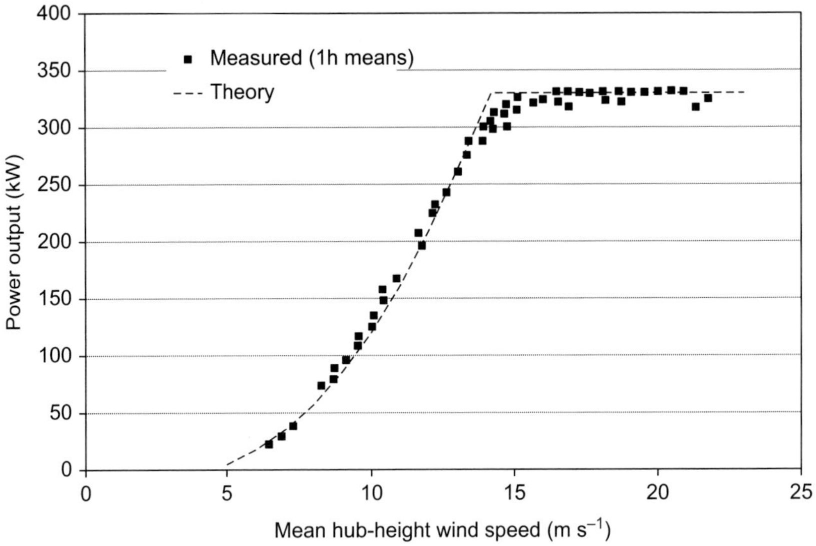

Figure 7.25 Measured power curve of the HWP26 wind turbine (hourly means, not binned).

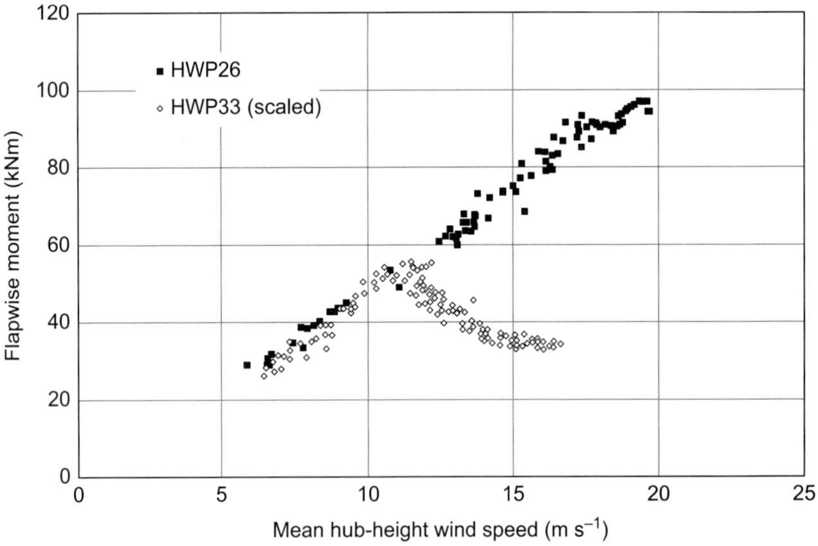

Figure 7.26 Mean blade root flapwise moment for the HWP26 and HWP33 wind turbines (the latter scaled to an equivalent rotor diameter). The curves illustrate the difference in loading for rotors with positive (HWP33) and negative (HWP26) pitch control.

non-standard hourly averaging was used.[7] Agreement with theory was otherwise quite good (note that the experimental data are here not binned), and BEM theory generally works well for the steep part of the power curve where the rotor is in attached flow and well below stall. Above rated wind speed the observed accuracy of the power curve is largely a function of the pitch control system and the measurements give less information about the rotor aerodynamic performance.

The blade flapwise mean loads are shown in Figure 7.26. These were based on strain measurements from the blade root, converted to bending moment via the results of a calibration pull-test, where the stationary blade was subjected to a known load. The bending moment shows a rising trend across the full range of wind speeds, with no significant change in slope at the onset of pitch control (around 15 m s^{-1}) or in high winds. As such, it contrasts with the pattern expected for an ideal pitch-regulated wind turbine, illustrated in Figure 4.1. This is because the HWP26 control system was based on negative pitch (active stall) so power control invoked a significant increase in drag rather than a reduction in lift (see Section 4.7.2). Howden experimented with both positive and negative pitch strategies, and the larger HWP33 was alternately configured with 'stalling' and 'flying' pitch control: the flapwise loading in the latter case is shown overlaid on Figure 7.26, where the HWP33 loads have been scaled to the diameter of the HWP26. The difference between positive and negative pitch control is clearly seen. In terms of power control Howden found no strong evidence to favour either method, with the mean power curve well maintained in both cases and comparable power excursions (30%–40% about the mean) seen in high winds. This level of control was fairly typical for CSVP wind turbines.

7.8.3 Load Spectra

More detailed insight into WEC structural response is gained by examining the frequency content of the measured loads. Strain gauge time histories for the HWP26 were processed via fast Fourier transform to yield plots of the kind shown in Figure 7.27, which shows the frequency spectrum of blade flapwise bending: the plot is based on an original analysis by SERI, one of the partners in the experimental programme (Wright et al., 1988). The distinct peaks seen at 1P, 2P, and 3P are due to a combination of deterministic (wind shear, tower shadow, gravity) and stochastic (rotationally sampled turbulence) influences. Other peaks are seen at the eigenfrequencies of dominant structural modes. The lowest flapwise mode appears at approximately 1.4 Hz and is close to the 2P excitation frequency, with consequently high amplitude: the plot reveals a potential design limitation of the HWP26 prototype. The spectral peak at 3.1 Hz is attributable to the first edgewise mode.

Similar spectral plots were produced from measurements made elsewhere on the wind turbine, and in this way the modal response of the complete structure was characterised for comparison with theory. Some of the predicted mode shapes and frequencies are shown in Figure 7.5; the Campbell diagram in Figure 7.28 is based on measured modal frequencies and shows

[7] The IEC standard averaging period for power curve measurements is 10 min.

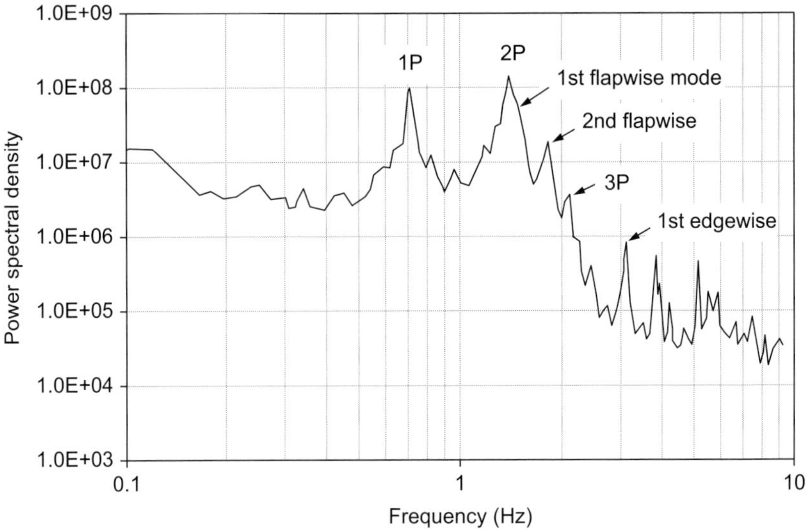

Figure 7.27 Spectral content of blade root flapwise loading for the HWP26 wind turbine, based on an original analysis by SERI (Wright et al., 1988). The first flap mode is close to the 2P excitation frequency.

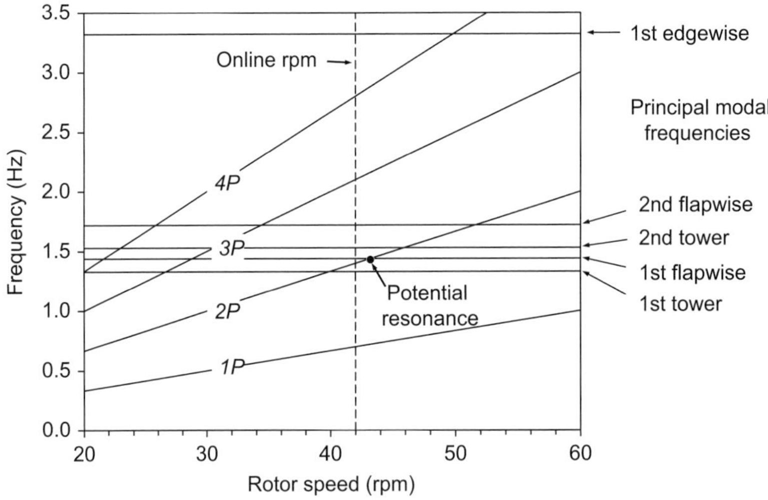

Figure 7.28 Campbell diagram for the HWP26 wind turbine, based on measured eigenfrequencies. At the nominal (fixed) rotation speed of 42 rpm, the first flapwise blade mode is close to 2P harmonic excitation.

the coincidence of 2P harmonic excitation with the first blade flap mode. This was commented on in test reports (Madsen et al., 1987). Avoiding such interactions during the design of a WEC can be something of an art, and field testing remains an important part of the process. In practice the HWP26 was not developed further, unlike the larger HWP33, which, despite an uncertain start (see above) ultimately enjoyed a long operational life.

7.8.4 Stochastic and Deterministic Loads

The harmonic peaks at 1P, 2P, etc., in the HWP26 blade load spectrum (Figure 7.27) are due to a combination of deterministic and stochastic excitation. To determine the relative magnitude of these two influences the technique of angular (or azimuthal) averaging was used. Deterministic loads occur predictably with blade position, whereas stochastic loads are quasi-random, so if a suitably long load time series is averaged on the basis of rotor angular position the stochastic content cancels out, leaving only the deterministic. Figure 7.29 shows a typical time history for blade flapwise loading: angular averaging is the equivalent of chopping this record into increments of one rotor revolution and overlaying them; the residual average is then the purely deterministic load cycle.

This technique was used to extract deterministic blade and tower loads from the HWP26 strain measurements for comparison with a theoretical dynamic model. The predictions were made using a time-stepping simulation incorporating a modal WEC model and BEM aerodynamics code incorporating the influences of wind shear, tower shadow, and yaw misalignment (Wehrey et al., 1988). The basis of the modal model is described in Section 7.2.3. Stochastic loading was not modelled at this time, and the output of the code was restricted to steady and deterministic response. Some results of these analyses are presented in Figure 7.30 and Figure 7.31. The former shows the increase in measured harmonic loading with wind speed; the latter is a comparison of the measured blade flapwise load waveform with prediction from the dynamic response code. The development of such codes (not to mention the computers on which they run) has progressed greatly since the results shown here were obtained, and a modern time-stepping simulation is far more capable in terms of modelling accuracy and computational power, while stochastic wind inputs are now incorporated as standard.

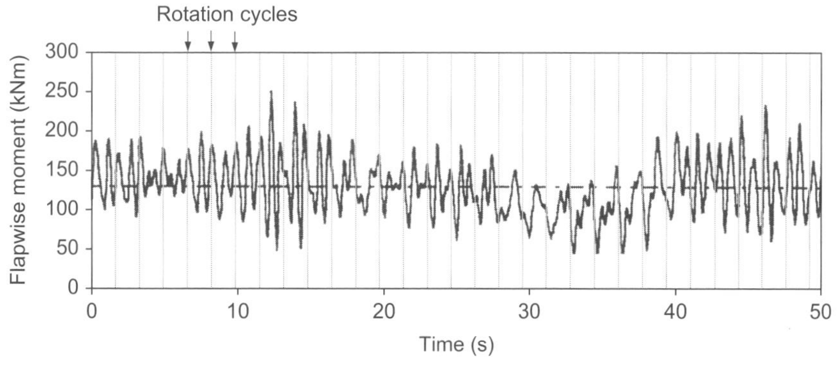

Figure 7.29 Time series of blade root flapwise moment. Angular averaging involves overlaying the loads from each rotor cycle to eliminate the stochastic (quasi-random) content, leaving only the deterministic.

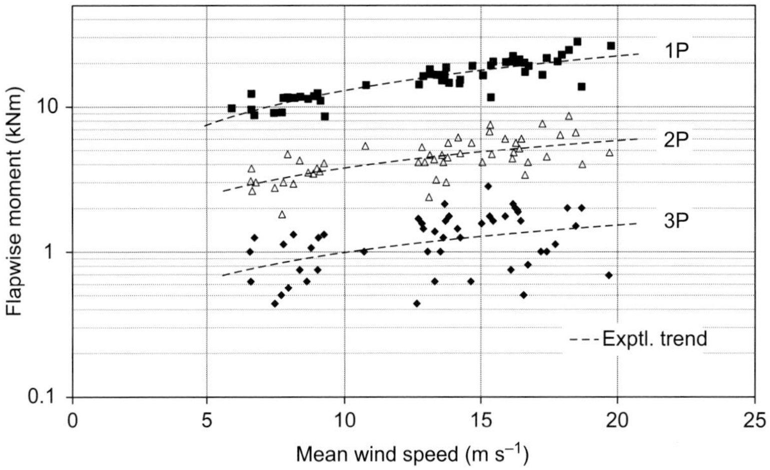

Figure 7.30 Deterministic content of HWP26 blade root flapwise loading, extracted by angular averaging (Wehrey et al., 1988). All the harmonic components exhibit a rising trend with mean wind speed.

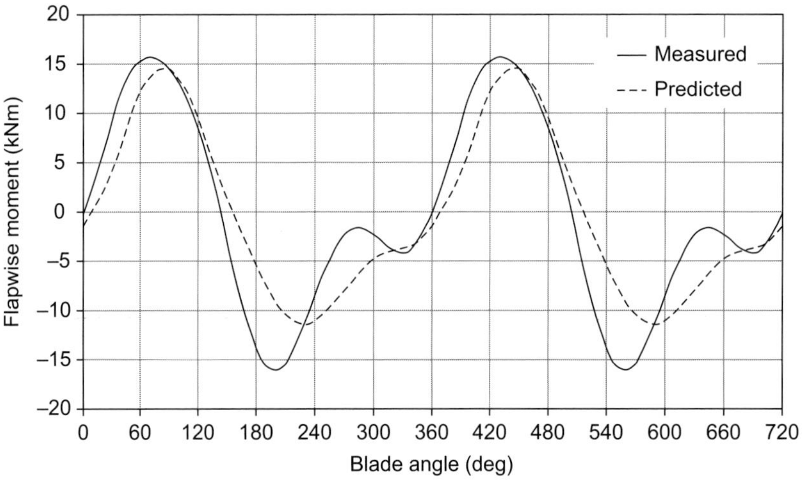

Figure 7.31 Deterministic content of HWP26 blade root flapwise moment: measured and predicted results for wind speed of 18 m s^{-1}. The measured data were obtained by angular averaging.

7.9 Complete Load Prediction

7.9.1 Modern Aeroelastic Codes

In the years since the HWP26 trials described above the capability of HAWT aeroelastic design codes has increased enormously, in part due to more complete mathematical descriptions of both the wind turbine structure and the interacting windfield, but also to the huge growth in computing power. Modern design codes are comprehensive and fast: good examples are the commercial package BLADED from DNV-GL, and NREL's public domain FAST suite. In addition to modelling the deterministic influences described above (see Section 7.3.3) these codes incorporate models for atmospheric turbulence (typically based on Kaimal or von Karman spectra) to allow prediction of stochastic loading; unsteady aerodynamic effects such as dynamic stall and dynamic inflow (see Section 3.9) are also included. A full description of the capabilities of a modern aeroelastic code is too detailed to be included here, but the reader is referred to the (continually updated) websites supporting BLADED (DNV-GL, 2019) and FAST (NREL, 2019).

These software packages can produce simulated time series (comparable to the real trace shown in Figure 7.29) for the loading and deflection at almost any location on the HAWT structure, including key areas along the blade, the blade root-hub connection, drivetrain, and tower. The designers can then simulate the response of the wind turbine in all foreseeable conditions, including online operation, starting and stopping, fault and emergency conditions, and extreme loading (both operational and stopped). One aspect in which aeroelastic simulation is now a powerful – and essential – tool is fatigue load analysis, the basis of which is outlined below.

7.9.2 Fatigue Prediction

Fatigue damage is caused when a structure experiences cyclic stresses that may be well within its ultimate strength, but which repeated many times can cause premature failure. The fatigue or residual strength of a material is characterised by the cyclic stress amplitude S_N that will cause failure after N load cycles, according to the relationship

$$S_N = S_0 N^{-1/x} \tag{7.14}$$

where S_0 is the ultimate (static) strength and x is an empirical factor (the Wohler coefficient). In logarithmic form the above equation is the S/N curve familiar to mechanical engineers:

$$\log S_N = \log S_0 - (1/x)\log N \tag{7.15}$$

Fatigue curves for materials commonly used in wind turbine construction are shown in Figure 8.1, where the characteristic reduction in strength with log cycle count is seen. The exponent x is a measure of the fatigue resistance of the material, and the plot shows why composites are superior to steel in fatigue. The way in which fatigue evaluation is incorporated in aeroelastic design codes can be described as a number of sequential steps:

1. A load time series (force or bending moment) is generated for a particular component or location on the structure under representative operating conditions.
2. The time series is converted to stress using a scalar multiplier based on the geometry of the part (for example the stress in a composite blade spar is proportional to the local bending moment).

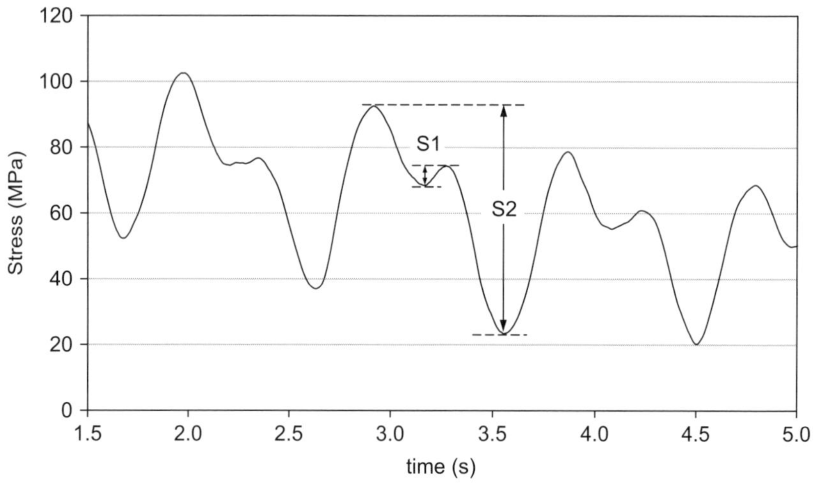

Figure 7.32 Synthesised stress history for a blade pitch system component. The Rainflow algorithm reduces the data to a set of turning points. Small half-cycles (S1) are then counted and removed before counting large ones (S2).

3. The stress time series is analysed using a routine that counts the number of fatigue cycles with given combination of mean stress s_m and range s_a. The 'Rainflow' cycle counting algorithm is commonly used for this (see below) and the result is a 2D matrix of cycle counts.
4. The influence of mean stress is accounted for using the Goodman relationship, and the 2D matrix reduced to a 1D array of cycle counts against equivalent fully reversing stress range $s_{a'}$.
5. Cumulative fatigue damage is evaluated by application of Miner's rule to the 1D array.

These steps can be illustrated with an example. Figure 7.32 shows part of a simulated stress history for a component of a blade pitch system: in practice this would represent a small part of a much longer simulation record. The trace contains cyclic activity due to stochastic and deterministic loading and the effects of pitch control activity. The Rainflow algorithm first identifies the turning points (peaks and troughs) in the time series and discards the intervening data so that fatigue cycles are defined only by their extremes. The algorithm then identifies small-amplitude half-cycles (example S1 shown) superposed on larger ones (S2), counts and removes them, then recursively processes the remaining data. The outcome is a 2D matrix containing the number of cycles with given stress range and mean.

The above is a somewhat simplified description of the Rainflow algorithm, which was originally devised by Matsuishi and Endo (1968); it is more fully described in ASTM Standard E-1049 (ASTM, 2017). By running simulations for each operating regime of the wind turbine, and extrapolating the number of cycle counts to represent the full design life, a 2D Rainflow matrix is obtained that represents the lifetime fatigue stress spectrum for the component under study, as shown in Figure 7.33.

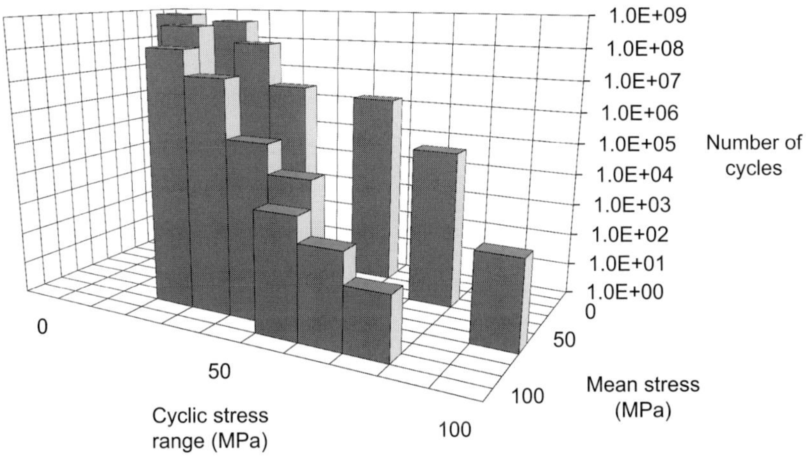

Figure 7.33 Rainflow matrix, showing the number of fatigue cycles with given stress range and mean. The data shown represent the lifetime fatigue spectrum for the component.

Each block in Figure 7.33 represents the number of cycles with mean stress s_m and cyclic range s_a. The influence of the mean stress is accounted for using the Goodman relationship:

$$s_a' = \frac{s_a}{(1 - s_m/s_{\text{UTS}})} \tag{7.16}$$

where $s_a\prime$ is the equivalent stress range for a fully reversing load (i.e. with zero mean) and s_{UTS} the ultimate tensile strength of the material. The outcome of Equation (7.16) is therefore material specific, and the influence of the mean stress is to increase the effective cyclic range. The Goodman relationship collapses the 2D Rainflow matrix into a 1D array of the number of fully reversing cycles at amplitude $s_a\prime$, as shown in Figure 7.34. Cumulative fatigue damage is then assessed using Miner's rule. Each stress range in the array has occurrence n_i and based on its S/N curve (see Equation (7.14)) an associated lifetime N_i; the cumulative lifetime fatigue damage C is then given by

$$C = \sum_{i=1}^{k} \left(\frac{n_i}{N_i} \right) \tag{7.17}$$

where k is the number of load blocks in the array. The criterion for fatigue survival is then

$$C < 1 \tag{7.18}$$

Fatigue analysis and design for wind turbines is a large subject, and the above is only a brief introduction. The rules for fatigue cycle counting and damage evaluation apply equally to simulated or measured load data, and Rainflow matrices derived from field measurements are

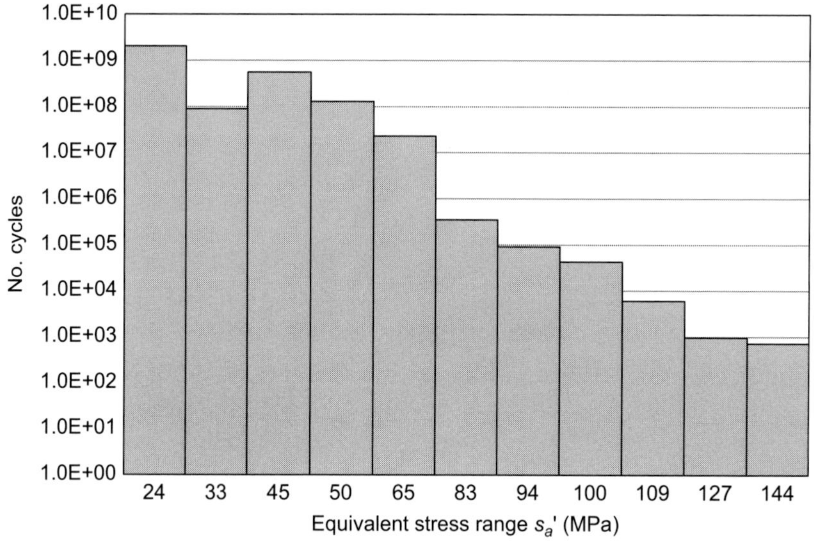

Figure 7.34 Fatigue stress distribution: 1D array of equivalent reversing stress cycles derived from 2D Rainflow matrix using the Goodman line relationship. Miner's rule can be applied to the data shown.

often used to inform subsequent WEC designs. Similarly, the basic relationships defined by the S/N curve and Miner's rule can be used to design accelerated fatigue test regimes for key components such as blades. In such cases a constant-amplitude load spectrum may be used that is predicted to cause the same lifetime fatigue damage C as the full fatigue spectrum, but with a cycle count sufficiently small to allow testing to complete in a few weeks: factory fatigue tests may still run to a few million load cycles, but will verify a 25 year design life with true cycle counts perhaps 100 times higher. An example is seen in Figure 7.35, which shows the final traces from a laboratory fatigue test on a blade root attachment fixture. A constant-amplitude spectrum was used, with loading applied by a hydraulic actuator. Failure is indicated by the sudden change in strain at around 4.5 million cycles. The component in question was designed on the basis of 1 million cycles at constant amplitude so the test result demonstrated a comfortable margin above the target design life.

7.10 Exercises

7.10.1 Fatigue Cycles

A wind turbine operates for 20 years, during which it spends 85% of the time online at an average rotor speed of 25 rpm. Estimate the number of (a) 1P blade gravity bending cycles and (b) 3P tower-top thrust cycles, accumulated during online operation.

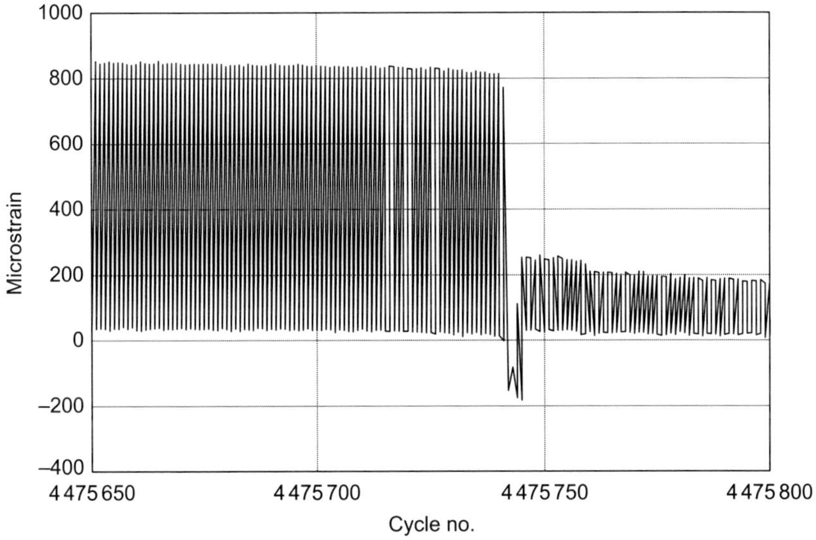

Figure 7.35 End of the road. The final traces from an accelerated constant-amplitude fatigue test on a full-scale blade: the strain measurements are from a root attachment stud. Failure is indicated by the sudden collapse in strain magnitude. (Based on lab test data kindly provided by NREL)

7.10.2 Thrust Load

The thrust load on a parked (stationary) wind turbine rotor is 160 kN in a wind speed of 55 m s^{-1}. What will the corresponding load be in a wind speed of 70 m s^{-1}?

7.10.3 Tower Shadow

The minimum clearance between the blade tip and the tower on a large upwind HAWT is 1.6 m. The tower is tubular with a diameter of 3.0 m. Calculate the percentage reduction in axial wind speed due to tower shadow when a blade passes directly in front of the tower.

7.10.4 Dynamic Magnification

A three-blade wind turbine rotor rotates at 62 rpm. Calculate the fundamental rotational frequency (1P) and its first two harmonics (2P and 3P) in Hz. If the lowest blade flapwise vibration mode is at 3.5 Hz, which of the above excitation frequencies will result in the highest dynamic magnification factor Q?

7.10.5 Blade Damping

The figures in Figure 7.36 represent the flapwise vibration response of an operational blade after being subject to a sudden impulsive load. The two traces correspond to different operating

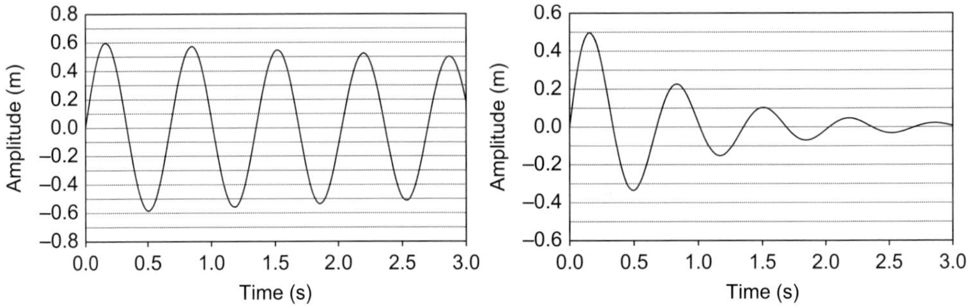

Figure 7.36 See Exercise 7.10.5.

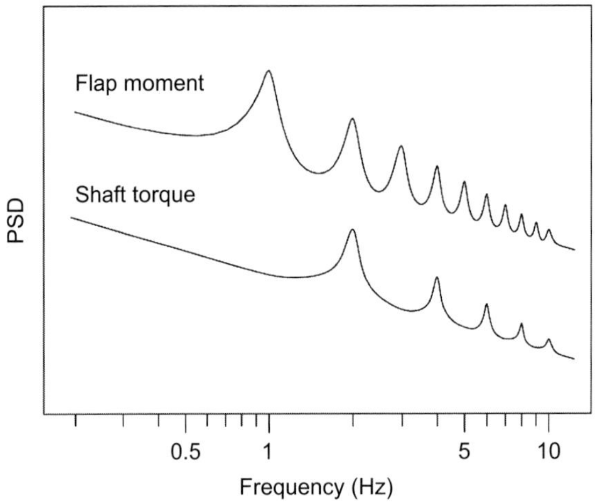

Figure 7.37 See Exercise 7.10.6.

conditions. Calculate the modal damping factor ζ in each case, and suggest a reason for the observed difference in the traces.

7.10.6 Harmonic Content of Shaft Torque

Figure 7.37 shows frequency spectra for the (a) blade root flapwise bending moment and (b) main shaft torque for a constant-speed pitch-regulated wind turbine. How many blades does it have? Explain your answer.

7.10.7 Unbalanced Blade

A three-blade WEC has rotor diameter of 48 m and hub height 55 m; the distance between the tower centreline and the rotor plane is 4.5 m, and the nominal rotor speed is 28 rpm. If one of the blades has an unbalanced mass of 0.5 kg in the tip, calculate the amplitude and frequency of the resulting cyclic (a) yaw moment and (b) tower base bending moment.

7.10.8 Smeaton's Windmill

Refer to the drawing of John Smeaton's experimental windmill apparatus shown in Figure 1.3. Why might his model have been subject to unrealistically high (a) aerodynamic yaw loads and (b) cyclic bending loads on the rotor shaft? Explain your reasoning.

CHAPTER 8 ROTOR BLADE TECHNOLOGY

8.1 INTRODUCTION

In 1958 the German aeronautical engineer Ulrich Hütter constructed a two-bladed, 100 kW, wind turbine. The design was innovative, optimised for high tip speed ratio with a lightweight downwind rotor and, significantly, featured blades made from glass fibre–reinforced plastic (GFRP). Other WECs of this era used more traditional materials – steel, aluminium, and in some cases, wood – but Hütter had a background in glider design where the requirements of lift, weight, and strength are similar to those of wind turbine rotors. His insight was sound and nowadays all wind turbine blades are manufactured from fibre-reinforced composites, with no metals to be found in their primary structure (wood, which is a natural composite, still plays a significant role in a few blade designs). The use of composites reflects their superior mechanical and manufacturing properties, and takes advantage of their *anisotropy*, the property that stiffness and strength are dependent on fibre orientation. Composites can thus be tailored to suit the direction of the principal stresses in a structure in the most material-efficient way. Anisotropy adds a degree of complexity to design calculations, but a wind turbine blade can still be analysed using elementary beam theory with good results. These topics are discussed in the present chapter.

8.2 PROPERTIES OF FIBRE-REINFORCED COMPOSITES

The composites most widely used in HAWT blade manufacture are glass fibre– and carbon fibre–reinforced plastic (respectively GFRP and CFRP). Their bulk properties are compared with other structural materials in Table 8.1. The figures for GFRP and CFRP are based on unidirectional laminates with high volume fraction (see explanation below). The first two columns in the table give tensile strength and stiffness (E-modulus); the final two columns are specific properties, i.e. strength and stiffness per unit weight, which are obtained by dividing the absolute values by density. The specific values have furthermore been normalised with respect to mild steel for convenience. The data show that the specific strength of GFRP and CFRP is significantly higher than steel; CFRP is stronger than mild steel in absolute terms. The table also indicates why wood-epoxy is still a useful blade material, with more than twice the specific strength of mild steel and comparable specific stiffness.

Table 8.1 Comparison of Material Properties: Tensile Strength and E-Modulus

MATERIAL	Ultimate tensile strength (MPa)	E-modulus (GPa)	Density ρ (kgm^{-3})	Specific strength* UTS/ρ	Specific stiffness* E/ρ
Mild steel	350	210	7800	1.0	1.0
High-strength steel	900	210	7800	2.6	1.0
GFRP (UD, $V_f = 0.5$)	600	38	1840	7.3	0.77
CFRP (UD, $V_f = 0.6$)	1675	130	1500	25	3.2
Wood-epoxy laminate	70	17	660	2.4	0.96

* Normalised with respect to mild steel.

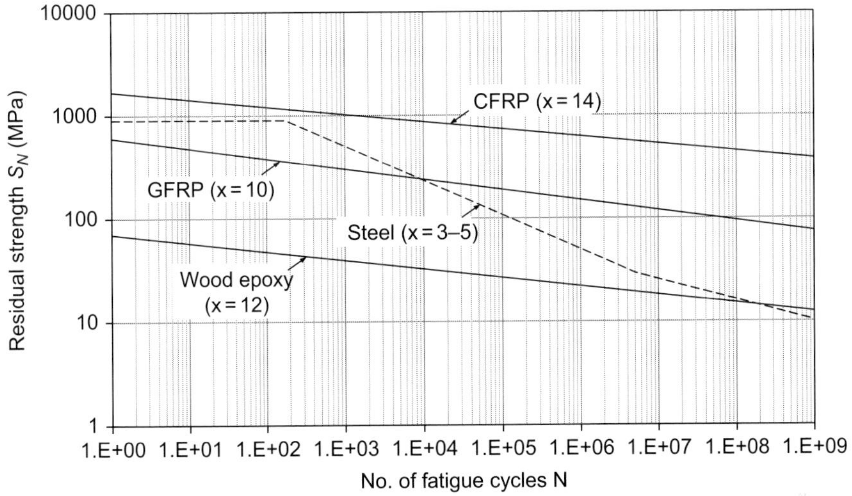

Figure 8.1 Fatigue curves for blade materials; see Equation (7.15); steel shown for comparison.

The above comparison shows that composites can provide lighter blades for given strength, but arguably their greatest advantage is superior fatigue resistance. Figure 8.1 shows the S/N curves (see Section 7.9.2) for the materials listed in Table 8.1, and it is seen that steel is relatively poor in fatigue compared with fibre-reinforced composites. Table 8.2 gives the residual fatigue strength for each material assuming 10^8 load cycles: for this duty GFRP and CFRP exhibit significantly higher strength than steel in absolute terms, and even wood-epoxy has comparable fatigue strength to steel. This is before considering specific (per unit weight) properties, when the composites' advantage becomes even greater. As noted above, the primary structure of large HAWT blades is nowadays made entirely from non-metal composites; steel blade spars were, however, used on some large prototype wind turbines. An example was the WEG LS-1, a 3 MW machine with a 60-m-diameter two-blade rotor; Figure 8.2. This machine operated for several years at the Burgar Hill test site on Orkney, but developed fatigue cracks on the rotor hub after a relatively short time (though its life was

Table 8.2 Comparison of Fatigue Strength

Material	Fatigue exponent x	Residual strength at 10^8 cycles (MPa)	Relative fatigue strength*
High-strength steel	3–5	16	1.0
GFRP (UD, $V_f = 0.5$)	10	95	5.9
CFRP (UD, $V_f = 0.6$)	14	449	28
Wood-epoxy laminate	12	15	0.9

Note. See Figure 8.1.

* Normalised with respect to high-strength steel.

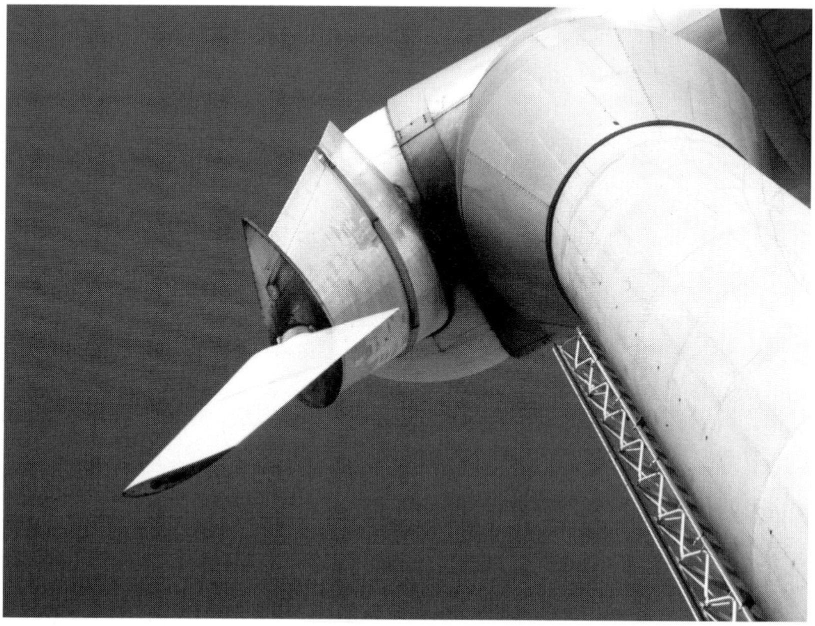

Figure 8.2 The WEG LS-1: the blades of this 3 MW prototype had a steel main spar.

successfully extended after on-site repairs). The weight of the LS-1 rotor was approximately 30 t: for the same diameter the weight of a GFRP rotor is around 30% less, and with significantly longer fatigue life.

The high strength of GFRP and CFRP derive from the microscopic properties of fibres, which in simple terms are too thin to harbour the flaws that would otherwise weaken the bulk material: glass fibres are much stronger than sheet glass, and carbon fibres stronger than solid graphite. This is a fundamental property of all fibres, including metallic ones (Harris, 1999), but GFRP and CFRP further benefit from low density, conferring the high specific strength and stiffness noted above. For both materials the composite is formed by embedding long fibres in a thermosetting polymer matrix: GFRP is manufactured using epoxy or polyester (or less commonly vinylester) resin; for CFRP epoxy resin is exclusively used. An important property of the finished composite is its fibre volume fraction V_f defined by

$$V_f = V_i / V_{\text{total}} \tag{8.1}$$

where V_i is the volume occupied by fibre type i in a total volume V_{total}, which includes the resin matrix and any other fibre types or orientations (see below). Up to a point the higher the fibre volume fraction, the greater the strength of the composite. There is a limit, however, as with too little resin to bind it the composite becomes 'dry' and cannot cohere. The volume fraction of fibre in GFRP varies with the method of manufacture: traditional hand layup achieves V_f of 30%–40%, but modern vacuum infusion techniques enable values of around 50% (see Section 8.5.1). With CFRP the volume fraction can be greater than 60%. The net modulus E_c of a fibre-reinforced composite may be calculated using the Rule of Mixtures, or Voigt estimate (Harris, 1999):

$$E_c = E_m V_m + E_f V_f \tag{8.2}$$

where E_m and E_f are the E-moduli for the resin matrix and reinforcing fibres, respectively, and V_m and V_f their volume fractions. The net properties are then the volume-weighted sum of the component properties.

Figure 8.3 shows the E-modulus of unidirectional glassfibre-epoxy composite as a function of fibre volume fraction: the modulus for pure glass fibres is 73 GPa and for cured epoxy 3 GPa, so for V_f of 0.5 the net modulus E_c is 38 GPa or roughly half that of the pure fibre. The contribution of the matrix to E_c is almost negligible, yet its presence is essential as without it the fibres are not bound and the composite cannot sustain any off-axis load. The solid line in Figure 8.3 indicates the practical range of fibre volume fraction for GFRP: values of V_f much below 30% would be considered resin-rich; values a little above 50% (dashed line) are above the practical limit for this material before becoming too dry. The compressive strength of GFRP peaks at V_f around 0.5 and begins to decrease with higher volume fraction (Harris, 1999).

Depending on its structural role a composite may contain fibres laid parallel to a single axis (unidirectional) or aligned in two or more directions (bi-axial or multi-axial). Glass fibre is commercially supplied as woven or stitched[1] fabrics in which a range of fibre orientation may be incorporated. Figure 8.4 illustrates some typical examples: unidirectional (UD) glass is used in the blade spar caps, which are subject to mainly longitudinal stress; biaxial fabric with fibres crossed at ±45° is used in the blade skins and the shear web for resisting shear and torsional

[1] Stitched fabrics are preferred to woven, as fibre orientation is better maintained. Light polyester stitching is used.

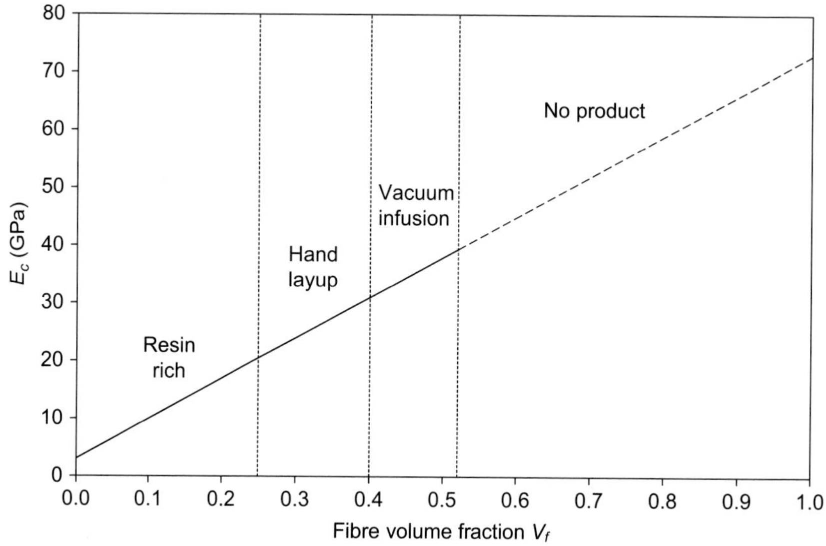

Figure 8.3 Net modulus E_c of unidirectional glass-epoxy composite as a function of fibre volume fraction, according to the Rule of Mixtures; see Equation (8.2). The volume fraction that can be achieved varies with the method of production, as indicated.

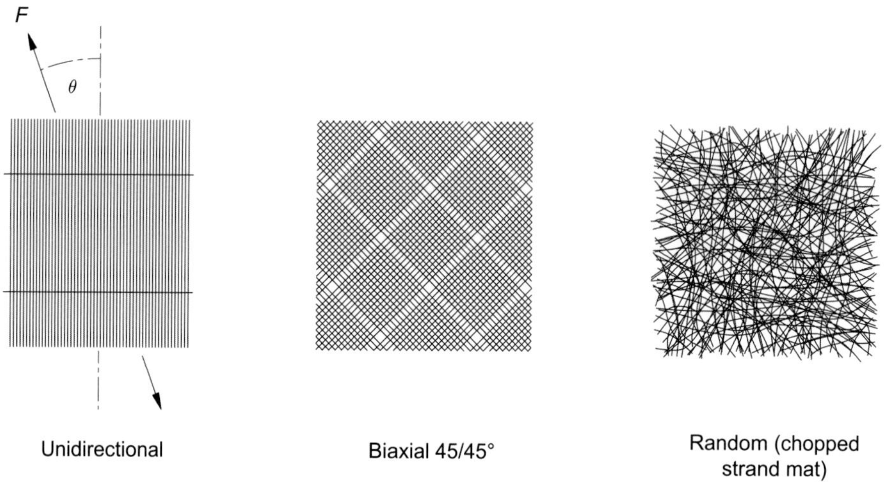

Figure 8.4 Fibre orientation in typical composite fabrics (illustrative). Strength and stiffness are functions of the angle θ made between applied load F and the fibre axis; see also Figure 8.5.

stresses (see Section 8.3.4); chopped strand mat has random fibre orientation and relatively modest strength: it is used in resin-rich surface layers for a smooth surface finish. Other fabrics are used elsewhere: at the blade root multi-axial composite layups are required as the loading is a complex combination of axial force, shear, and torsion.

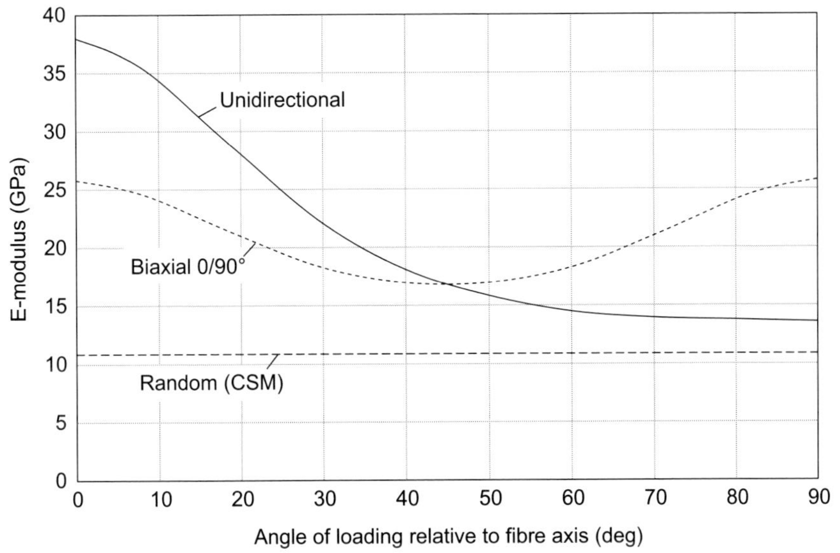

Figure 8.5 Variation in *E*-modulus of different types of GFRP composite with direction of applied loading. Based on published data (Harris, 1999) with an assumption of 50% glass volume fraction.

As noted above, fibre-reinforced composites are anisotropic, i.e. their engineering properties vary with the direction of loading. Stiffness and strength are greatest parallel to the fibre axis, but diminish with off-axis loading. This seen in Figure 8.5, which shows the *E*-modulus of GFRP as a function of the applied load angle. The three fabric types shown correspond to those in Figure 8.4 (note that 0/90° biaxial is equivalent to ±45° with the loading axis rotated by 45°); their net moduli are based on experimental results (Harris, 1999) scaled to represent GFRP with 50% glass volume fraction. When calculating the properties of multi-axial composites the Rule of Mixtures is again used, with the net modulus found from

$$E_c = E_m V_m + E_1(\theta_1)V_1 + E_2(\theta_2)V_2 + E_3(\theta_3)V_3 + \cdots \tag{8.3}$$

where $E_i(\theta_i)$ is the modulus of fibres lying at angle θ_i to the loading direction, and V_i their corresponding volume fraction; the matrix properties apply as before. Equation (8.3) can also be used where fibres of different type (e.g. glass and carbon) are combined in the same structural component. The way in which composites of different fibre orientation are used to achieve high structural efficiency is described in more detail in Section 8.3.4.

8.3 Blade Structure

8.3.1 Cantilever Beam Model

A HAWT blade can conveniently be modelled as a one-dimensional cantilever beam, as shown in Figure 8.6. A hollow section is shown (solid blades are too heavy for all but the smallest wind turbines) with aerofoil cross section of chord c and thickness (depth) t. The dominant aerodynamic

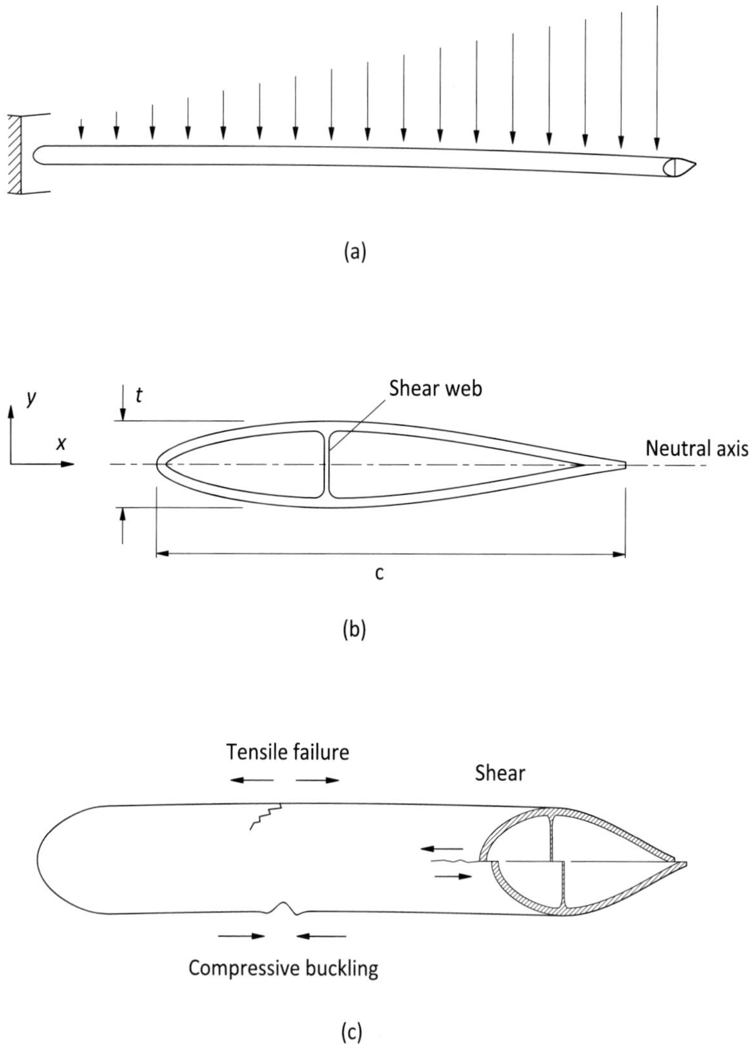

Figure 8.6 Cantilever beam model of wind turbine blade: (a) flapwise aerodynamic load distribution under optimum conditions; (b) beam cross section; (c) principal modes of failure due to flapwise bending: the upwind blade surface is in tension and the downwind in compression.

loading under most conditions is axial thrust, acting normal to the rotor plane: on an untwisted blade this corresponds to flapwise loading, and for simplicity this assumption is made here. At optimal aerodynamic efficiency the flapwise force per unit span increases linearly with radius,[2] and a typical load distribution (calculated from BEM theory) is seen in Figure 8.7, together with the associated flapwise bending moment. Under this form of loading three potential failure modes are identified, namely

[2] For explanation, see Section 3.5.

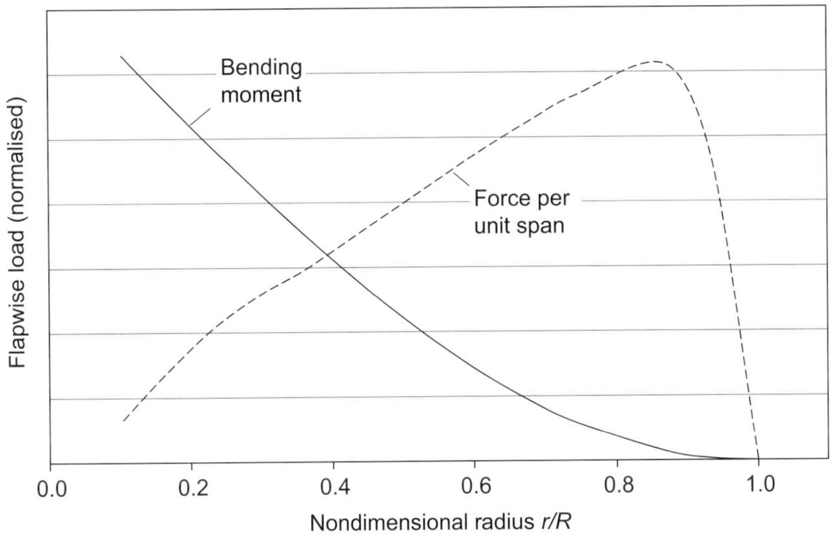

Figure 8.7 Radial distribution of flapwise force and bending moment calculated from BEM theory for a blade operating near $C_p^{\max}$. The flapwise bending moment dominates blade structural design.

- tensile/compressive failure (longitudinal stress)
- buckling
- shear

These modes are illustrated in Figure 8.6(c). Tensile failure is most likely to occur on the upwind (high pressure) blade surface, as it bends out of the rotor plane (for illustration see Figure 3.8). On the downwind surface compressive failure can theoretically occur but elastic buckling is a much likelier possibility. The locations most susceptible to shear failure lie on the flapwise neutral axis, at the leading and trailing edges of the blade or on the shear web. The structure of the blade must be designed to efficiently resist each of these failure modes and this is achieved by tailoring the cross-sectional geometry, and by selective application of composite materials.

From elementary beam theory, the longitudinal stress σ at an arbitrary cross section is found from

$$\sigma = \frac{My}{I_{xx}} \tag{8.4}$$

where M is the local bending moment, y the distance from the flapwise neutral axis, and I_{xx} the second moment of area of the section. Referring to Figure 8.6, the neutral axis of the aerofoil profile lies on (or close to) the chord line and σ is tensile on the upwind (high pressure) surface of the blade, and compressive on the downwind (suction) side. Maximum stress occurs for $y = t/2$, hence the local flapwise rigidity is proportional to I_{xx}/t; furthermore in the case of a section with constant skin thickness rigidity becomes directly proportional to section depth t. Thicker aerofoil profiles are therefore desirable for structural efficiency. Optimum aerodynamic performance,

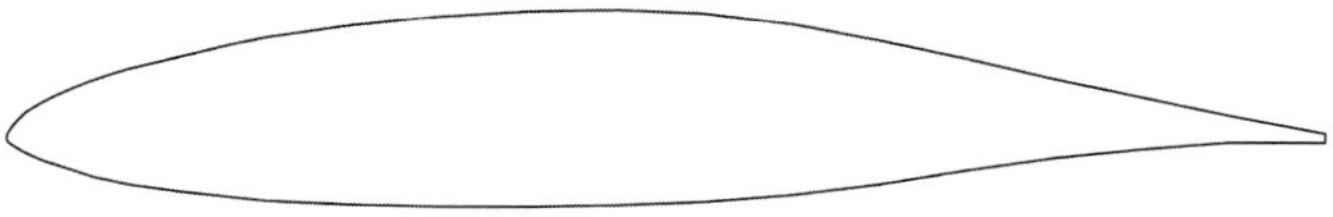

Blade tip: S813 (15%)

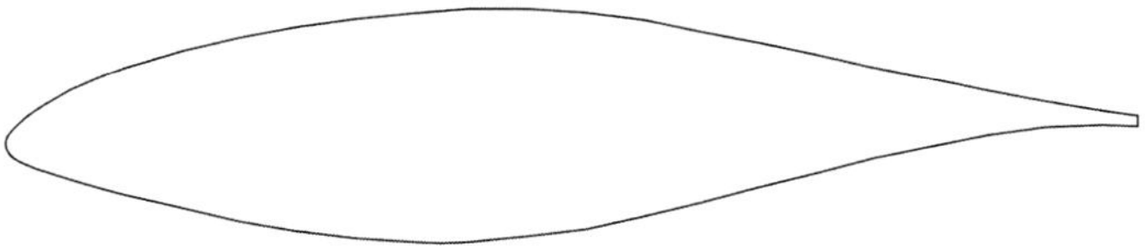

80% span: S812 (20%)

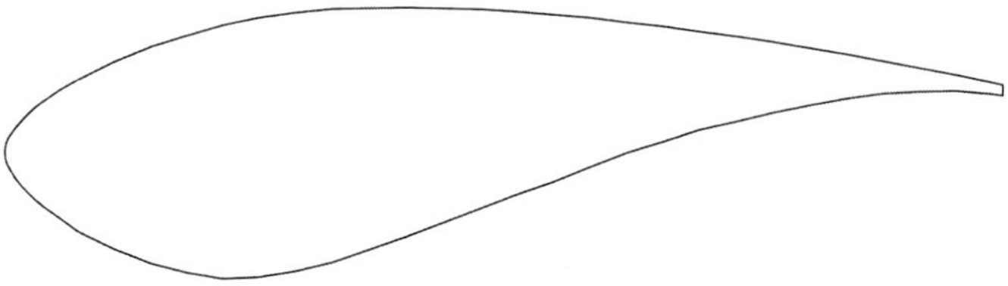

45% span: S814 (26%)

Figure 8.8 Aerofoil profiles developed by NREL for large wind turbines, showing the variation in thickness with radial position (Tangler et al., 1995). Thickness/chord ratios are in brackets.

however, favours thin profiles, so a key design challenge is the trade-off between aerodynamic performance and structural rigidity.

Development of thick aerofoils has been pursued by groups in several countries, including NREL[3] in the USA (Tangler et al., 1995), Delft University of Technology in the Netherlands (Timmer et al., 1991), and FFA in Sweden (Björck, 1990). Figure 8.8 shows a series of profiles developed by NREL with thickness ratios ranging from 15% for use at the blade tip, where aerodynamic efficiency is paramount and bending loads are small, to 26% at mid span. Thicker sections still may be used towards the blade root, where aerodynamic efficiency is less important (FFA have demonstrated a 36% thick cambered aerofoil profile). The flapwise stiffness of the NREL profiles is compared in Figure 8.9; the data are normalised by the stiffness of a solid rectangular section with 2% thickness, i.e. with approximately the same wetted cross-sectional area as the given aerofoil profiles. The plot illustrates the linear relationship between flapwise stiffness and thickness ratio, and shows the NREL profiles to be 10–16 times stiffer than a solid rectangular bar of the same weight. This trend will be similar for most aerofoil families.

[3] National Renewable Energy Laboratory, formerly known as SERI (Solar Energy Research Institute).

The internal blade structure plays an equally important role in achieving the required flapwise stiffness. For a given external profile the ratio I_{xx}/t can be optimised by judicious arrangement of material, as illustrated in Figure 8.10, which shows two sections with the same profile but different

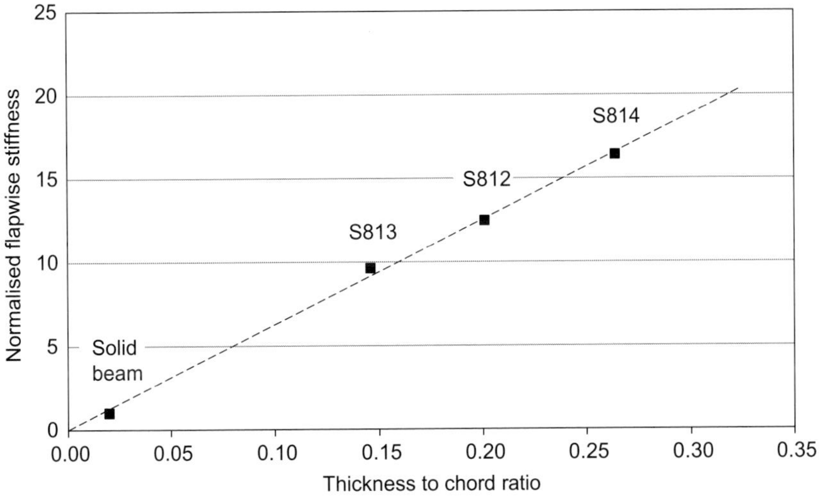

Figure 8.9 Comparison of flapwise stiffness (I_{xx}/t) for the NREL profiles shown in Figure 8.8. Hollow sections are assumed, with wall thickness equal to 1% of chord and stiffness normalised relative to that of a solid rectangular beam of the same chord and cross-sectional area.

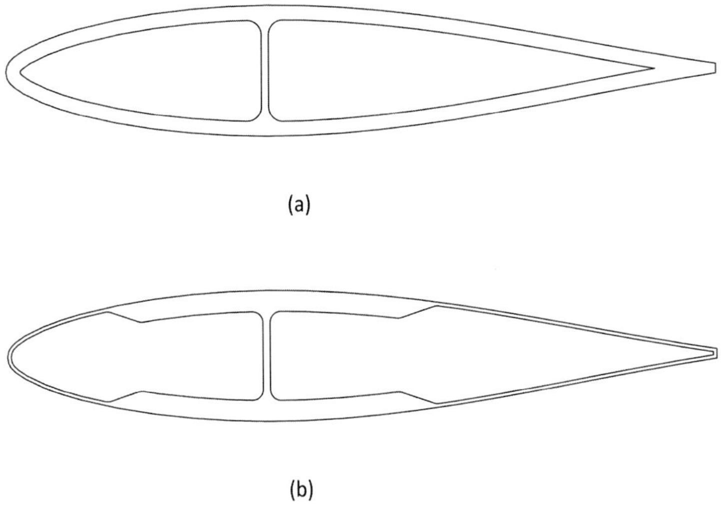

Figure 8.10 Blade cross sections with the same flapwise stiffness (I_{xx}/t) but different weight. Section (a) has uniform wall thickness in the shell; in section (b), the spar caps are thickened and the shell is reduced elsewhere, with 25% less material used overall.

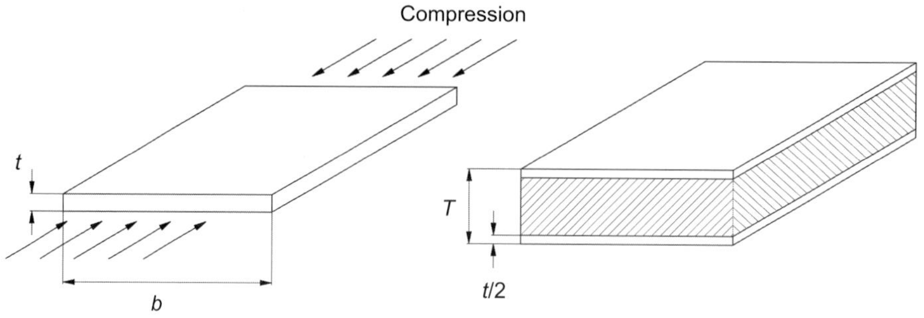

Figure 8.11 Sandwich structure is used for blade skin panels and shear webs to improve buckling resistance. The sandwich panel (right) has the same amount of skin material as the plain sheet (left), but the critical compressive stress increases as T^2. The core adds minimal weight.

internal dimensions. Section (a) has uniform wall thickness; on section (b) the skin is selectively thickened in the regions of maximum longitudinal stress and reduced elsewhere; the thickened area corresponds to the spar cap or girder. The two sections have the same I_{xx}/t ratio so are equally resistant to flapwise bending, but section (b) uses 25% less material, and would result in a correspondingly lighter blade: its cross section can be thought of as a combination of a traditional I-beam to support the bending load, with a thin shell to provide the required aerodynamic form. This kind of structural optimisation is further enhanced by the use of composite materials, as discussed in Section 8.3.4.

8.3.2 Compressive Buckling

Under flapwise loading the downwind (suction) surface of the blade experiences longitudinal compression, which can potentially cause the skin to buckle. This is an elastic stability phenomenon, and thin blade skins may buckle under stress levels that are well within the compressive strength of the material. Figure 8.11 shows a rectangular skin panel subject to compressive loading; the critical stress σ_{crit} at which buckling will occur is given by

$$\sigma_{\text{crit}} = kE(t/b)^2 \tag{8.5}$$

where E is the elastic modulus, t the panel thickness and b its width; the constant k is a function of the panel aspect ratio and edge constraints (Megson, 1972). To increase the buckling strength of the panel sandwich construction is employed, where the skin is bonded to a lightweight core of low-density material and the effective panel thickness is increased without a significant weight penalty. Assuming a given thickness t in the skins and an overall thickness T the skin buckling strength (σ_{crit}) in a sandwich panel is proportional to T^2. A core of relatively modest thickness can thus increase skin buckling resistance by an order of magnitude.

Sandwich construction may also be used on the shear web, an important function of which is to maintain the blade profile geometry. Under flapwise loading there is a tendency for the section to deform, potentially altering its aerodynamic characteristics, and reducing its flapwise stiffness.

The shear web acts as a strut keeping the upper and lower blade surfaces apart, and as such comes under compression. By providing the web with a lightweight core similar to that used in the blade skins its effective thickness and buckling resistance are increased. On large blade sections there are usually two or more shear webs: this reduces the loading on each web and also divides the unsupported blade surface into narrower panel widths, increasing the ratio t/b and (following Equation (8.5)) further improving the resistance to buckling.

8.3.3 Shear

Flapwise bending causes shear stresses both in the plane of the blade cross section and orthogonal to it. In the latter case the stress acts as though to make the upper half of the blade slip lengthwise relative to the lower, as shown in Figure 8.6(c). This tendency is resisted primarily by the material in the blade leading and trailing edges and by the shear web. The analytic treatment of shear stress is less straightforward than longitudinal stress, but in sizing the shear web a conservative approach is to assume that it supplies all the shear resistance, neglecting any contribution of the leading and trailing edges. For a thin web of the type shown in Figure 8.10 the maximum shear stress τ occurs on the neutral axis, with value

$$\tau = \frac{3F}{2A} \tag{8.6}$$

where F is the shear force and A the cross-sectional area of the web. If the web is of box section with a foam core (see below) the contribution of the core is ignored and area A defined only by the vertical walls. If a section has multiple shear webs – large blades may have two or even three at some stations – their area can be summed. As noted, use of Equation (8.6) to determine the web thickness is quite conservative. It is in any case unusual for shear stress to be a determining factor in blade strength, which is mainly governed by longitudinal fibre stress and/or buckling resistance.

8.3.4 Elements of a Composite Blade

The main principles of HAWT blade design – the importance of profile thickness, the function of the shear web and spar caps, and the use of sandwich skin construction – have long been understood by aircraft designers.[4] From the time of Ulrich Hütter (see above) onwards, however, the wind turbine industry arguably took the lead in the application of high-strength composites to primary structure and HAWT blades up to 80 m long are now manufactured almost entirely without metal components. Fibre-reinforced composites enable the stresses arising in the blade to be resisted in a targeted fashion and thus achieve high structural efficiency. This can be illustrated with reference to Figure 8.12, which shows a section of a typical composite blade with its main structural components identified.

---

4 See e.g. Megson (1972).

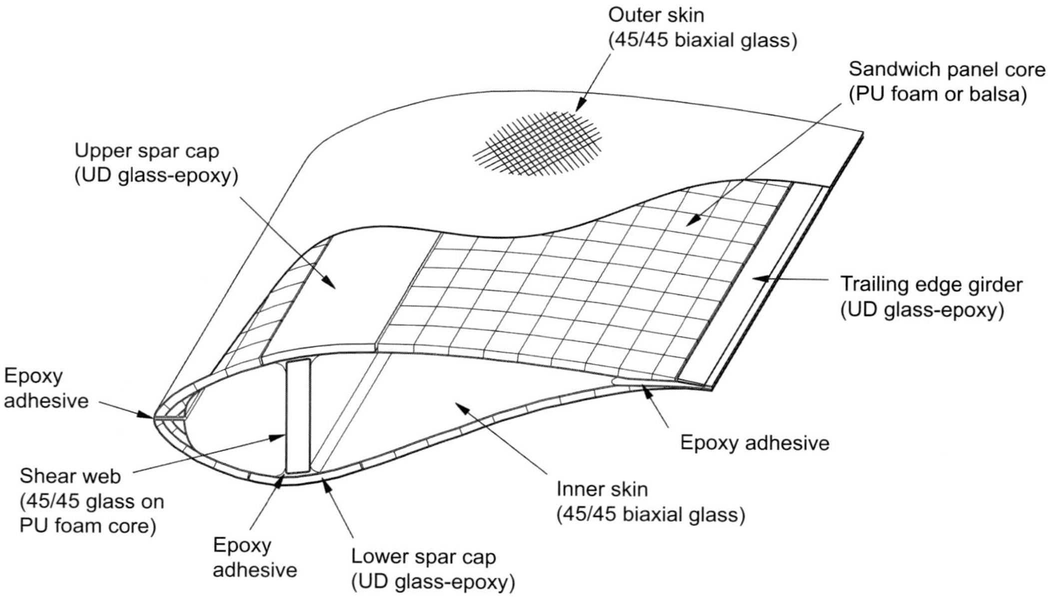

Figure 8.12 Structure of a typical composite blade. The design shown is based on an NREL thick profile (see Figure 8.8), with unidirectional glass-epoxy spar caps and a single shear web.

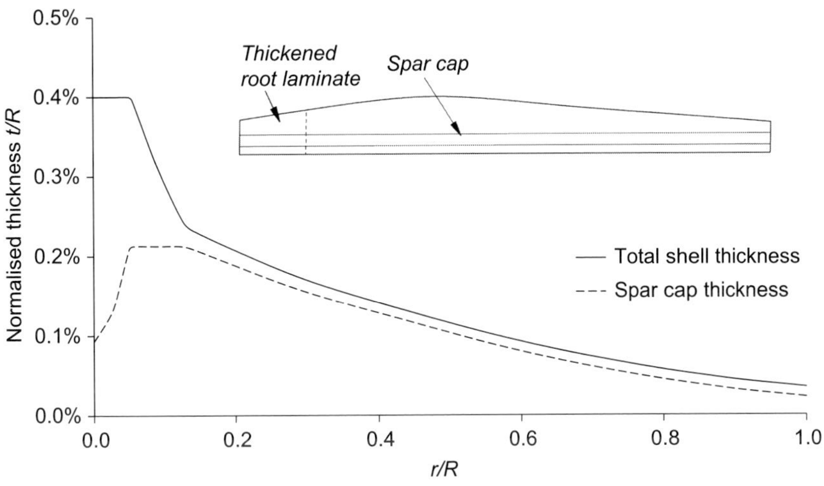

Figure 8.13 Variation in spar cap thickness (normalised by tip radius) along the blade length. In this case, the spar cap has constant chord.

The spar caps (aka girders) are made from unidirectional composite with fibres parallel to the blade long axis and located at the thickest part of the blade profile. This is the most efficient cross-sectional geometry for flapwise stiffness (see Figure 8.10) and as the stress developed in the spar caps is primarily longitudinal, so UD material offers the highest efficiency. GFRP is most commonly used; CFRP is superior in strength and stiffness (and particularly good in fatigue) but it is unusual to find it on blades below about 50 m length due to its high cost, which is several times that of GFRP; carbon is, however, used on some of the very largest blades (see Section 8.7).[5] The spar cap becomes thinner along the blade, reflecting the decrease in flapwise bending moment with radial location (Figure 8.7). Spanwise variation of the spar cap thickness is illustrated in Figure 8.13; the chordwise dimension may also decrease, though on some blades, such as that illustrated here, a constant chord spar cap is used. Towards the blade root the loads increase significantly, and become more complex due to a combination of flapwise, edgewise, and torsional components; to support the resulting 3D stresses the shell is thickened in this region using multi-axial composites and there is no sandwich material towards the blade root (see Figure 8.14). On very large blades carbon fibre may be incorporated in the most highly stressed areas, e.g. in the laminate around the root attachment bolts.

The principal functions of the shear web are (a) to resist flapwise shear loading and (b) to provide shape stability. Maximum shear stress occurs at the flapwise neutral axis with stress lines developed at ±45° to the blade long axis; accordingly the shear web skins are made from ±45° biaxial GFRP, providing maximum resistance in the required plane. High tensile strength is not as important here as longitudinal stress in the web is negligible. It must, however, support compressive (crushing) forces arising between the upper and lower blade surfaces and be resistant to

[5] Though CFRP is notably absent from the 154-m-diameter rotor of the Siemens 7 MW wind turbine.

Figure 8.14 Root of a 16 m GFRP blade being finished, NOI Scotland factory, circa 2003. Note the thick root shell where multi-axial glass fabrics are used to resist complex stresses; also note the T-bolt root attachments.

buckling in this direction; a box section with lightweight core, as shown in Figure 8.12, is often preferred for this reason. The core material may be rigid polymer foam (polyurethane or PVC) or balsa wood, similar to that used in the skin sandwich panels (see below).

The blade skins are of sandwich construction, formed from ±45° biaxial GFRP on a lightweight core. The biaxial laminate provides shear strength and torsional rigidity; the core material carries little load, but its thickness gives the sandwich panels buckling resistance. This is particularly important on the downwind blade surface, which experiences high compressive loads. At the trailing edge of the blade UD material similar to that in the spar caps may be incorporated for stiffness in respect of chordwise (edgewise) bending. This component is sometimes known as the chordwise girder, but it requires significantly less UD material than the flapwise girder as the blade is inherently stiffer in the chordwise sense due to the aerofoil thickness/chord ratio.

The majority of blades are still manufactured as two half-shells, which are bonded together with epoxy adhesive at the leading and trailing edges and along the shear web surfaces. The adhesive joints must support shear force, and the required contact area is dictated by the adhesive lap shear strength (which may be as little as 3 MPa after material safety factors are taken into account). There

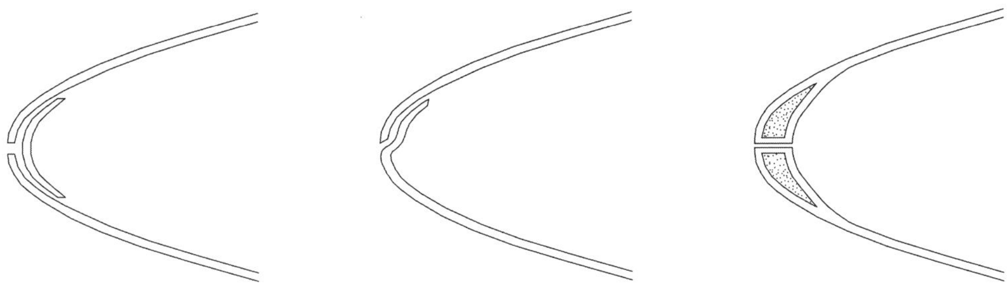

Figure 8.15 Options for the leading-edge joint of a two-part blade: (a) profiled joining angle; (b) slip joint; (c) developed flange using core material. In all cases, the joint is completed with epoxy adhesive when the mould halves are brought together.

are different methods for the leading-edge joint, as seen in Figure 8.15. In case (a), a prefabricated joining angle is bonded to the inside of the leading edge; in case (b), the lower blade shell has a return flange formed on its leading edge to make a 'slip joint' when the two halves are brought together; in option (c), a wide flange is developed on both shells using core material. The last of these options is also shown in Figure 8.12, and mimics the joining detail of a wood-epoxy blade (see Figure 8.22) which has inherently thick leading-edge shells. In all cases the leading-edge joint is completed using epoxy adhesive paste; similarly at the trailing edge, where the section profile inherently lends itself to a wide bond line. On some large GFRP blades bond lines and adhesives have now been eliminated by moulding the blade as a single piece, achieving a continuous thin shell at the leading edge: the manufacture of GFRP blades is described in more detail in Section 8.5.

8.3.5 Bending Analysis

The stress and strain in a blade due to bending can be calculated from classical beam theory modified to allow for composite materials. The cantilever beam model still applies, but the bending stress formula of Equation (8.4) now becomes

$$\sigma_i = \frac{My_iE_i}{\sum_i EI_{xx}} \tag{8.7}$$

As previously M is the bending moment at a given radial station; σ_i is now the longitudinal stress in component i of the composite, y_i the applicable distance from the net neutral axis,[6] E_i Young's modulus for the material in question, and $\sum_i EI_{xx}$ the net section modulus summed over all structural elements. The corresponding strain is given by

[6] Calculating the neutral axis for a non-symmetric composite section is slightly complex; for the governing equations for composite beams, see e.g. Chapter 7 of Crandall et al. (1978).

$$\varepsilon_i = \frac{\sigma_i}{E_i} = \frac{My_i}{\sum_i EI_{xx}} \tag{8.8}$$

Equations (8.7) and (8.8) are a generalised version of the beam equations applicable to composites; when there is only one material present they reduce to the more familiar beam equations for stress and strain. The engineering theory of composites assumes that under an external load the different component materials are subject to a common strain but develop stress σ_i proportional to their individual E-moduli. The stiffer components thus experience a higher stress or 'take more share of the load'. An analogy is two dissimilar springs in parallel being compressed by a common load: the springs undergo equal deflection (strain) but the stiffer spring takes a greater share of the load (stress).

To illustrate such a calculation, consider the GFRP blade in Figure 8.12. The structure is divided into two component groups of dissimilar E-modulus, namely (1) the spar caps and trailing edge girder and (2) the blade skins and shear web walls. The analytic geometry is shown in Figure 8.16: note that sandwich core material is neglected in this analysis as its low modulus contributes little to the net section stiffness.[7] Group 1 components are made from UD glass composite with high longitudinal stiffness ($E_1 = 38$ GPa) while Group 2 are made of ±45° biaxial glass with high shear strength but lower modulus ($E_2 = 17$GPa). The second moments of area I_1 and I_2 are calculated for the two groups with respect to the flapwise neutral axis, and from Equation (8.7) the stress in the Group 1 (UD) material is

$$\sigma_1 = \frac{My_1E_1}{E_1I_1 + E_2I_2} \tag{8.9}$$

Maximum stress occurs at the outer fibres where $y_1 \cong t/2$, where t is the profile thickness. Maximum stress in the spar cap is then found from

$$\sigma_{1(\max)} = \frac{MtE_1}{2(E_1I_1 + E_2I_2)} \tag{8.10}$$

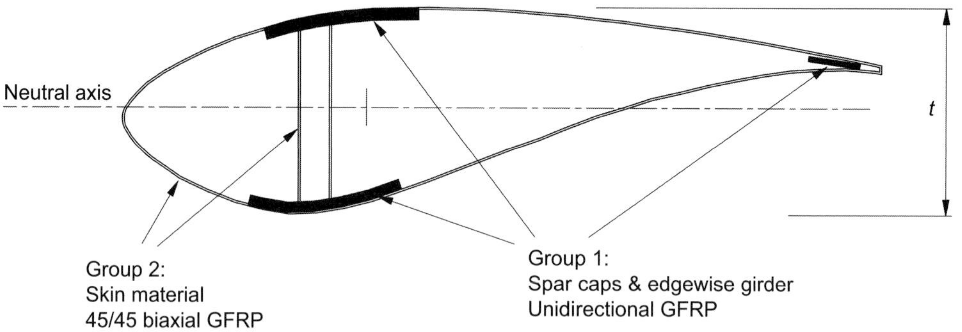

Figure 8.16 Bending moment analysis of composite blade. The key structural elements are separated into component groups of the same E-modulus. Stress and strain in each component are found using Equations (8.9) through (8.13).

[7] Typically, E≤0.05 GPa for polyurethane (PU) foam.

Maximum stress in the skins, again assuming a location at the outer fibre, is given by

$$\sigma_{2(\max)} = \frac{MtE_2}{2(E_1I_1 + E_2I_2)} \tag{8.11}$$

These calculations are carried out at each radial station along the blade, with stress maxima in each case checked against the allowable strength of the materials. For composite materials the allowable strength is based on a characteristic test value with a conservative material safety factor. In the Germanischer Lloyd recommendations the material safety factor γ is the product of several partial safety factors, with

$$\gamma = C_1 \times C_2 \times C_3 \times C_4 \times C_5 \tag{8.12}$$

The partial safety factors are themselves based on a range of environmental and manufacturing criteria, and the following values are typical for GFRP (Germanischer Lloyd, 1993):

PSF	Factor	Value
C_1	General safety factor	1.35
C_2	Creep strength factor	1.5
C_3	Temperature effect factor	1.1
C_4	Production factor	1.2 (hand layup), 1.1 (vacuum infusion)
C_5	Heat treatment factor	1.0 (controlled cure), 1.1 (uncontrolled)

Based on the above figures a GFRP laminate produced by hand layup without temperature control during the resin cure would have material safety factor γ of 2.94. If vacuum infusion was used with temperature control the corresponding value of γ would be 2.45. The admissible strength of the composite is then of the order 35%–40% of its characteristic strength depending on the manufacturing process, indicating a relatively conservative design margin. An alternative to the stress calculation procedure above is to calculate maximum strain at each blade section. Equation (8.8) applies, and as maximum strain occurs at the outer fibre we again set $y = t/2$, hence

$$\varepsilon_{\max} = \frac{Mt}{2(E_1I_1 + E_2I_2)} \tag{8.13}$$

The admissible strain for composites is largely independent of fibre orientation so the same limiting value or 'cracking limit' would apply to the different structural components referred to in the preceding analysis. For GFRP this is typically 0.4%–0.5% (4000–5000 $\mu\varepsilon$) including material safety factors. On this basis, the admissible stress in an arbitrary GFRP composite may be taken as approximately 0.5% of its E-modulus.

8.4 Root Attachment Methods

At its root the entire thrust and bending loads developed by a blade are transferred to the rotor hub. This is a critical interface between materials of different properties, e.g.

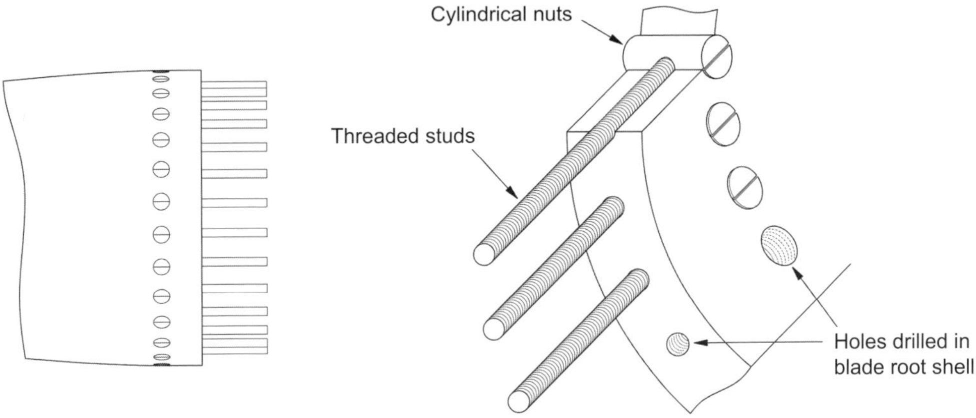

Figure 8.17 T-bolt root attachment. Widely known as the 'IKEA' fixture for obvious reasons.

a composite blade and cast iron hub, and an area of potentially high stress concentration. Development of reliable blade attachment methods was for many years a preoccupation of the wind turbine industry. Some early solutions were subject to premature fatigue failure, but as a result of significant R & D blade root attachments are now highly reliable. One of the most widely used is the T-bolt, which can be seen on the blade in Figure 8.14, and is illustrated in Figure 8.17. In this system a ring of threaded studs are located in sockets in the blade root shell and secured by cylindrical nuts. The system is (inevitably) known as the 'Ikea joint' for its similarity to the well-known Swedish self-assembly furniture. The T-bolts are secured to the hub by conventional nuts, which put the bolt heavily into tension, and the composite root shell correspondingly into compression.

The T-bolt is thus a good example of a pre-loaded joint,[8] where the external blade load is shared by the bolts and composite shell according to their relative stiffness. In practice the bolts may be tensioned to 60% or more of their yield stress, but they experience only a fraction of the external cyclic blade loading and are thereby protected against fatigue. The bolt pre-load maintains the laminate in compression under the highest bending load to prevent joint separation and the dimensions of the cylindrical nut are designed to ensure that the contact stress at the root shell does not exceed the compressive strength of the composite (multi-axial laminate is used in this region). Although drilling a ring of transverse holes in the most highly loaded part of a cantilever beam seems like a recipe for stress concentration, the T-bolt joint has an excellent track record and failures in service are almost unknown. For an overview of design rules and experimental test results see Martinez et al. (2011).

An alternative root attachment method utilises profiled steel studs bonded with epoxy adhesive into the blade shell, as shown in Figure 8.18. This technique was historically developed

[8] The principles are described in many good textbooks, e.g. Shigley (1981).

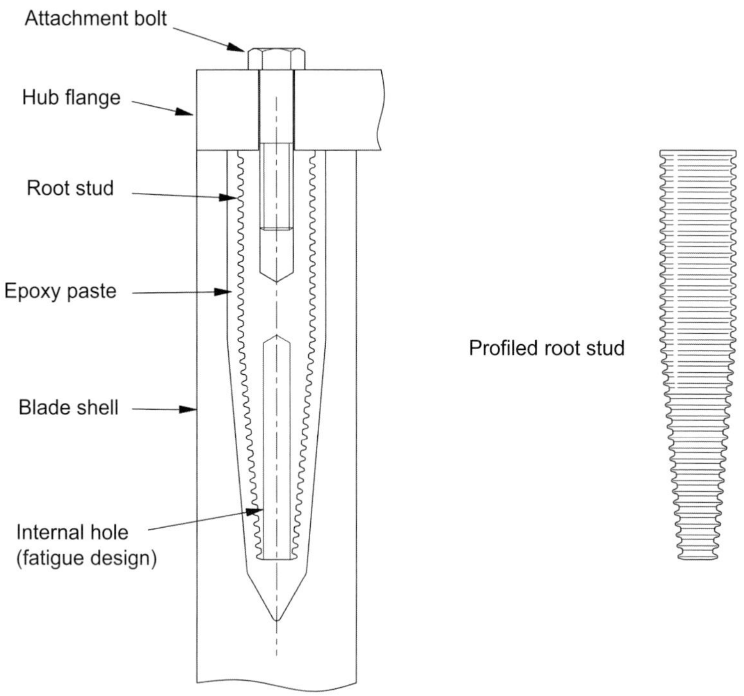

Figure 8.18 Adhesive-bonded root stud with internal thread. This concept was originally developed for wood-epoxy blades, but similar designs are used with GFRP.

for wood-epoxy blades (see Section 8.5.2) but is also widely used for GFRP. The studs may be internally threaded to accommodate an attachment bolt (as shown) or have a threaded extension to mate with the rotor hub flange, similar to the T-bolt. The cross-sectional dimensions and external profile of the stud are designed to minimise stress concentration, with tapered diameter and an internal counter-bore to achieve a smooth reduction in cross-sectional area. The parallel grooves on the stud surface give a strong bond with the epoxy adhesive, and their profile is formed by rolling (rather than machining) to avoid small radii that may potentially cause stress concentration. The limiting pull-out strength of the stud is dictated by the adhesive strength and dimensions of the bond interface.

A more recent development is embedded root studs, which are bonded into the root shell during the moulding process. The external profile of the stud is shaped so as to key into the composite and the stud pull-out strength is then determined by the composite shear strength rather than an adhesive bond, facilitating a smaller and lighter stud. The blade manufacturer Aerpac introduced keyed studs on a commercial 50 kW blade in 1999, with laboratory tests demonstrating a static pull-out strength in excess of 34 t. The Danish company SSP Technology has more recently developed a system using prepreg composite materials to embed the stud, and this method has been employed on very large blades (SSP Technology, 2018). Keyed studs may represent the most structurally efficient method of blade

Figure 8.19 Flanged blade root. No longer used, this attachment method was susceptible to fatigue due to stress concentration at root–flange interface. (Windmaster 300 kW blade at Blyth Harbour, 1995)

attachment, allowing closer circumferential spacing and hence more bolts for a given root circle diameter. Their principal disadvantage is that a damaged stud is hard to remove: the T-bolt is superior in this respect, being a purely mechanical fixture.

Older blade root attachment methods based on steel 'trumpet' flanges bonded or bolted to the blade laminate are now rarely seen, and were more susceptible to fatigue due to high stress concentration. An example is shown in Figure 8.19. The more modern solutions described above have the advantage over a flanged design that the bolt pitch circle occupies the full diameter of the blade root shell, resulting in a much stiffer geometry with higher load-carrying capacity, and ultimately lighter blades; see Section 8.7. For a review of historic blade root attachment methods see (Burton, 2011).

8.5 Blade Manufacture

Most GFRP blades are nowadays manufactured by vacuum resin infusion moulding (VRIM), which has largely superseded the open-mould 'wet layup' techniques originally adopted from boatbuilding and general purpose glass fibre manufacture. In the moulding process sheets of glass fabric are laid up (laminated) and impregnated with a thermosetting resin; when fully cured the resulting composite has the high strength and stiffness properties described earlier (see Section 8.2). Carbon fibre is harder to infuse than glass due to its smaller fibre diameter and when carbon is used it is usually supplied in the form of 'prepregs', or fibre sheets pre-impregnated with uncured epoxy resin, which are heated in the mould to effect the polymer reaction. The following is a short description of the basic VRIM manufacturing process for GFRP blades.

8.5.1 GFRP Blade Manufacture (VRIM)

In the commonest VRIM applications the blade is manufactured as two half-shells, which are subsequently bonded together with structural adhesive. Separate moulds are therefore required for the upper and lower surfaces, with each mould being the full length of the blade. The procedure for laying up the dry blade materials and preparing for infusion can be explained with reference to Figure 8.20, which shows the key elements used in the VRIM process (note that some details are omitted for clarity).

Layers of dry glass fabric are laid into the mould starting with the outer skin (the blade is effectively laid down from the outside inwards) and adding successive layers of fabric and/or foam

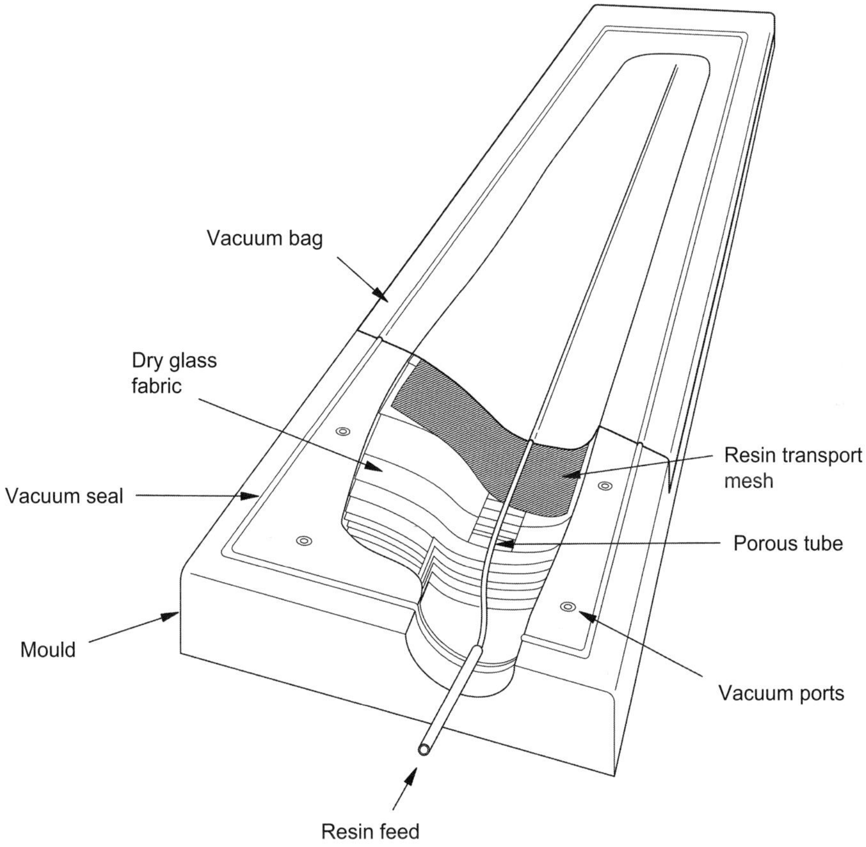

Figure 8.20 Vacuum resin infusion of a GFRP blade shell. Dry glass fabric is first laid in the mould and covered by an impermeable bag. Vacuum pressure compresses the laminate and draws the resin into it.

sandwich material as required. The spar cap is laid down as unidirectional (UD) fabric oriented along the blade axis, with the number of layers decreasing smoothly from root to tip (see Figure 8.13). Towards the blade root the laminate becomes relatively thick, as dictated by the high bending strength needed in this region and to accommodate the root attachment studs or T-bolts. At the tip the laminate is much thinner, typically 10% of that at the root.

Once the layup of dry fabric into the mould is complete, it is covered by the resin transport mesh: this is a porous plastic sheet (similar in texture to fine wire mesh) that aids the flow of resin across the upper surface of the glass fabric. Finally an impermeable vacuum bag is laid over the complete assemblage, and sealed around the mould edges. Vacuum is then applied to the mould contents. This serves two purposes, namely (a) to compress the laminate under 1 bar of atmospheric pressure, forcing it to conform to the mould profile, and (b) to draw the liquid resin into the mould and make it flow into the evacuated spaces in the glass fabric. This is the infusion process. It takes typically 2 hours to fully wet out the dry fabric, during which time the mould is heated to speed the flow of resin (by reducing its viscosity). Heating is achieved by thermostatically controlled electric panels built into the mould. Heat accelerates the epoxy cure, but as the chemical process is exothermic care must be taken to avoid excessive heat build-up (and in the worst case fire) in areas of thick laminate. Programmed control of the mould heating cycle is therefore required. Some moulds incorporate water circulation systems that can provide heating or cooling at different times in the infusion process.

The resins mainly used in GFRP blade production are epoxy or polyester (or, less commonly, vinyl acetate). Epoxy requires no solvent: the polymer reaction is activated by mixing a resin and hardener, whose proportions can be varied to accelerate or slow the cure. Epoxies give high-dimensional stability in the final product, whereas polyester resin may suffer from up to 2% shrinkage on demoulding; polyester also incorporates a volatile and flammable solvent (styrene) and fume extraction may be required in the production environment. The main advantage of polyester over epoxy is its significantly lower cost.

Once the resin has cured the two blade moulds are stripped of vacuum bags and other disposables, and internal components are fitted to the blade shells: in addition to the shear web these may include lightning conductor parts (see Section 5.7), tip brake components (for stall-regulated blades), and prefabricated joining pieces to facilitate the bonding of the two halves. Thickened epoxy adhesive is then applied to the leading and trailing edge bond lines and the shear web, and the two mould halves then brought together and securely clamped. Accurate alignment is achieved by registration pins at key locations around the mould perimeter. The epoxy paste takes several hours to fully cure, and again heat may be applied to shorten the time. Following this the blade is demoulded, after which it may be post-cured in an oven (typically 'soaking' at 70°C for 6–7 hours) to develop maximum strength in the composite. Some areas of the blade may have additional wet laminate added, for example at the join lines or in the thickened root region. Root attachment fixtures (T-bolts or root studs) are then installed and the completed blades are then dressed, smoothed, and sanded before painting and balancing (see below).

Figure 8.21 Manufacture of 23 m glass-epoxy blades via resin infusion moulding. Dry glass fabric is being laid into the two half moulds to facilitate the vacuum resin infusion process. Partly finished blades can be seen in the background. (Aerpac UK factory, Kirkcaldy, circa 2000)

The VRIM process was originally introduced in the manufacture of GFRP boats, and was adopted by the wind industry around the time that blades started to exceed 20 m in length. An example of a typical two-part mould for a 23 m blade is seen in Figure 8.21, where the dry glass fabric has been laid up prior to infusion. Blades up to 80 m long are now manufactured in two-part moulds but a recent development is a single-piece blade using the Siemens 'Integral blade' moulding process, in which the mould is closed while the laminate is dry and an internal vacuum bag applied to the complete inner surface prior to infusion (Stiesdal et al., 2006). The result is a seamless blade with no adhesive joints, and as a result higher structural efficiency (see Section 8.7). Blades up to 75 m long have been manufactured using this technique, and the scale of the product helps facilitate production as workers can gain internal access to the closed mould in a way not possible with smaller blades.

8.5.2 Wood-Epoxy Blades

For many years wood-epoxy laminate was used as primary structural material on commercial wind turbine blades, and is still found in some types. Wood is a natural composite with favourable engineering properties (see Table 8.1) and is also an inherently sustainable raw material, if sourced from managed plantations. The first wood-epoxy HAWT blades were manufactured in the US

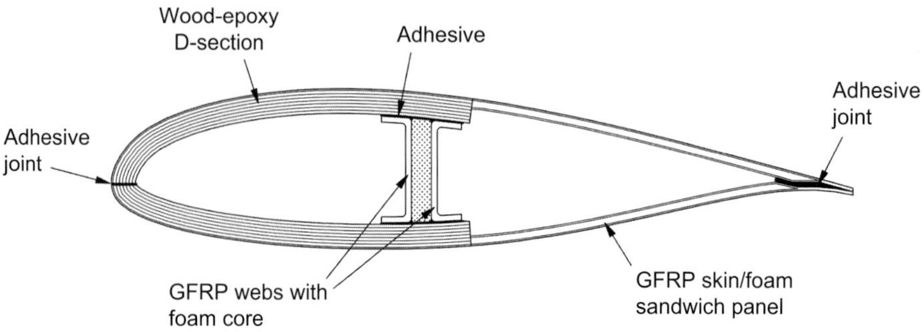

Figure 8.22 Section of a blade with wood-epoxy primary structure; the trailing edge and shear web structure are similar to a GFRP blade. This design was used by Howden on wind turbines up to 1 MW.

(Spera, 1990) using methods evolved from commercial boat building (Gougeon Brothers, 2018). Although arguably more difficult to manufacture than GFRP, wooden blades have some interesting advantages. The relatively low density of the parent material results in thick blade skins with inherently high buckling resistance, so sandwich construction is unnecessary. Wood-epoxy laminates have thin glue lines and require much less epoxy resin than GFRP blades, in which half of the composite by volume is epoxy. For blades up to a certain size wood laminate is thus a potentially cheaper and more environmentally benign technology.[9] This last advantage extends to end-of-life disposal: wood will biodegrade, whereas there are as yet no entirely satisfactory solutions for disposal of GFRP or CFRP. The preferred wood species for wind turbine blades are birch, poplar, and Douglas fir. At one time mahogany (*Khaya ivorensis*) was used but fell out of favour due to issues of sustainability.

As blade size grew the engineering properties of wood become less favourable, and the largest wood-epoxy blades were ultimately hybrid designs with some GFRP, and in some cases CFRP, components. Figure 8.22 shows the design favoured by the Howden company for a range of wind turbines culminating in a 1 MW machine with 26.5 m blades (Milborrow et al., 1988). The blade has a D-section main spar of wood-epoxy occupying the forward 60% of chord; the remainder of the section is similar to a GFRP design with sandwich skin construction over the rear chord. The aerofoil profile is NASA LS1(Mod) with 21% thickness ratio. Larger wood-epoxy blades were subsequently developed by UK manufacturer Taywood Aerolaminates and later NEG Micon, who introduced pre-cured carbon fibre strips into the wood shell using a patented vacuum infusion technique (Gunneskov et al., 2002): a typical cross section is shown in Figure 8.23. Blades up to 40 m long were produced using this technique and saw service on the NM82 1.65 MW wind turbine.

[9] Wood-laminate blades are arguably more aesthetically pleasing too: they are traditionally surfaced with polyester gel coat so the blade demoulds with a high-gloss finish.

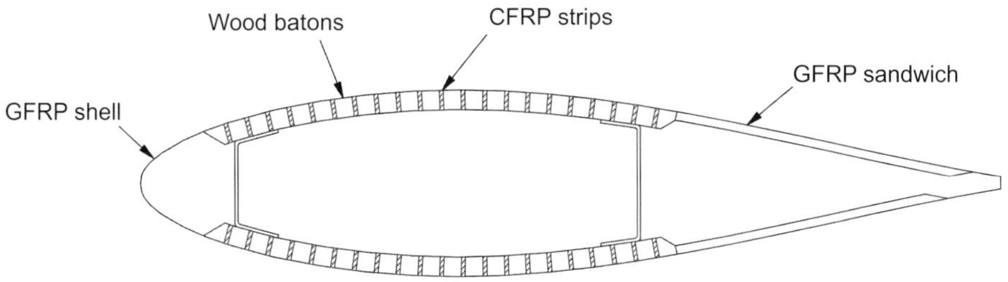

Figure 8.23 Wood-epoxy blade incorporating CFRP strips and manufactured by vacuum infusion. This technique was patented by NEG Micon (Gunneskov et al., 2002) and used for blades up to 40 m long.

Figure 8.24 Joining the halves of a wood-epoxy blade. The shear web has been pre-bonded into the lower blade shell and adhesive paste applied to the leading and trailing edge bond lines. (Aerpac UK factory circa 1997)

At the other end of the scale the original blade for the Atlantic Orient AOC15/50 wind turbine was almost 100% wood, including full blade shell and shear web. Its manufacture was typical of the US-evolved WEST system, with veneers of Douglas fir typically 3–4 mm thick coated with liquid epoxy and laid up in a two-part mould, then covered with plastic sheeting to allow vacuum compression (as in VRIM). The two cured shells were trimmed and, after installation of the shear web, bonded with thickened epoxy paste: Figure 8.24 shows the mould halves being brought together. Cold-curing epoxies were used, with around 8 hours required before the mould could be opened and the blade released.

8.5.3 Blade Balancing

The final step in production is to balance the rotor blades in matched sets, in order to prevent excessive vibration arising in operation. An unbalanced rotor can excite tower transverse and yaw vibration at the rotor rotation frequency, with potentially dangerous consequences; see Section 7.5. The blades in a set must be matched to the same weight moment, which is a more important criterion than their absolute weight (blades can have slightly different weights but still be accurately moment-balanced). A factory procedure for blade balancing is explained with reference to Figure 8.25, which shows the setup used to establish the weight moment M about the rotor centreline. The finished blade is first weighed, then bolted at its root to a narrow board, which rests freely on the floor and acts as a knife-edge pivot. The tip of the blade is supported beneath an electronic load cell.

The blade absolute weight W and the 'tip weight' w measured by the load cell are both recorded. The weight moment M about the rotor centreline is given by

$$M = Wc \tag{8.14}$$

where c is the radius of the centre of mass. From Figure 8.25 then

$$M = wL + Wh \tag{8.15}$$

where w is the weight recorded by the load cell (aka the tip weight), L is the distance between the blade root support board and the load cell, and h is the distance from the rotor centreline to the blade root.

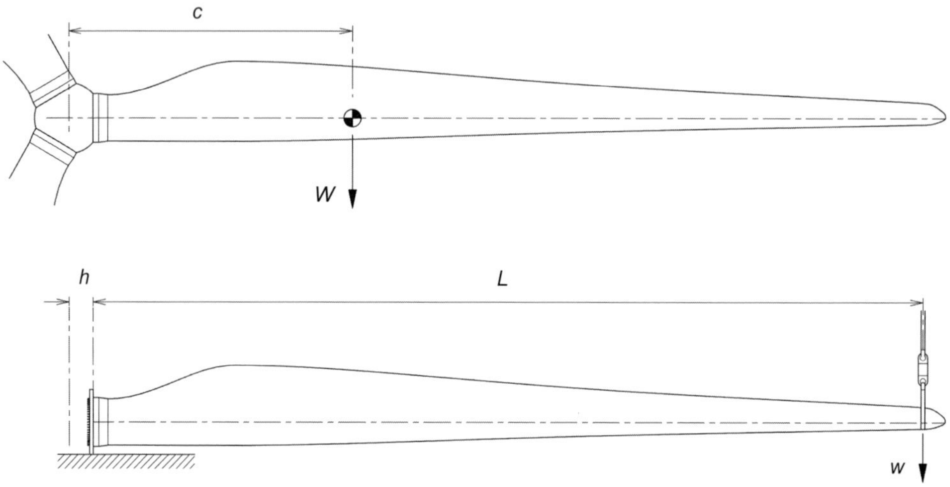

Figure 8.25 Mass balancing. To prevent vibration, blade sets must be matched in the factory to the same weight moment $M = Wc$. Equation (8.15) applies. The 'tip weight' w is measured by a load cell and adjusted by adding mass in the tip of the blade.

To produce a matched set, three blades of similar weight moment are initially selected; each of the two 'lighter' blades then has mass added near the tip until the value of M calculated from Equation (8.15) matches that of the 'heavy' blade. This is achieved by adding iron or lead shot in epoxy to a small tip cavity, which is subsequently sealed. Using this procedure blade weight moments can be matched to within a fraction of a per cent. The term Wh in Equation (8.15) accounts for the offset of the blade root from the true rotor centreline: depending on the balancing accuracy required this term may potentially be neglected, in which this case the criterion for balanced blades is simply that they have the same tip weight w.

8.6 Blade Testing

Testing full-scale blades on the ground requires specialised facilities, but is the most reliable way to verify the strength and dynamic characteristics of a new design before it goes into production. For small blades a simple static test may suffice, with a single point load applied at around two-thirds of the tip radius by a hydraulic ram or winch cable. Although single-point loading can provide the correct bending moment and shear force at the blade root, however, the load distribution elsewhere along the blade will not be representative. Referring to Equation (3.27) the axial thrust per unit length on an optimal blade is proportional to radial position, and the resulting distributions of axial shear force and bending moment are

$$F_{ax}(x) = F_0\left(1 - x^2\right) \tag{8.16}$$

$$M_{ax}(x) = M_0\left(1 - \frac{3}{2}x + \frac{1}{2}x^3\right) \tag{8.17}$$

where $x = r/R$ and F_0 and M_0 are, respectively, the shear force and bending moment at the blade root. The difference between the above load distributions and those achieved with single-point loading are illustrated in Figure 8.26: while the bending moment is reasonably well modelled over the inboard blade by a single point load the shear force is significantly overestimated (the sections outboard of the load point are generally of less concern as strength margins are inherently much higher).

For testing larger blades a more representative load distribution is achieved using multi-point loading. There are various methods, one of which utilises a 'whiffle tree', an arrangement of spreader bars that distribute the applied load in a realistic radial pattern, as shown in Figure 8.27; whiffle trees were traditionally used in the aircraft industry for wing strength testing. An alternative is to hang distributed weights from the blade as seen in Figure 8.28, which shows a 45 m blade under test at the NWTC test site in Colorado. The test stand here supports the blade angled upwards from the horizontal to allow for the large flapwise deflection under loading. In other test setups the blade is mounted edgewise on the stand and flapwise loading is applied horizontally via winch cables, which again facilitates the large deflections involved.

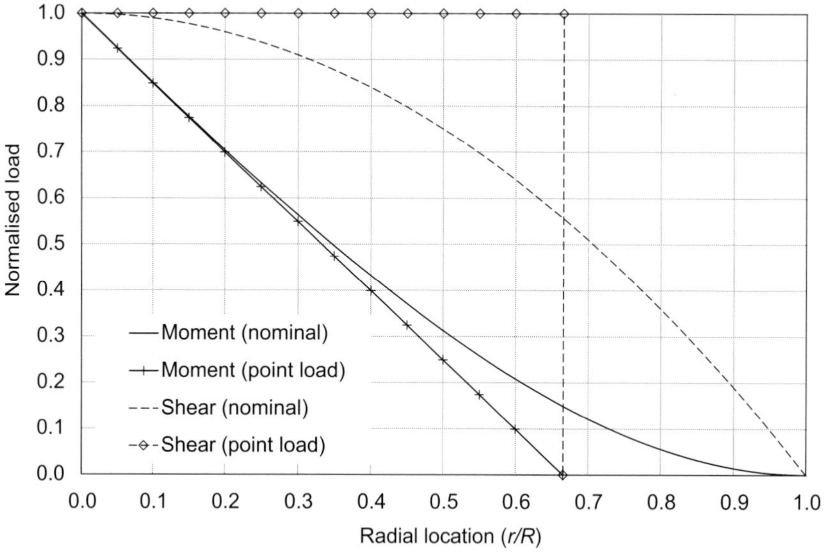

Figure 8.26 Comparison of axial bending moment and shear force distributions for an optimal HAWT blade with those obtained from a single-point load applied at two-thirds of tip radius. The moment distribution is a reasonable approximation, but the shear force is overestimated.

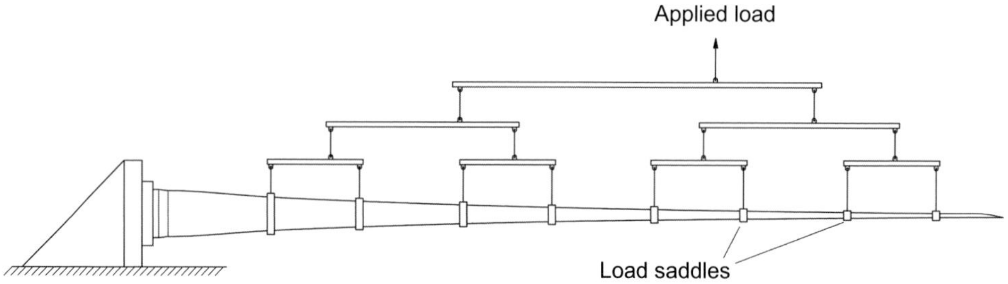

Figure 8.27 The 'whiffle tree' apparatus for blade static load testing. By suitable design of the spreader bar geometry, the applied load produces realistic shear force and bending moment distributions.

For fatigue testing the loading is applied dynamically, using computer-controlled hydraulic or electric actuators. Single-point loading may again be sufficient for small blades, but large ones require multi-point loading. The most sophisticated test rigs use multiple actuators to deflect the blade in representative mode shapes, with biaxial loading to simulate the combined response to flapwise and edgewise forces. Large-scale blade test facilities include the US National Wind Technology Centre in Colorado, the DTU Large Blade Facility at Risoe Campus at Roskilde, Denmark, and the ORE Catapult (Narec) test centre at Blyth in the UK; the last of these is capable of taking blades up to 100 m long (Williamson, 2012).

Figure 8.28 Static testing of a 45 m blade using distributed weight loading at the National Wind Technology Center, Colorado, in 2006. The blade is angled upwards to allow for static bending deflection. (Reproduced with permission from the National Renewable Energy Laboratory. Original image from NREL/PR-500-48898, available at https://www.nrel.gov/docs/fy10osti/48898.pdf)

Short-term tests to establish the strength, stiffness, and natural frequencies of a new blade typically take a few weeks; an accelerated lifetime fatigue test may take 3–4 months. In some cases blades are tested to destruction: an example is seen in Figure 8.29, which shows a 7 m GFRP blade following a destructive flapwise load test at the NWTC Colorado facility. The blade structure is similar to that shown in Figure 8.12, and the photograph illustrates several failure modes. The upper spar cap and adjacent skin panels have buckled in compression; there is transverse cracking on the lower surface of the blade, indicating tensile failure, and the blade leading-edge joint has parted. The test report concluded that the blade ultimately failed due to buckling (Musial et al., 2001) and with its cross-sectional shape compromised the other failure modes would have quickly followed. The blade had previously passed its extreme load 'proof test', and the flapwise load at failure was around 150% of the proof load.

8.7 Weight Trends

According to the laws of simple scaling the structural efficiency of wind turbine blades decreases with size. Aerodynamic bending moments increase as R^3, and Equation (8.4) predicts that direct scaling of a section geometry is needed to maintain constant stress: this leads back to the 'square-

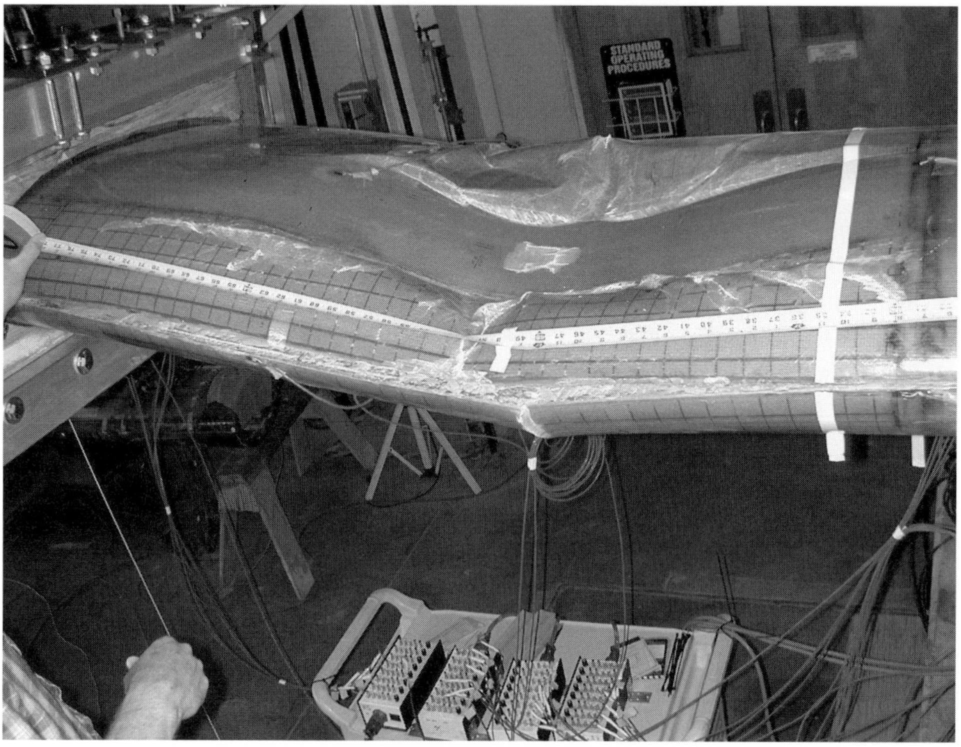

Figure 8.29 Destructive testing of a GFRP blade. The test article has been subjected to flapwise loading beyond its design limits; the upper surface has buckled, and there is evidence of tensile failure on the lower surface and at the leading-edge joint. (Reproduced with permission from the National Renewable Energy Laboratory)

cube' paradox (see Section 1.5) whereby the ratio of the structural weight (hence cost) of a wind turbine to its power output should theoretically increase as R. Today's largest blades appear to buck this trend, and as their size has grown the cost of wind energy has continued to fall. In practice the scaling laws are sound, but via a series of changes in design philosophy the blade weight curve has been continually reset to allow blades to grow in size without becoming unfeasibly heavy and uneconomic.

This process can be illustrated with reference to historic designs of the Netherlands blade manufacturer Aerpac BV (and its predecessor Stork). Figure 8.30 shows two Aerpac blade types: the WPX series dated from the mid 1980s and were glassfibre–polyester blades produced by hand layup; the aerodynamic design was relatively conservative with a maximum profile thickness of 24% at the widest chord, and the blades had a flanged root attachment incorporating a steel collar (Figure 8.19 shows a WPX blade). Around a decade later the company introduced their APX series of glass-epoxy blades, which differed in several key aspects. The planform was more optimally tapered and maximum thickness increased to 36%; the structure changed from a 'stressed-skin' to a 'concentrated

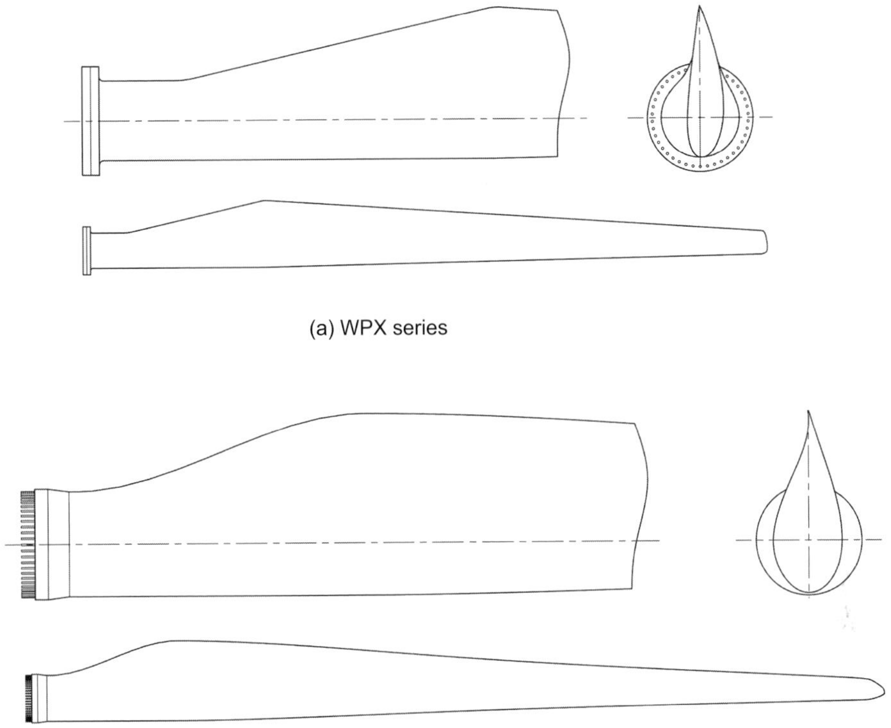

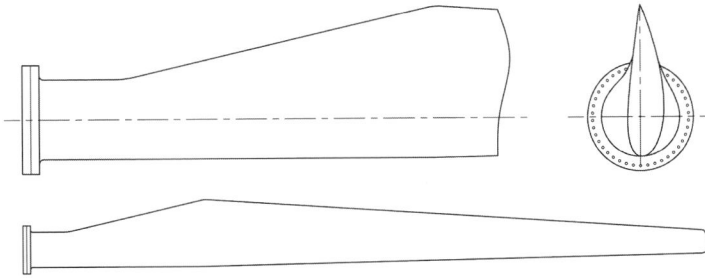

(a) WPX series

(b) APX series

Figure 8.30 Blade design trends. The Aerpac WPX series (*top*) had a steel flange root attachment and 24% thick profile at the widest chord; the later APX blades (*below*) had T-bolt root attachment and maximum thickness of 36%. The APX blades were also manufactured using VRIM rather than hand layup.

girder' design (see Figure 8.10) and exploited the advantages of oriented fibre composites to maximise spar efficiency. A T-bolt root attachment replaced the heavy steel flange, and the series culminated in the 34 m APX70, shown in Figure 8.31. The APX blades were in addition manufactured by vacuum resin infusion moulding (VRIM), enabling higher volume fraction laminates than the hand-laminated WPX. Taken together, these design changes led to a significant increase in structural efficiency, as evidenced in Figure 8.32. Whereas the weights of both blade series follow roughly cubic[10] scaling laws, the APX curve is shifted bodily to the right, indicating a significant weight reduction for given blade length. The smallest APX blade at 16 m length was only half the weight of the corresponding WPX 16 m design.

[10] The curves in Figure 8.32 actually have dependency of $R^{2.6}$.

Figure 8.31 The Aerpac APX70 incorporated a number of design improvements that helped to reduce the weight of large blades; see also Figure 8.30. (Taken at Aerpac BV Almelo factory, 1997)

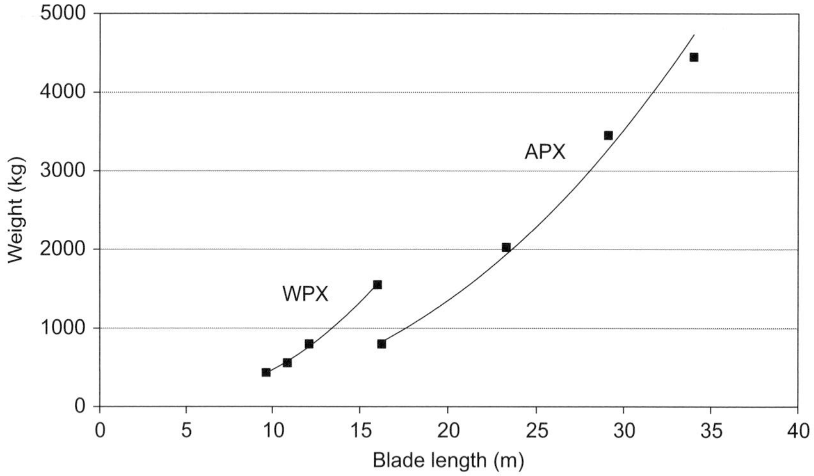

Figure 8.32 Blade weight trends. Changes in design philosophy between the Aerpac WPX and APX series (see Figure 8.30) allowed the designers to subvert the 'square-cube law'. At 16 m length, the smallest APX blade was only half the weight of the comparable WPX design.

The internal structure of the APX series was typical of today's very large blades, exploiting the optimisation techniques described in Section 8.3.4. More recent design innovations have allowed the scaling laws to be challenged further still, as seen in Figure 8.33: the LM88 blade

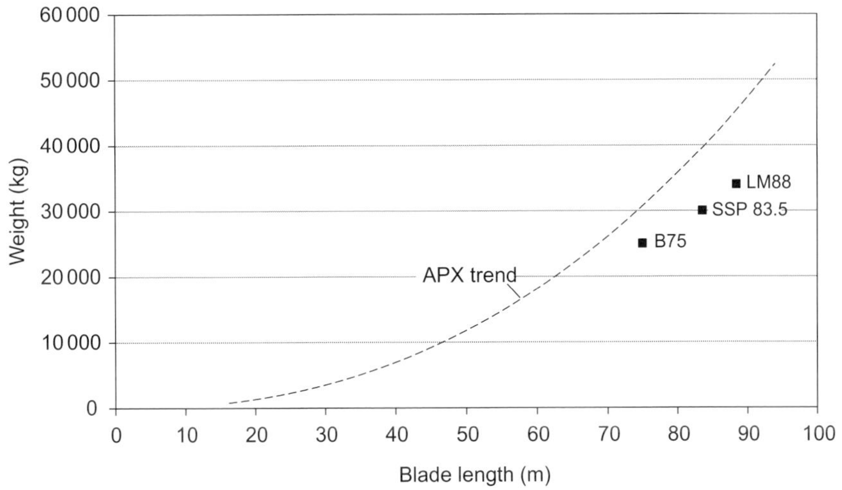

Figure 8.33 Very large blades. The LM88 and SSP blades include carbon fibre in the spar caps; the Siemens B75 is all glass but moulded as a single piece without joins. Innovation helps to beat the weight trend.

Figure 8.34 Samsung S7.0 wind turbine at Methil, Fife, on the east coast of Scotland. The rotor diameter of 171 m was the world's largest when installed in 2014 but has now (inevitably) been superseded.

introduces a proportion of carbon fibre in the spar caps, taking advantages of its high specific strength; the Siemens B75 blade is glass–epoxy, with no carbon, but moulded as a single piece using patented 'Integralblade' technology (Stiesdal et al., 2006): this reduces the weight associated with an adhesive join between blade halves. The SSP 83.5 blade has some carbon and uses keyed root studs: an example of a wind turbine using this blade is shown in Figure 8.34. These blades have successfully pushed the weight curve for HAWT blades farther to the right. One day a size limit will be reached for HAWT blades, but it is a brave person who would predict what it will be.

8.8 Exercises

8.8.1 Composite Fatigue Strength

The composite material used to form the spar caps (girders) of a wind turbine blade has ultimate tensile strength of 600 MPa, and an equivalent fatigue strength of 177 MPa at 2×10^5 cycles. Calculate the logarithmic fatigue exponent x for this material, and suggest what the material might be. What is its fatigue strength at 2×10^8 cycles?

8.8.2 Volume Fraction

The spar cap of a large wind turbine blade is 900 mm wide (chordwise) and 60 mm thick, with the thickness comprised of 40mm of unidirectional GFRP, and 20 mm of unidirectional CFRP. Using the material properties in Table 8.1 (Chapter 8) calculate (a) the net tensile modulus E_c and (b) the net density for the composite material in the spar.

8.8.3 Material Safety Factor

A particular GFRP blade laminate is moulded using vacuum resin infusion (VRIM) and post-cured for several hours in a temperature-controlled oven. Calculate (a) the appropriate material safety factor to apply when calculating the admissible tensile strength of the laminate, and hence (b) the admissible strength assuming the characteristic strength (UTS) of the material is 600 MPa.

8.8.4 Spar Cap Buckling

A blade spar cap is made from unidirectional GFRP with E-modulus of 38 GPa and ultimate compressive strength 500 MPa; the cross section (normal to the blade long axis) is 150 mm wide by 15 mm thick. Calculate (a) the critical buckling stress σ_{crit} of the spar cap and (b) the maximum width b of the cap above which buckling becomes a more likely cause of failure than compressive strength. Assume a value of $k = 1.3$ for the panel buckling geometry constant.

8.8.5 Sandwich Panel

The outer skin of a blade is made from sandwich construction, comprising two skins of GFRP bonded to a lightweight polyurethane foam core. Each skin is 1.2 mm thick, the core thickness is 10 mm, and the critical skin buckling stress (σ_{crit}) for the panel is 77.6 MPa. Calculate the critical buckling stress assuming the core thickness is increased to (a) 15 mm and (b) 20 mm.

8.8.6 GFRP Skin Thickness

The glass fibre fabric used in the skin of a HAWT blade has a dry weight of 1200 g m^{-2}. If the density of pure glass is 2550 kg m^{-3}, and the fabric is impregnated with epoxy resin to give a resulting fibre volume fraction V_f of 0.45, what is the final thickness of the laminated glass skin?

8.8.7 Blade Balancing

The table below contains factory measurements made on a set of 30 m blades after production and finishing, but before balancing. The data are the blade weight and the distance of the centre of gravity from the rotor axis (note that the root offset h is already accounted for). To balance the set additional weight must be added at an internal location 28.5 m from the axis: calculate (a) the additional weight needed for each blade to achieve a balanced set and (b) the final weight of each blade.

Blade	Initial weight (kg)	C.G. distance from rotor axis (m)	Added weight (kg)	Final weight (kg)
A	3530	10.40		
B	3510	10.50		
C	3490	10.48		

8.8.8 Blade Test Load Distribution

Equations (8.16) and (8.17) in Chapter 8 are expressions for, respectively, the axial shear force and bending moment distribution on an aerodynamically ideal HAWT blade. The constant F_0 represents the shear force at the blade root, and M_0 the corresponding blade root bending moment. Using the relationships developed in Chapter 3, find simple expressions for F_0 and M_0.

Chapter 9 Siting and Installation

9.1 Introduction

There are many steps leading to the successful installation of a wind energy project, whether it be a single machine or a large array. Wind conditions at the prospective site must be fully understood, as these will dictate which turbines are most suitable – not all types may be certified for the site characteristics – and how much energy they are expected to produce over the project lifetime. Terrain and access surveys are necessary to establish how easily the project can be delivered and constructed, and what type of foundations are best for the ground conditions and prevailing wind regime. Where large arrays are planned the efficiency loss and increased fatigue due to wake interactions must be assessed and taken into account in the overall project economics. The background to some of these topics and the methods employed to address them are discussed in the present chapter.

9.2 Site Wind Assessment

Assessing the wind characteristics at a prospective site is needed for two reasons:

1. to predict the energy output (hence economic outcome) of the project
2. to verify that a proposed wind turbine type is technically suitable for the site

The common aim is to minimise risk. Energy production is highly sensitive to wind speed and over-predicting the resource will lead to a shortfall in revenue: if the project economics are finely balanced this can have serious financial implications. Similarly, wind turbines must be operated within their design envelope if technical problems are to be avoided, and not all types are suitable for sites with extreme wind speeds or high turbulence. For these reasons some form of site wind assessment should be carried out for all projects, although the scope and detail of the exercise should normally reflect the scale of the project and/or level of risk involved.

9.2.1 Wind Resource

Site selection often begins with a desktop wind speed assessment. In many countries the regional wind resource is mapped at large scale, giving developers an initial guide to local variation. In the

UK the NOABL database gives average wind speeds at heights of 10 m, 25 m, and 45 m above ground at 1 km horizontal resolution.[1] The data are interpolated from long-term weather station records, corrected for terrain height (Burch et al., 1988). The NOABL database should be used with caution, however, as the estimated wind speeds are accurate only to within ±10% and relate to the average terrain height in each 1 km square; they do not take account of local differences in surface roughness, and in complex or hilly terrain there may be wide variation in wind speed that can only be resolved with finer scale terrain modelling, on-site measurement, or in some cases both.

For a single-turbine project a met mast can be installed at exactly the proposed location. Where large arrays are planned, however, it is uneconomic to erect separate masts at each turbine position; in practice several full-height masts may be installed at representative locations across the development area, with wind speeds between these positions interpolated using digital terrain flow models. These are available in commercial software packages such as Windfarm, Windfarmer, and WindPro, all of which provide the same essential features (some packages incorporate the Danish WAsP algorithm for terrain flow modelling, others use inbuilt routines).

The minimum wind speed at which a project will be economic is not a fixed figure, but a function of the overall project cost. With low infrastructure costs a project can be viable at a lower wind speed. In general, however, it is unusual to install wind turbines at sites with hub height wind speeds below 6 m s^{-1} (in the UK at least). In addition to establishing the mean wind speed a site assessment exercise must also verify the characteristic turbulence intensity, shear profile, and maximum gust speed (see Chapter 2 for definitions). These parameters are then compared with the design limits for candidate wind turbines. For this purpose the International Electrotechnical Commission has compiled the IEC 61400 standard, which was first issued in 2001 following the harmonisation of several national standards, and contains a site wind classification system now universally adopted by the wind industry (IEC, 2005). The key IEC wind class parameters are given in Table 9.1.

Wind Classes I-IV are defined by the average speed at hub height and associated maximum 50 year extreme gust speed; sub-classes A and B are accorded depending on turbulence intensity, and the same limiting shear index applies to all cases. This system is used to categorise wind sites on the basis of measurements (referred to hub height) and also as the basis for wind turbine design specifications. A particular machine can then be sold, for example, as 'Class IA' or 'IIB', indicating the limiting wind class for which it is suitable.[2]

9.2.2 Site Measurements

The standard method of wind measurement involves one or more anemometers installed on a met mast of height ranging from 10 m to 80 m, depending on the application and the level of accuracy required.

[1] At the time of writing, the database can still be downloaded from the UK national archives website.

[2] Not all WEC designs fall within the limits in the table, and the IEC Class 'S' designation is reserved for site-specific designs, e.g. with average speed above Class I. Specific wind characteristics accompany the certification.

Table 9.1 The IEC Wind Classification System

Wind class	I	II	III	IV
Vav average wind speed at hub height (m s^{-1})	10.0	8.5	7.5	6.0
V50 extreme 50 year gust at hub height (m s^{-1})	70	59.5	52.5	42.0
I15 characteristic turbulence Class A		18%		
I15 characteristic turbulence Class B		16%		
α wind shear exponent		0.20		

Note. Data from IEC (2005).

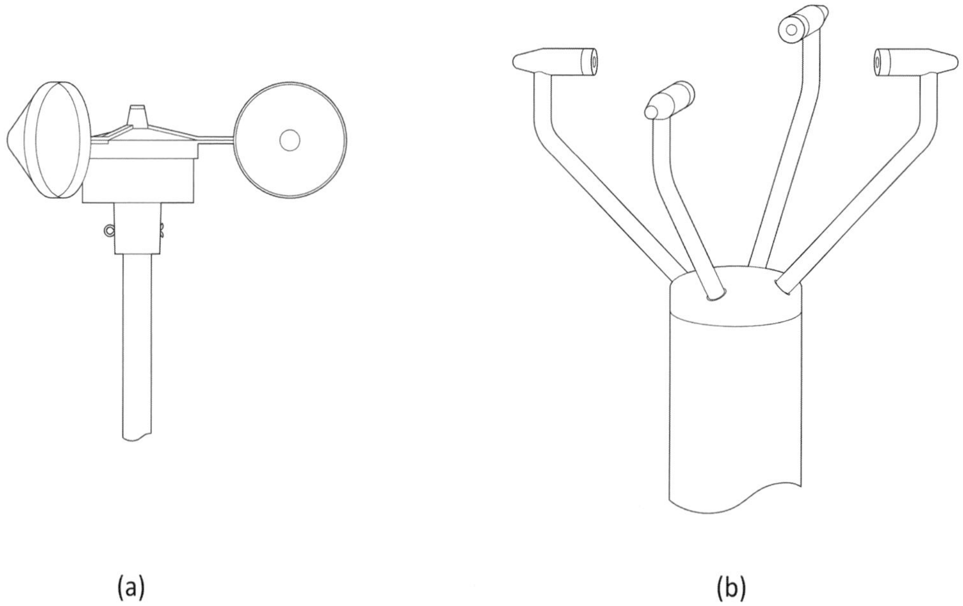

Figure 9.1 Anemometers: (a) some traditional cup types do not require a power supply; (b) 2D ultrasonic anemometers measure horizontal speed and direction: 3D units are also available, with six pickups to allow vertical wind component to be measured.

Ideally wind speed measurements are made at hub height, but this may be expensive and shorter masts are acceptable if equipped with several anemometers to enable the shear profile to be extrapolated upwards. Some turbine manufacturers recommend measurement at a minimum of two-thirds of hub height. In cases where the terrain is simple and the technical risk low, single anemometer measurements at 10 m may be acceptable. In all cases wind direction is measured by a wind vane alongside (or just below) the top anemometer on the mast; tall masts may be equipped with more than one.

The most common wind measuring instrument is the traditional cup anemometer; see Figure 9.1. These are accurate and relatively cheap; the output may be an analogue voltage, or a frequency pulse proportional to wind speed. Some models use electro-optical encoders and require a power supply; others incorporate a rotating permanent magnet and are effectively mini-generators requiring no external power source. A more modern instrument is the ultrasonic anemometer (see Figure 9.1) which has no moving parts and simultaneously measures wind speed and direction; accuracy is very high, but they are expensive and require more power than cup anemometers. Ultrasonic anemometers are more commonly seen as control sensors on wind turbine nacelles. The chief disadvantages of all types of anemometer, however, are the need for the supporting mast, and the ability to measure at only a single point in space.

These limitations are overcome by remote measurement techniques such as sodar (sonic detection and ranging) and lidar (light detection and ranging). Both are ground-mounted and enable wind speed and direction to be measured up to heights of several hundred metres and over wide areas. Comparisons of lidar with conventional cup anemometry indicate a high level of agreement; lidar can, however, be sensitive to weather conditions as rain tends to clean the atmosphere of the dust particles on which the system depends for scattering (Albers, 2009). In addition lidar has in some cases been found to overestimate turbulence intensity (Westerhellweg, 2010). Remote sensing anemometry remains expensive and in onshore applications is used primarily over short periods to resolve specific micrositing issues; the equipment power requirements are also much higher than for conventional anemometry. Where lidar and sodar come into their own, however, is offshore, where the cost of conventional anemometry may be extremely high, and remote sensing enables a much wider area to be assessed from a single location.

On-site wind measurements are frequently used in conjunction with digital terrain modelling to allow wind speed characteristics from the measurement location to be extrapolated to prospective turbine positions further away. This is a feature of most commercial windfarm software packages, and the results are generally dependable for sites in relatively smooth terrain. These models are, however, less appropriate for analysing complex terrain where flow separation and high turbulence occur, and in such cases computational fluid dynamics (CFD) models are increasingly used. Experimental studies have been carried out to compare the results of CFD modelling with high-resolution wind measurements made in complex terrain, demonstrating how CFD can reveal significant flow separation phenomena not predicted by simple linear models (Abiven et al., 2009).

9.2.3 MCP Analysis

In principle at least 10 years of wind measurements are needed to establish the long-term average wind speed and inter-annual variability at a site to the level of confidence demanded by prospective owners or investors (see Section 2.3). This timescale allows the project lifetime output to be accurately forecast and the variation between 'good' and 'bad' wind years to be bracketed (which is important for financial modelling). For most projects such long-term wind measurements would, however, be impractical. To overcome this a technique has been devised whereby short-term measurements from a target site are correlated with simultaneous records from an existing weather station, and historic trends at the latter then extrapolated to the target site. The procedure is known as measure-correlate-predict (MCP) and the essential steps are as follows:

1. Measurements of hourly[3] wind speed and direction are taken at the proposed site over a period of at least 12 months; simultaneous hourly records are obtained from a reference met station in the same geographical area.
2. The joint hourly data are co-plotted as a scatter diagram of target site wind speed V_{site} against reference wind speed V_{ref} and a best-fit straight line found with $V_{site} = AV_{ref} + B$, where A and B are constants.
3. The inter-site linear relationship is applied to the historic (10+ years) average wind speed at the reference location to yield a corresponding long-term mean for the target site.
4. The correlation is also applied to individual annual speeds at the reference site over the historic period to give corresponding inter-annual variation at the target site, indicating the highest and lowest expected speeds, and a wind speed probability distribution.

The procedure can be illustrated with an example, based on a comparison of two sites 20 km apart in northeast Scotland. Figure 9.2 shows simultaneous hourly wind speeds for the target and reference sites over a 10 day period, sampled from a 6 month record; the two locations experience broadly the same wind variation, but with higher speeds at the target site. The joint hourly wind speeds are co-plotted in Figure 9.3, yielding a best-fit straight line with coefficients $A = 1.12$ and $B = 1.35$; the correlation coefficient (R^2) of 0.78 indicates a reasonably strong correlation. This linear relationship is then used to scale long-term data from the reference site, as illustrated in Figure 9.4, where a 21 year record of historic wind speeds is shown with corresponding MCP-derived estimates for the target site; the dashed lines indicate the long-term average speeds at the two locations.

The MCP technique is also useful to characterise inter-annual variation. In this case the hindcast annual wind speeds are presented as an exceedance curve, showing the probability that the speed at the target site will exceed a certain figure, as shown in Figure 9.5. Marked on the plot are the annual wind speeds that will be exceeded in 90% and 50% of years: these values are denoted P90 and P50, respectively, and are a common benchmark in project economic appraisals. Translated into annual energy capture, P50 represents the expected median[4] output (or central estimate) of the project; P90 is a conservative figure representing the yield that should be exceeded in 9 out of every 10 years (in layman's terms the lowest yield expected in a 10 year period). When a more conservative appraisal is required P95 may be used instead of P90.

The foregoing is a simplified version of the MCP procedure, neglecting directional variation. In practice the inter-site correlation will vary with wind direction due to sheltering or other topographic effects at either the target or reference site. If the directional variation over the short-term measurement campaign is not representative of the long term, then the outcome may be

[3] Hourly records are standard with the UK Met Office.

[4] The median of a long-term wind distribution is usually close to the mean, in which case P50 also indicates the long-term average wind speed (or energy yield).

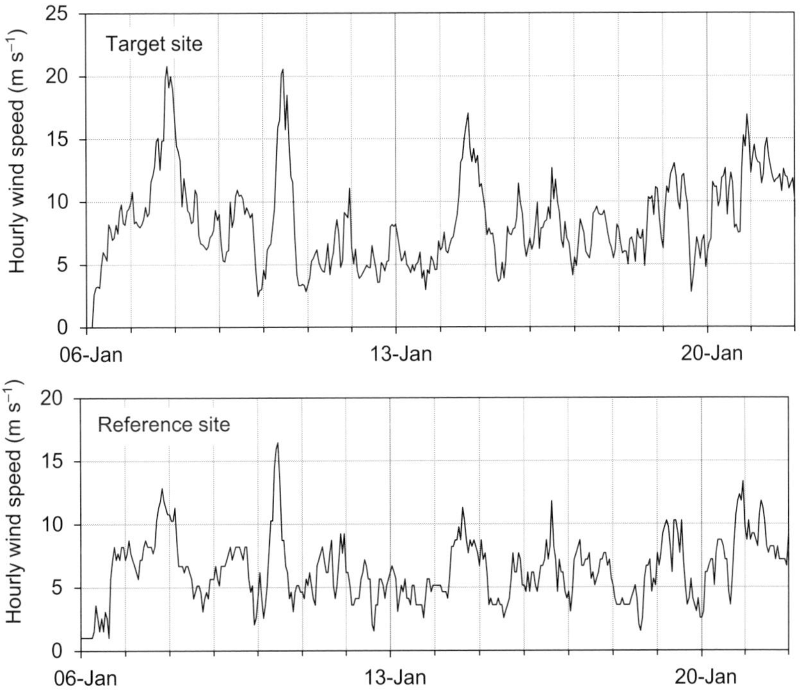

Figure 9.2 Simultaneous hourly wind records from MCP target and reference sites. Shown is a 10 day joint record from a 6 month measurement campaign; the locations were approximately 20 km apart.

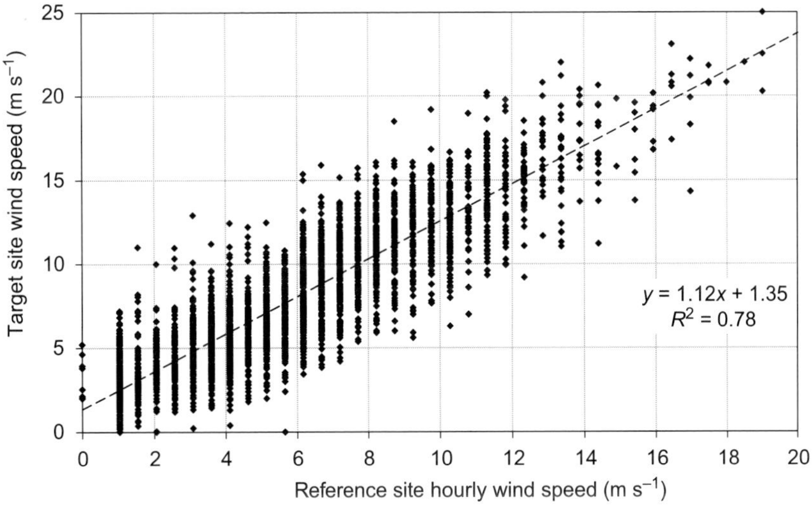

Figure 9.3 Scatter plot of MCP hourly wind speeds, with best-fit straight line.

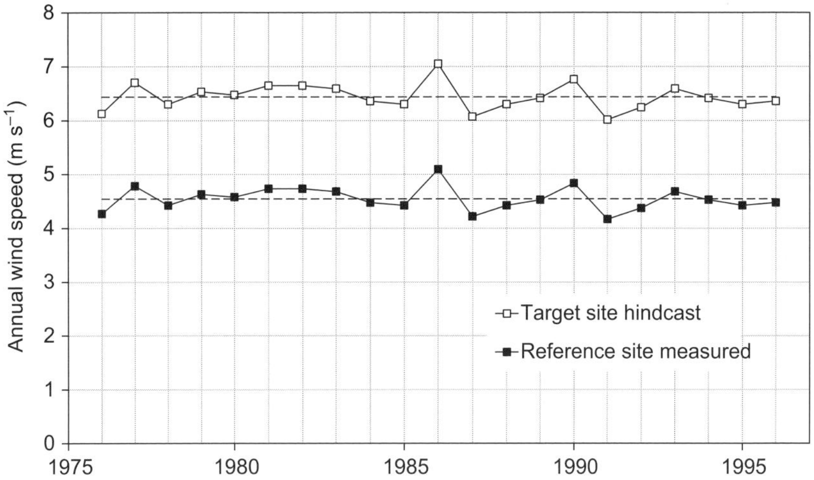

Figure 9.4 Inter-annual wind speed variation. Target site wind speeds are estimated by applying the linear relationship from the MCP plot (Figure 9.3) to long-term means at the reference site.

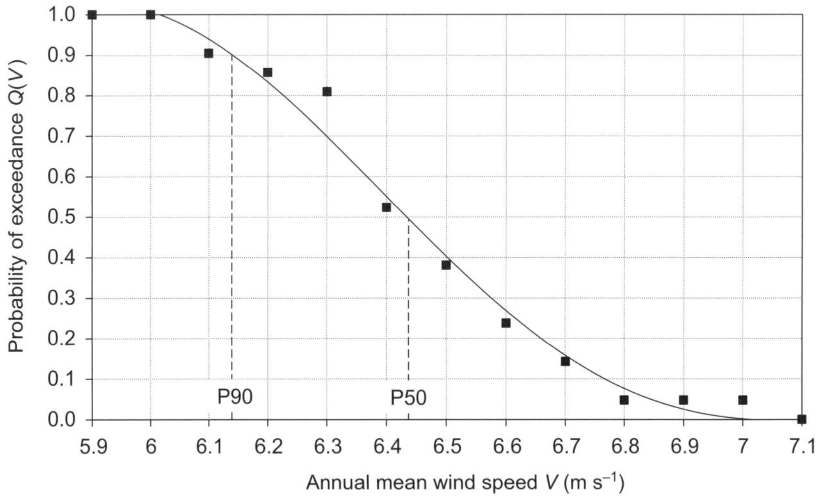

Figure 9.5 Exceedance curve for annual mean speeds at the target site. P90 and P50 represent the wind speeds that will be exceeded in 90% and 50% of years, respectively.

biased. To account for this, the joint hourly data set is divided into 30° sectors based on the reference wind direction and a separate correlation carried out for each sector. Long-term directional statistics are meanwhile obtained for the reference site, and the overall average wind speed at the target site is found by weighting the sectoral means by the proportion of time spent in them, i.e.

$$\overline{V}_{\text{site}} = \sum_{\theta=30,60\ldots} P_\theta \left(A_\theta V_{\text{ref}}(\theta) + B_\theta \right) \tag{9.1}$$

where P_θ is the proportion of time historically spent in sector θ at the reference site, $V_{\text{ref}}(\theta)$ is the associated sectoral mean speed, and A_θ and B_θ are the regression line constants for that sector.

MCP is the industry norm for long-term wind estimation, although there are variations on the method described above. In some cases the inter-site correlation is not linear and other best-fit functions may be preferred; similarly, although 12 months is normally recommended for the inter-site measurement campaign, periods down to 6 months may give satisfactory results. The distance between target and reference sites is not a fixed quantity: usually the reference will be sought within 30 km or so of the target site, but this may not always be possible. Sites at greater separation may still correlate well depending on the influence of intervening terrain: for instance two island sites 50 km apart may see very similar wind patterns; conversely, sites in close proximity but on different sides of a mountain range may not correlate well. The strength of the correlation can ultimately be judged by the scatter plot R^2 value. For a comprehensive review of the MCP method, see Derrick (1992).

9.2.4 Turbulence and Shear

Local turbulence and shear characteristics are best established by on-site measurement. This can be achieved over a relatively short period, again less than 12 months, so long as the full range of potential wind speeds and directions is sampled. Ideally measurements are made at hub height, but if not, both shear and turbulence can be extrapolated from measurements at lower height. In the case of shear either a logarithmic or power law relationship can be assumed (see Section 2.2) with a best-fit line plotted through the mean speeds from two or more anemometer stations and extrapolated to hub height. When extrapolating turbulence intensity the log law may be used – see Equation (2.10). In some instances (e.g. on a low budget) wind measurements may be taken at a single height. The wind shear profile is then estimated by inspecting the local terrain, and allocating roughness length z_0 or shear index α from lookup tables (see Table 2.1). To account for differences in terrain a directionally weighted shear profile should be used. When using measurements from an anemometer on a short (e.g. 10 m) mast it is advisable to bracket the shear estimate to give low and high estimates of hub height wind speed, where the former can be used for energy yield calculations, and the latter to characterise the IEC Wind Class. This approach avoids overestimating the resource, or underestimating the Wind Class. Turbulence intensity can again be extrapolated using Equation (2.10).

9.3 Array Interactions

9.3.1 Array Losses

In Section 3.7 it was seen how the wake of a wind turbine rotor is characterised by reduced wind speed and increased turbulence. One machine situated downwind of another will therefore experience reduced energy capture and higher fatigue loading. Wake influence is illustrated in Figure 9.6,

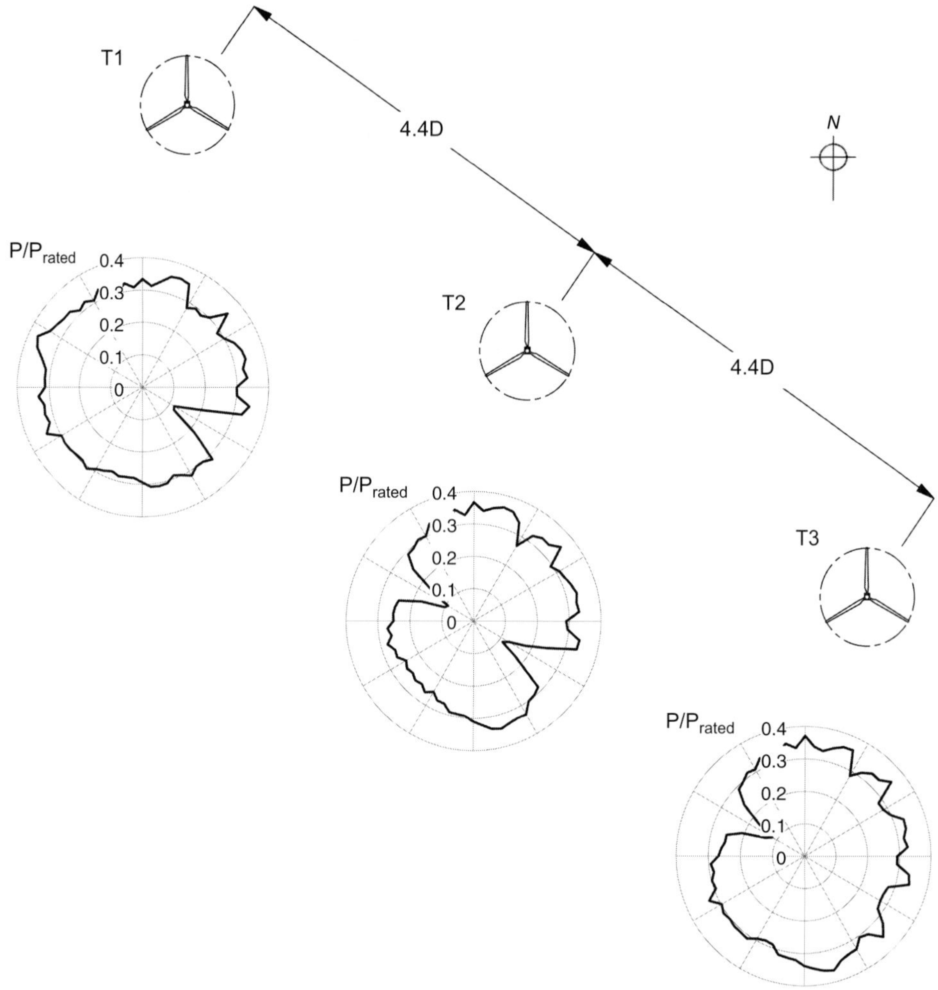

Figure 9.6 Wake influence in a three-turbine array. Polar plots show the average power for each turbine as a function of wind direction for $U = 8\ \text{m s}^{-1}$. Turbines T1 and T3 are each wake affected from one direction, while T2 is affected from two. Overall array efficiency measured over 12 months exceeds 97%. (Data from Balquhindachy windfarm, supplied by kind permission of Greenspan Energy Ltd)

which shows the arrangement of three wind turbines spaced 4.4D apart in a linear array at Balquhindachy in Aberdeenshire: the windfarm is shown in Figure 3.21. The accompanying polar diagrams show the measured power output for each machine as a function of wind direction in a freestream wind speed of 8 m s^{-1}. The 'notches' in the polar plots reveal how the output of a turbine drops when it is operating in the wake of another. The outer machines, T1 and T3, experience a power deficit from only one direction, while the central turbine (T2) is wake affected from two directions. The wake effect persists over a wind sector of about 30° with output falling by as much as 70% at the midpoint; outside this sector the power is unaffected (although there is some inherent variation due to terrain influences).

Wind turbines in arrays therefore experience losses, and array efficiency is defined as the ratio of the long-term output of a windfarm to that (hypothetically) achieved by the same turbines in isolation, i.e. without wake interference. Efficiency is a function of the array geometry and the statistical distributions of wind speed and direction. The directional influence is illustrated by Figure 9.6. The dependence on wind speed is more subtle: array losses are highest in light winds when the ambient power is low and the turbines operate at peak extraction efficiency. In high winds the turbines go to full power, so above a certain wind speed threshold the array operates at 100% of its rating and wake influence no longer affects the output. The overall efficiency of the windfarm in Figure 9.6 was measured at 97% over a 12 month period, which is a typical figure for a small array. Although the velocity deficit can be as much as 40% when a turbine is wake affected (actual wake profiles for Balquhindachy T2 can be seen in Figure 3.20) the overall length of time spent in this condition is relatively small, so the net array loss is modest.

Very large windfarms, however, containing tens or hundreds of turbines in 2D arrays, experience a multiplicity of wake interactions and, as a result, higher losses. Predicting array efficiency is therefore of some economic importance. The aerodynamic structure of the near-wake is complex (see Section 3.7), but most interactions occur in the far wake region at least 3D downstream of the nearest rotor, and simple analytic models have been developed to account for this regime. Foremost is the Jensen–Katic model, originally based on negative-jet expansion theory (Jensen, 1983) and later extended to account for rotor aerodynamic loading (Katic et al., 1986). In this treatment the wake diameter is assumed to expand linearly with downstream distance, with a uniform cross-wake velocity profile.[5] The analytic geometry is shown in Figure 9.7. At distance X downwind of the rotor the ratio of wake velocity V to freestream velocity U is given by

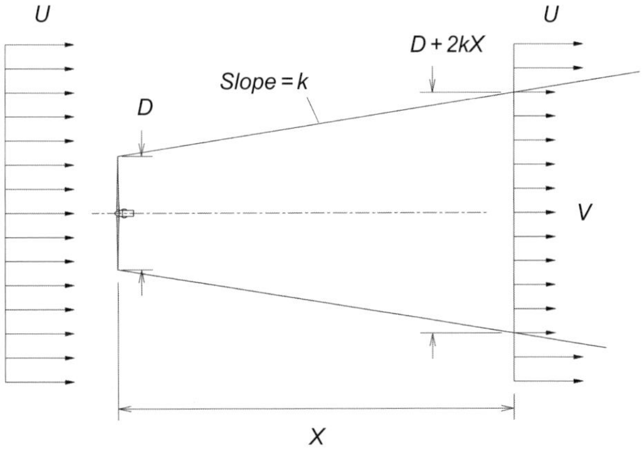

Figure 9.7 Geometry assumed in Jensen–Katic wake deficit model; see Equation (9.2).

[5] The true velocity profile is quasi-Gaussian (see Section 3.7.3); the Jensen–Katic model uses an averaged deficit.

$$V/U = 1 - \left(1 - [1 - C_T]^{1/2}\right)\left(\frac{D}{[D + 2kX]}\right)^2 \tag{9.2}$$

where C_T is the thrust coefficient of the upstream rotor and D its diameter. The horizontal extent D_w of the wake at the downwind position is given by

$$D_w = D + 2kX \tag{9.3}$$

The parameter k is an empirical constant related to surface roughness, which dictates the wake expansion ratio and velocity decay with downstream distance. Typical values for k are 0.075 for onshore arrays and 0.04 for offshore.

Figure 9.8 shows the wake velocity ratio based on the Jensen–Katic model plotted against non-dimensional downstream distance for $k = 0.075$ and $C_T = 0.89$, i.e. behind an onshore rotor operating at the Betz limit. As the wake expands its velocity recovers asymptotically; at 1D downstream the velocity is 50% of the freestream value, rising to nearly 80% at 5D and 95% at 20D. The Jensen–Katic model can be further extended to cover multiple wake interactions: referring to the linear array in Figure 9.9, the total kinetic energy deficit at the Nth turbine downwind is assumed to be equal to the sum of the energy deficits due to the individual wakes, hence the local velocity ratio is found from

$$\frac{V_N}{U} = 1 - \sqrt{\sum_{i=1}^{N} (1 - V_i/U)^2} \tag{9.4}$$

The development of wake velocity along a linear array is shown in Figure 9.10 for turbine spacing of 3D, 5D, and 7D. As might be expected, the greater the separation the less the velocity deficit. The wake velocity converges quite rapidly, however, to a minimum by around the third or

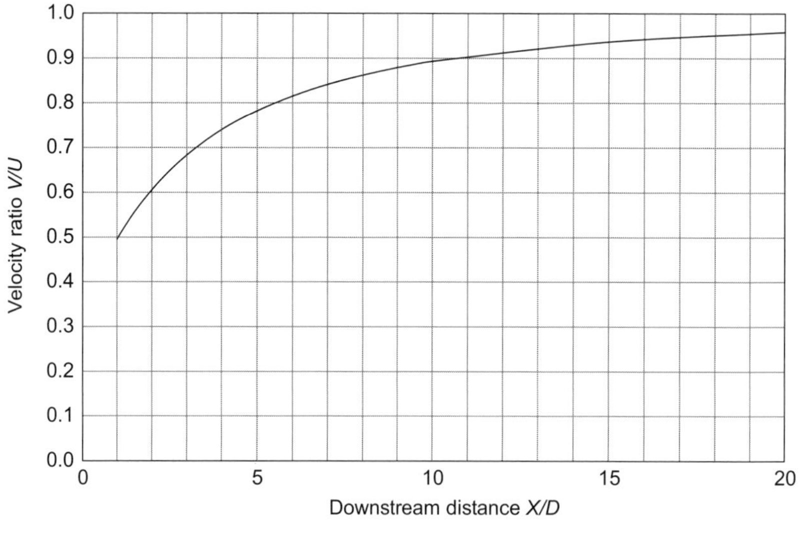

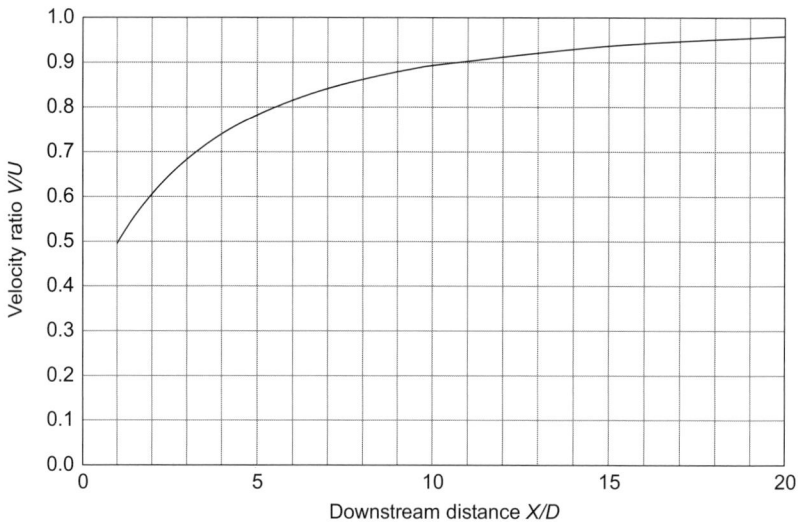

Figure 9.8 Wake velocity ratio V/U as a function of non-dimensional downwind distance X/D; the upwind rotor is here assumed to be operating at the Betz limit with $C_T = 0.89$, and the applicable roughness factor $k = 0.075$.

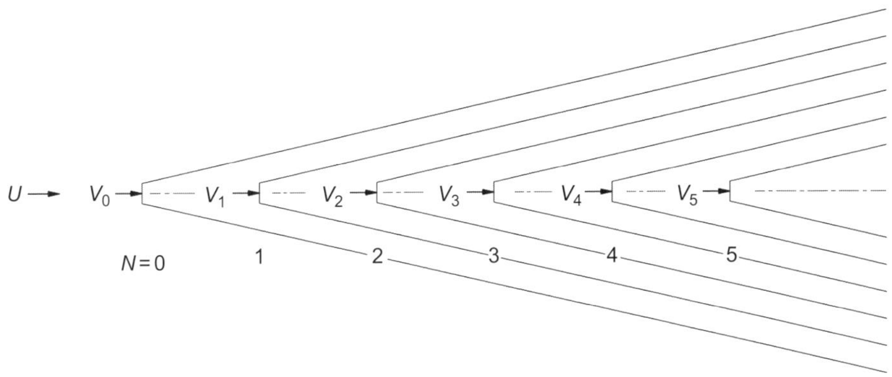

Figure 9.9 Multiple wake array loss model; see Equation (9.4).

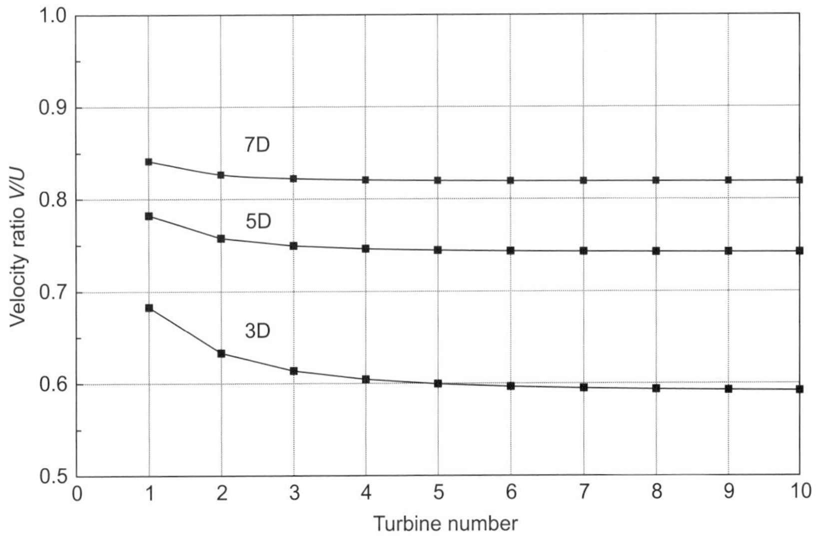

Figure 9.10 Wake velocity ratio in a linear array, calculated for inter-turbine spacing of 3D, 5D, and 7D. Operation at the Betz limit is assumed (C_T = 0.89) for all turbines. Turbine 1 is the first wake-affected machine.

fourth turbine, with little further change thereafter. This pattern is broadly observed in practice. Note also that the curves shown in Figure 9.10 assume a uniform value of C_T, which is not the general case: turbines in an array experience a range of local wind speeds and operate at different power levels and thrust coefficient. More sophisticated analyses (still based on the Jensen–Katic model) can be run to account for the local variation within the array. The patterns in Figure 9.10 also apply to worst-case conditions, i.e. with the wind blowing directly along the row of turbines.

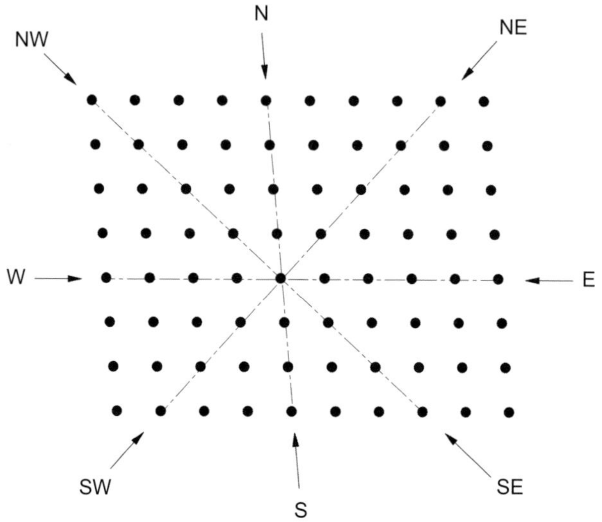

Figure 9.11 Geometry of the Horns Rev I offshore array: 80 wind turbines on a 10 × 8 grid, with 7D spacing in both directions. The arrows indicate wind directions causing maximum wake loss.

In reality this situation occurs only a small percentage of the time, and when the wind direction changes sufficiently the wake influence is reduced. The overall efficiency of a linear array can still be quite high, particularly if there is a strongly prevailing wind direction that the array is designed to avoid (e.g. the line of turbines is installed at right angles to the prevailing wind).

Most large arrays are, however, two-dimensional and wake interactions arise from multiple wind directions. This is well illustrated by the Danish Horns Rev I offshore array depicted in Figure 9.11, comprising 80 × 2 MW turbines on a near-rectangular grid with 7D spacing. Using the Jensen–Katic model, researchers at DTU have modelled the wake interactions at Horns Rev I to produce polar diagrams of park efficiency (Pena et al., 2013). A typical result is seen in Figure 9.12 for freestream wind speed $U = 8$ m s^{-1}. The highest losses are evident when the wind blows directly along the axes identified in Figure 9.11. For N–S or E–W winds the efficiency drops to around 50%; on the diagonal axes (NW–SE and NE–SW) the efficiency is around 70% as the inter-turbine spacing is greater (9.6D on average). Figure 9.12 represents a worst case, however, with the wind turbines operating below rated power. In a more comprehensive analysis based on year-round SCADA measurements the Horns Rev I operator Elsam compared total array output to that achieved by the 'free standing' turbines at the exposed corners, and calculated an overall efficiency of 87.6% (Sørensen et al., 2006). This figure is probably typical for a large-scale 2D array.

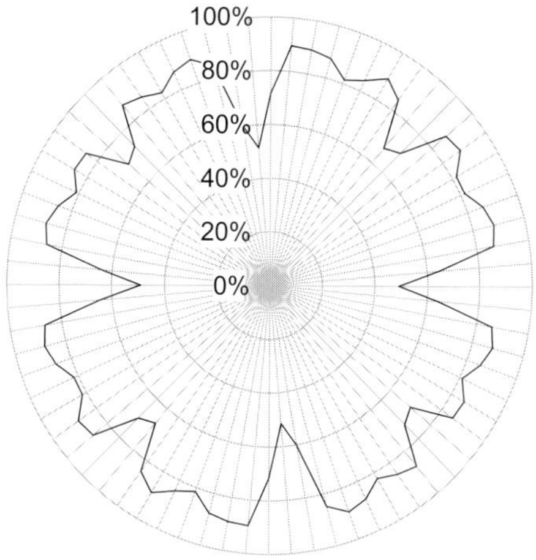

Figure 9.12 Calculated array efficiency for Horns Rev I as a function of wind direction, for freestream velocity $U = 8\ \mathrm{m\ s^{-1}}$. Higher losses correspond to the principal wind directions shown in Figure 9.11. Plot based on data presented in Pena et al. (2013).

9.3.2 Wake Turbulence

The net turbulence intensity I_w in the downwind wake of a HAWT rotor is the sum of the prevailing ambient turbulence I_0 and the added contribution due to the rotor I_a according to

$$I_w = \sqrt{I_0^2 + I_a^2} \tag{9.5}$$

The magnitude of I_a is a function of the rotor loading, the downstream distance, and to a minor extent the ambient turbulence level. Various empirical formulae have been proposed for I_a and a detailed review is given in Vermeer et al. (2003). Some of the models are compared in Figure 9.13, which shows I_a as a function of non-dimensional distance (X/D) downstream of a highly loaded rotor, for ambient turbulence I_0 of 0.15. The models all predict an asymptotic decline in turbulence with distance, with varying rates of decay. In all cases, however, wake turbulence decays more slowly than velocity deficit (see Figure 9.8), indicating that turbulence persists after the velocity has recovered. The following formula for I_a was proposed by Hassan on the basis of wind tunnel tests (Hassan, 1992):

$$I_a = 5.7{C_T}^{0.7}{I_0}^{0.68}(X/X_N)^{-0.57} \tag{9.6}$$

where I_a and I_0 are expressed in per cent, C_T is the upstream rotor thrust coefficient and X the downstream distance; X_N is the length of the near-wake, which is typically of the order 1–3D.[6] The bold line in Figure 9.13 is based on the following formula in IEC standard 61400-1, which is used to estimate the influence of wake turbulence on fatigue (IEC, 2005):

6 A formula for X_N is given in Chapter 2 of Burton (2011), from which Equation (9.6) is also quoted.

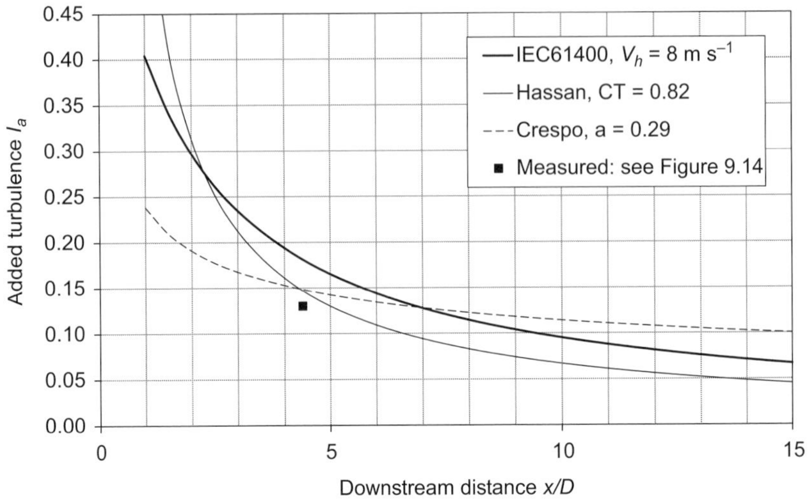

Figure 9.13 Added wake turbulence I_a as a function of non-dimensional downstream distance X/D, assuming a highly loaded upstream rotor and ambient turbulence $I_0 = 0.15$. For details of all the theoretical models referred to, see Vermeer et al. (2003).

$$I_a = \sqrt{\frac{0.9}{\left(1.5 + 0.3(X/D)\sqrt{V_h}\right)^2}} \tag{9.7}$$

Note that in Equation (9.7) hub height wind speed V_h is strictly in m s^{-1}; X/D is non-dimensional downstream distance, and I_a is a non-dimensional ratio (not per cent). The formula does not explicitly take account of rotor thrust loading, but its effect is accounted for in the dependency on V_h. Turbulence estimated using Equation (9.7) should be conservative when used in regard to fatigue design.

The single data point in Figure 9.13 is derived from SCADA measurements taken at 4.4D downwind of one of the turbines at Balquhindachy windfarm (see above). Figure 9.14 shows measured cross-wake turbulence profiles for freestream wind speeds of 8 m s^{-1} and 16 m s^{-1}, corresponding to operation of the upstream rotor below and above rated wind speed. In the former case the turbulent wake boundary is clearly defined and within it net turbulence I_w rises to approximately 0.20. Outside the wake boundary $I_0 = 0.15$ and application of Equation (9.5) gives added turbulence $I_a = 0.13$. This is reasonably in line with the predictions in Figure 9.13 for a highly loaded rotor. In the high-wind case the turbulent wake is ill-defined and extracting a value for I_a is not possible; in this case the applicable thrust coefficient is low ($C_T \approx 0.2$) and the predicted value of I_a is around 0.07, so its contribution to net turbulence is small.

The array layout of large windfarms is highly site specific, but a balance has to be struck in terms of inter-turbine separation. In simple terms,

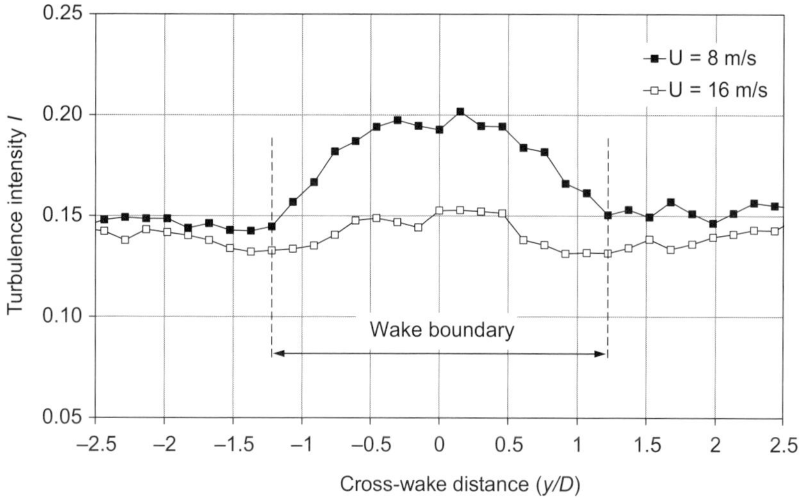

Figure 9.14 Wake turbulence profiles measured at 4.4D behind a V52 wind turbine in ambient wind speeds of 8 m s^{-1} and 16 m s^{-1}. In the former case, added turbulence I_a is calculated as 0.13, based on $I_W = 0.20$ and $I_0 = 0.15$. (Using data from Balquhindachy windfarm, kindly provided by Greenspan Energy Ltd)

- the smaller the turbine separation, the lower the costs of site infrastructure (including site tracks and underground cables) and the smaller the overall land imprint, which can also mean lower environmental impact.
- the greater the turbine separation, the higher the array efficiency, and the less fatigue damage due to added turbulence.

In modern windfarm design codes these constraints (plus others based on environmental and planning considerations; see Chapter 10) are weighted, and the optimum turbine placement sought using multi-rule based algorithms. The above formulae for wake deficit and added turbulence apply, though for developments in complex terrain these simple models may not be sufficiently reliable and CFD studies are sometimes required. For small arrays in smooth terrain a rule of thumb is to use 5D turbine spacing if there is no strongly prevailing wind direction; otherwise spacing may be increased (typically to 7D) in the prevailing direction, and in the crosswind direction reduced to as little as 3D.

9.4 Construction

9.4.1 Introduction

No two wind energy projects are exactly alike, and the challenges and costs of construction can vary widely depending on scale and location. The best wind resource is often found in remote or hilly areas where there is little transport or electricity infrastructure, and providing these elements – collectively known as balance of plant – increases project costs compared with developments on 'easy' sites. Wind turbine foundation designs are also site specific, having to reflect local ground conditions and prevailing wind characteristics. Electricity networks are stronger and better developed near large population centres, but many projects are sited in remote locations and major grid reinforcement may be needed to accommodate their output. In all these aspects planning is key, and a project of any size must be prefaced by numerous surveys and technical assessments. Some of the requirements are discussed below.

9.4.2 Foundations

The majority of onshore wind turbines are mounted on gravity bases, which are reinforced concrete plinths onto which the tower is securely attached. The principle of the gravity base is simple, as illustrated in Figure 9.15: the foundation must be sufficiently heavy that under maximum wind loading conditions there is no tendency for the wind turbine to tip over. More formally, the weight moment M_w taken about the edge of the foundation must exceed the rotor thrust moment M_t by a sufficient margin to prevent any possibility of movement, where

$$M_w = WB/2 \tag{9.8}$$

$$M_t = TH \tag{9.9}$$

The applicable value of weight W includes the weight of the foundation slab and wind turbine, plus the overburden soil and rock (backfill) that cover the plinth; B is the slab horizontal dimension (the side length for a square plinth, or the diameter for a circular one), T is the rotor thrust load, and H the hub

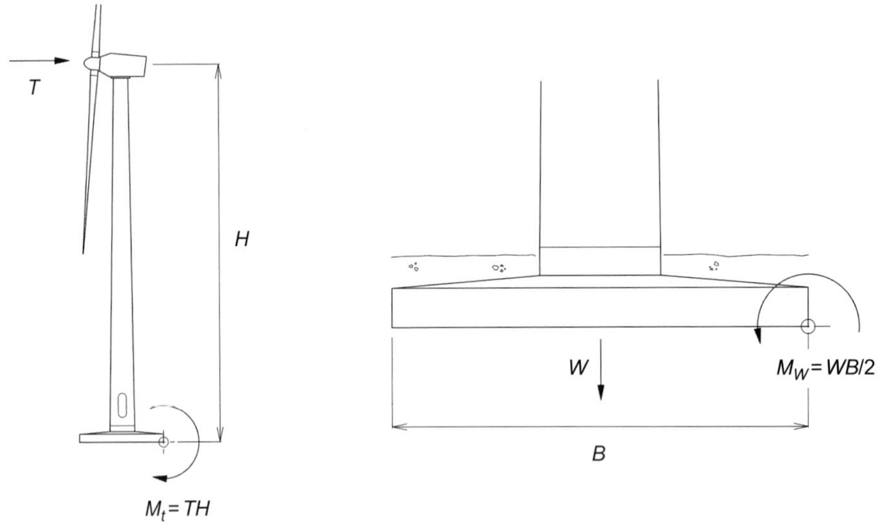

Figure 9.15 Principle of the gravity base foundation. To prevent overturning, weight moment M_w must exceed thrust moment M_t by a suitable margin: GL recommend a safety factor of 3 (Germanischer Lloyd, 1993).

height. The Germanischer Lloyd rules for foundation design stipulate that the maximum overturning moment due to rotor thrust must not exceed $WB/6$, which effectively gives a safety factor of 3 on stability, or

$$M_w > 3M_t \tag{9.10}$$

The higher the weight moment M_w, the more stable the foundation. With reference to Equations (9.8) and (9.9) the following are noted:

- For a given weight W, foundation stability increases with base dimension B, for which reason concrete plinths tend to be much wider than they are deep.
- For a given rotor thrust T, the overturning moment M_t is proportional to H; hence the same turbine type in terms of rotor diameter and power rating will require a larger foundation if a taller tower is used.
- Because the extreme rotor thrust load is dependent on local wind characteristics, foundation designs are site specific, and related to the IEC Wind Class (see Table 9.1). The higher the maximum wind speed, the larger the foundation.

There are variations on the basic concrete plinth: some are rectangular, others circular, but the construction process is similar. A pit is first excavated, typically about 2 m deep, to accommodate the foundation. A geotechnical survey then establishes the strength of the soil at its base. If the ground is too soft, with clay, peat, or sand, the excavation continues until a suitable load-bearing stratum is found; stone is then laid in and compacted to bring the floor height back to the datum. A thin (50 mm or so) layer of 'blinding' concrete is then poured to provide a smooth level surface off which construction of the plinth can proceed. A cage of steel reinforcing bar is constructed that defines the three-dimensional envelope of the plinth; this is the most time-consuming exercise as the re-bar is fixed by hand and literally hundreds of pieces may be required (steel reinforcement accounts for around 10% of the final slab weight). Circular bases are more challenging than square ones due to the number of different bar radii involved.

Achieving a high-integrity connection between the foundation and wind turbine tower is critical, and two different methods are commonly used. One is the 'stub tower' in which a tubular steel can, essentially a short tower section, is integrated into the steel reinforcing cage prior to concrete pour. A flanged bolt ring at the top of the can provides the interface to the bottom tower section. Vestas have favoured this design for several of their medium-scale wind turbines and an example is shown in Figure 9.16: the reinforcing bar is threaded horizontally through the can and rises vertically inside it, with concrete poured both inside and out.

An alternative design is the 'slab and pedestal' in which the tower holding bolts are incorporated into the concrete base. The long bolts are secured deep in the foundation reinforcement by a captive ring, and sleeved to prevent their threads adhering to the concrete: this allows the bolts to stretch when tensioned. An example of a slab-and-pedestal foundation for an Enercon E48 is shown in Figure 9.17, at the stage where concrete is being poured. In contrast to the re-bar

Figure 9.16 Stub tower foundation under construction for a Vestas V52. The steel 'can' is threaded with reinforcing bars and forms an integral part of the base once the concrete is poured. The bottom tower section is bolted to the flange ring at the top of the can.

Figure 9.17 Construction of a slab-and-pedestal foundation for an Enercon E48. Concrete is being poured into the reinforcing cage; note the captive tower bolt ring rising out of the central pedestal. Once the foundation has been back-filled, only the pedestal will remain visible.

construction, the concrete pour is relatively quick, and ideally accomplished in a continuous process over half a day or so; this ensures the concrete cures monolithically with no internal discontinuities.[7] Concrete may be ready-mixed and delivered to site by a succession of wagons, or (particularly in remote locations) batch-mixed on site. During the pour concrete samples are continuously extracted for 'cube testing': they are set aside to cure with their strength tested at timed intervals. The strength of concrete develops non-linearly with time typically to

- 20% of final strength after 1 day
- 60% after 1 week
- 80% after 28 days

The design value is taken as the 28 day figure, although the final strength may be achieved only after a year or more (Mosley et al., 1999). Erection of the wind turbine may begin within the 28 day timescale, but commissioning and operation must wait until cube tests verify the concrete strength. Of the two foundation types described above, the stub tower design has greater simplicity in regard to the tower interface, but precise levelling of the 'can' is essential prior to the concrete pour – there is no second chance afterwards. In contrast, levelling adjustment of the slab-and-pedestal bolts can be carried out once the bottom tower section is in place, with the interface then filled with a thin layer of cement-based grout.

The disadvantages of the conventional gravity base are mainly a consequence of the large amounts of material required. The foundation for a 2–3 MW turbine typically requires 500 t of concrete and 50 t of reinforcing steel. In addition to the cost, many vehicle movements are needed: supplying the above volume of concrete ready-mixed would involve 40–50 truck deliveries, plus those required for the steel formwork and concrete pump. Concrete also attracts an environmental penalty in terms of the energy required in its manufacture and the accompanying release of CO_2. It is estimated that cement production is responsible for 5% of all man-made carbon dioxide emissions, half due to the chemical process and half from direct energy use (IPCC, 2007).

An alternative foundation type that avoids most of these disadvantages is the rock anchor, which does not rely on gravity, but essentially pins the WEC structure to the ground. Wind turbines both large and small have been successfully installed using rock anchors, and a system developed for MW-scale turbines is illustrated in Figure 9.18. The anchors are threaded steel rods typically 6 m in length, which are inserted into pre-drilled holes in bedrock and bonded with grout (high-strength cement) at their lower end. The rods pass through a steel adapter ring or plate, which is mounted almost directly on to the rock surface: a thin layer of reinforced concrete is laid to level the adapter, but otherwise none is used. The upper part of the anchor is sleeved to prevent adhesion and secured by a nut on the adapter plate. The bottom tower section is bolted to a flange ring on the adapter plate. Prior to installing the wind turbine the anchors are heavily post-tensioned: this stretches them, and exploits the compressive strength of the bedrock to protect the anchors against fatigue (in a similar way to a T-bolt blade root attachment, with rock taking the place of blade composite; see Section 8.4).

[7] If necessary, the pour time can be extended by incorporating retardants in the concrete mix.

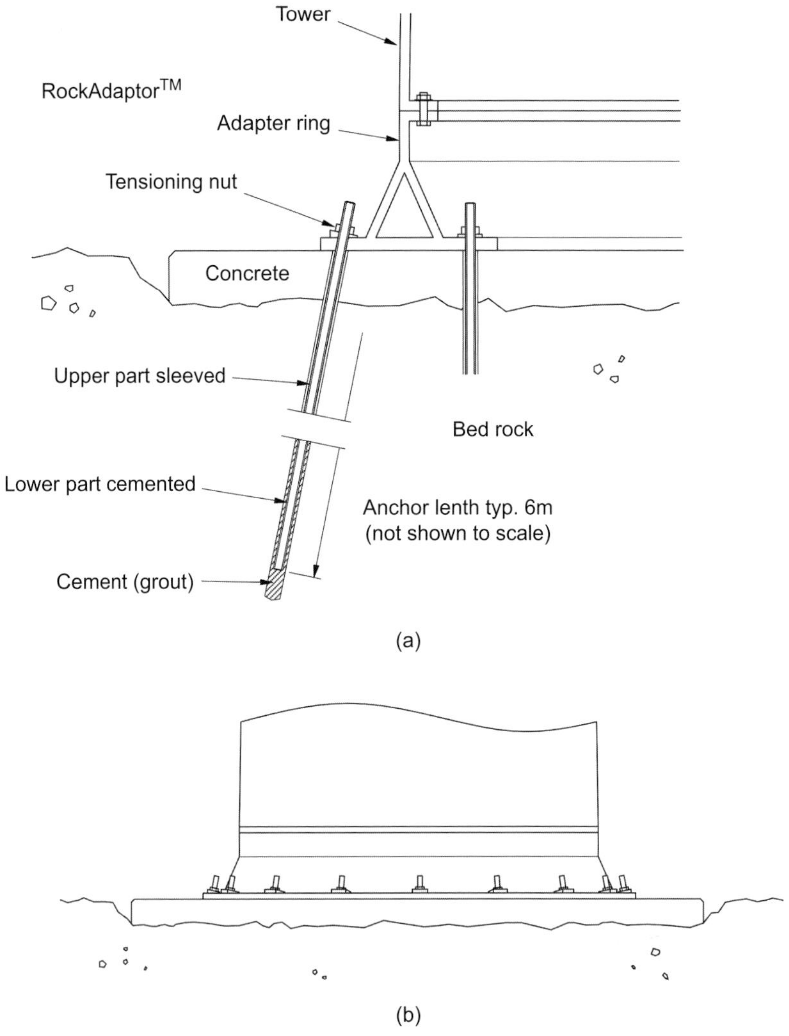

Figure 9.18 The RockAdapter™ rock anchor solution: (a) threaded steel anchors secure the adapter ring onto bedrock; the lower part of the anchor is grouted into the rock, with the upper part sleeved to allow post-tensioning; (b) completed foundation. The concrete requirement can be less than 5% of a comparable gravity base. (Based on drawings provided by kind permission of Windtechnique A/S)

The system shown in Figure 9.18 was devised by Scandinavian civil engineering firm Windtechnique A/S under the trade name RockAdapter™ and has been used on several sites in the Faroe Islands and Northern Europe. The anchors are splayed outwards to achieve the required load capacity and maintain the integrity of the bedrock. The concrete requirement is limited to that filling the gap between the adapter ring and a shallow depression chipped out of the bedrock, and typically less than 5% of the concrete in an equivalent gravity base.

Figure 9.19 Hybrid rock anchor foundations. Each leg of the hinged lattice tower attaches to an individual reinforced concrete plinth, which is pinned to bedrock by two 6 m rock anchors. Demonstration project on the Island of Luing, Scotland. (Photo courtesy Shane Cadzow)

Rock anchors may also be applicable in situations where bedrock lies at greater depth, in which case hybrid foundation designs are used. The wind turbine tower is mounted on a concrete plinth of reduced volume, which is pinned to subsurface bedrock by the anchors. An example is seen in Figure 9.19: this shows a demonstration project on the Island of Luing, Scotland, using a 50 kW wind turbine on a lattice tower. Each leg of the tower is mounted on a small reinforced concrete block pinned into slate bedrock by two 6 m anchors. The total concrete requirement was 5.4 m^3, or approximately 10% of that required for a comparable gravity base. A project report comparing the two solutions concluded 'the total costs are similar, and there is potential for cost reduction in the rock anchored version through repetition and competitive tendering' (Robb, 2003, p. 9). The wind turbine was winched into place, with an additional rock-anchored block to support the winching equipment (see Figure 9.27).

The main disadvantage of rock anchors is that not all rock strata are suitable to accept them: specialised geotechnical surveys must be carried out in each case, and foundation designs are highly site specific. Where they are used the material advantages of rock anchors are, however, significant, and there seems no good reason why they should be restricted to remote sites. The currently limited application of rock anchors compared with gravity bases may be partly due to industry conservatism. If the cost of concrete were to rise significantly in the future, or its environmental penalty be given greater weight, then rock anchors may become more common. A summary of the advantages and disadvantages of the two foundation types is given in Table 9.2.

Table 9.2 Comparison of Gravity Base and Rock Anchor Foundations

Type	Slab (gravity base)	Rock anchors
Advantages	Well-established design and construction method, widely used in other fields. Simple and dependable. Many contractors are capable of design and construction; competition reduces costs.	Significant reduction (up to 95%) in concrete requirement. Few vehicle movements needed: better for remote or inaccessible sites. Low environmental impact. Potentially more cost-effective (depending on take-up by the industry).
Disadvantages	Uses large quantities of concrete; construction requires many vehicle movements to/from site. Environmental penalty: cement production estimated to produce 5% of global CO_2 emissions.	Not all rock strata are suitable. Design and installation are site specific: requires specialist contractors, hence currently more expensive.

9.4.3 Transport and Access

Transporting a large wind turbine from factory to site requires careful planning, and consultation with many parties. Delivery routes initially involve the public road, in some cases over distances of several hundred kilometres, but ultimately purpose-built tracks are needed within the boundaries of the development. At an early stage the complete route must be assessed, taking account of vehicle weights and dimensions, road widths, turning radii, gradients, and load-bearing capacities. Surveys of bridges and culverts, overhead lines, and other infrastructure along the route are needed and a single omission can potentially cause serious delays and unforeseen costs. Specialist transport companies now offer complete end-to-end route surveys, including port and ferry capabilities, and 'swept-path' analyses that enable the footprint of a delivery vehicle to be superimposed on critical bends; where necessary, temporary road improvements are recommended to enable successful delivery. On this basis the transport company can contractually guarantee delivery of all components, helping to remove a major source of project risk.

The longest single components to make the journey are the rotor blades; the widest and tallest are the bottom tower sections or (especially with direct-drive turbines) the generator; and the heaviest vehicles are the construction cranes required to erect the turbines. Figure 9.20 shows typical vehicle configurations for delivery of a 2–3 MW wind turbine: the blade is on an extendable-bed ('trombone') trailer which is shortened for the return journey, and its independent rear-wheel steering allows tight corners to be negotiated; the tower section is on a low loader to minimise overall height and maintain clearance under bridges and power lines. A typical swept-path analysis is shown in Figure 9.21, for transit of a long blade through an urban area; Figure 9.22 illustrates the capabilities of an extended-bed trailer with steerable front and back axles.

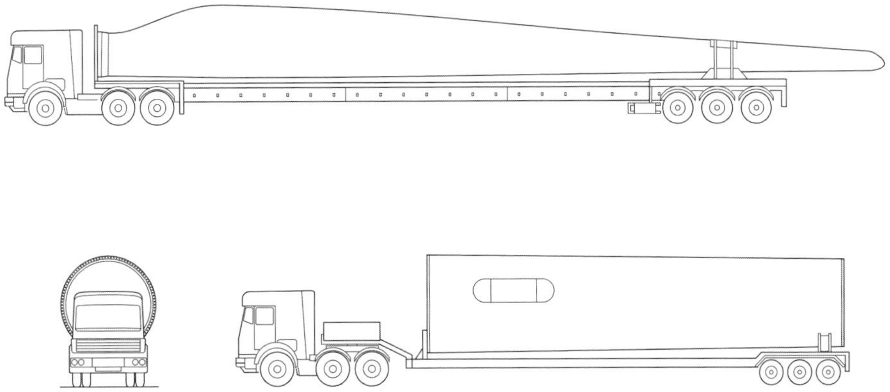

Figure 9.20 Typical transport for a MW-scale wind turbine. (*Top*) extendable-bed trailer for blade delivery, with independent rear-wheel steering. (*Bottom*) low loader with tower section.

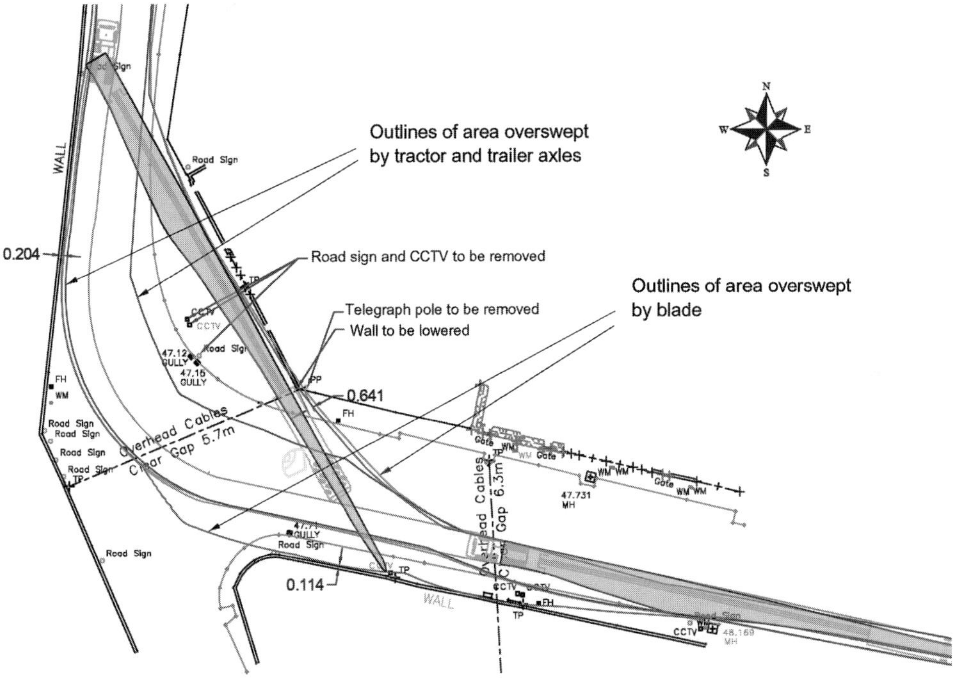

Figure 9.21 Swept-path analysis. Sophisticated desktop surveys identify 'pinch points' on the public road network where temporary widening may be needed or obstacles may require to be removed. (Reproduced with kind permission of Collett Transport Ltd)

Within a windfarm access tracks to each turbine base are constructed from compacted stone topped with gravel. A typical cross section is shown in Figure 9.23. Tarmac and concrete are rarely used due to cost and environmental factors: stone tracks offer better drainage and

Figure 9.22 One good turn. Delivery of a 52 m blade to an upland UK windfarm using an extended-bed trailer; the access track has been locally widened for clearance. (Reproduced with kind permission of Collett Transport Ltd)

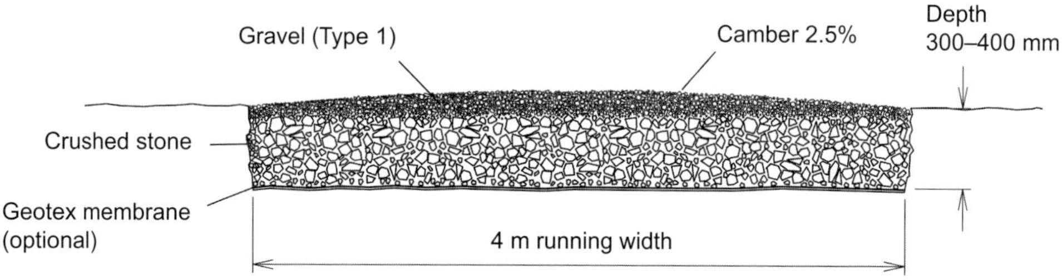

Figure 9.23 Cross section of typical site access track.

blend back visually into the landscape after a year or two. Stone is imported from quarries or won on site: the distance to the nearest supply may be the determining factor, particularly for projects in remote locations. The maximum vehicle weight the track must support can be up to 200 t – the weight of a large all-terrain crane – but it is distributed, and the local limit is dictated by standard axle weights of 10–12 t. Track dimensions depend on the size of wind turbine to be delivered, but a maximum running width of 4-5 m is typical, as vehicle wheelbase is ultimately limited by the public road dimensions. Maximum track gradients of 6°–8° are recommended, though steeper sections are permissible with assisted traction. Minimum bend radii are dictated by vehicle and load length, and will be specified by the turbine manufacturer and/or transport specialist.

A single wind turbine may require an access track a few hundred metres long to connect to the nearest public road. Local contractors can usually be found for the work, or in some cases the developer will carry it out themselves. On a large project, however, track construction becomes a major contractual item: the 539 MW Whitelee windfarm near Glasgow has 130 km of site access tracks (averaging around 600 m per turbine) built from stone won on site. In contrast, a number of windfarms in the UK have been developed on disused airfields whose old runways and taxiways provided ready-built access roads: examples of such developments include Haverigg in Cumbria, Lissett in Yorkshire, and Boyndie in Aberdeenshire. These projects benefited from lower balance of plant costs as a result.

9.4.4 Crane Operations

Onshore wind turbines are usually erected using two cranes, with a large one to carry out the main lifts and a smaller 'tailing' crane for ground assembly and handling work and to steady the lift.[8] Crane operations impose high ground loading, and the vehicles themselves may weigh up to 200 t, so a strengthened hardstanding area is developed beside each turbine base. A typical layout is shown in Figure 9.24. The crane pad is of similar stone construction to the access tracks (Figure 9.23) but with greater depth of stone due to the higher loading imposed. The arrangement of the cranes on the pad is carefully designed to facilitate (a) the main lifting operations and (b) efficient offloading of the turbine components and ground assembly of the rotor; once in place, the positions of the cranes remain fixed throughout the entire operation.

Cranes are generally of all-terrain type with hydraulically extendable boom; their operating envelope is given by a load chart relating lifting capacity to jib height and horizontal radius: an example is shown in Figure 9.25. The crane's maximum lift is achievable only at small radius but for practical reasons it must operate at 15–30 m (depending on turbine size) from the foundation centreline; consequently the heaviest lift is only a fraction (typically 10%) of the crane's maximum lift capacity. For example, a crane of 80 t capacity may be needed to lift the 8 t nacelle of a 225 kW wind turbine into place; to lift the 50 t gearless generator of a 3 MW wind turbine requires a 500 t crane. Wind turbine erection typically requires four to six lifting operations depending on the tower height (hence number of sections) and turbine type; for a conventional geared wind turbine the following are typical:

- tower sections: three lifts
- nacelle, complete: one lift
- rotor, pre-assembled: one lift (see Figure 9.26 for an example)

[8] Some turbine manufacturers (e.g. Enercon) have devised systems to lift the rotor with a single crane: a mechanism under the hook turns the rotor through 90°, transitioning from the horizontal to the vertical plane once suitably clear of the ground.

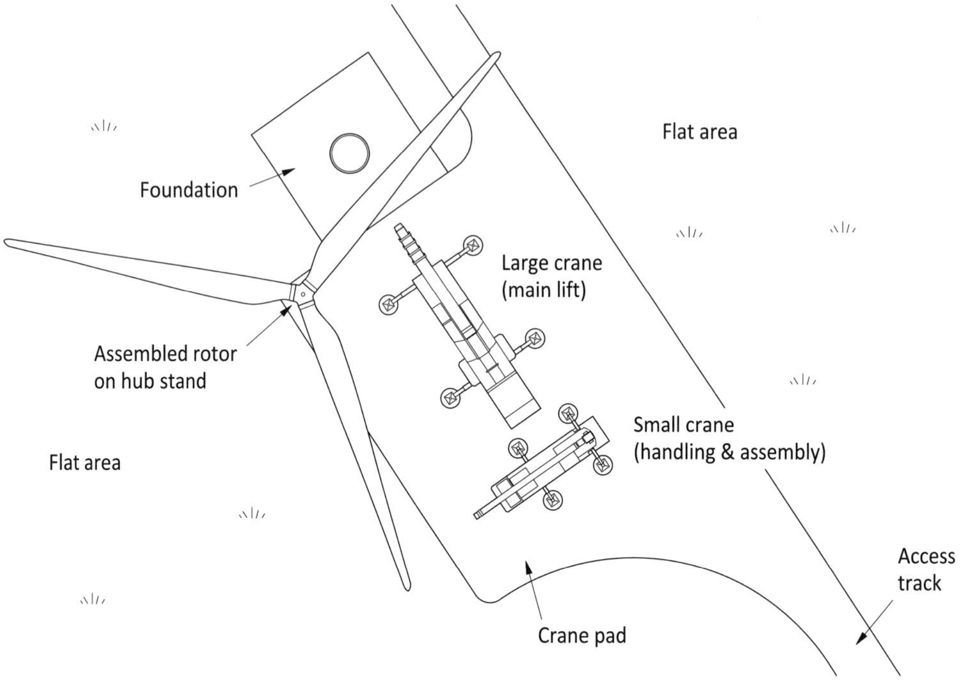

Figure 9.24 Crane pad layout for wind turbine assembly and erection. In this arrangement, the rotor is fully assembled before lifting into place. Usually two cranes are employed, with the smaller used for low-level handling and assembly work.

A gearless wind turbine may require an extra operation, with the nacelle and generator lifted separately. Strict wind speed limits must be observed during crane work, and the rotor lift is the most critical operation; an anemometer on top of the crane jib monitors the hub height wind speed and lifting work is suspended if it is too high. This can be frustrating and costly, and it has been known for a final rotor lift to be interrupted and delayed for a week or more due to high winds.

It is possible to erect wind turbines without using cranes but instead winching them into place. The technique is illustrated in Figure 9.27, showing a 50 kW machine on the Island of Luing in Scotland. The specially adapted tower has hinged feet, and a steel A-frame temporarily attached to its base; the powered winch is located on an adjacent plinth. The wind turbine is fully assembled on the ground and the A-frame provides a lever arm during the early stages of the lift, when the weight moment about the hinge is greatest; a multi-pulley block and tackle is used to limit the cable tension. The winch load reduces progressively as the turbine nears the vertical and when its centre of gravity crosses the hinge line a restraining cable (back stay) comes into play. The turbine is then gently lowered into place under gravity. The example shown was a project to demonstrate low-cost infrastructure solutions for remote sites, with no requirement for a crane, gravity plinth (rock anchors were instead used; see above) or site access tracks (Robb, 2003).

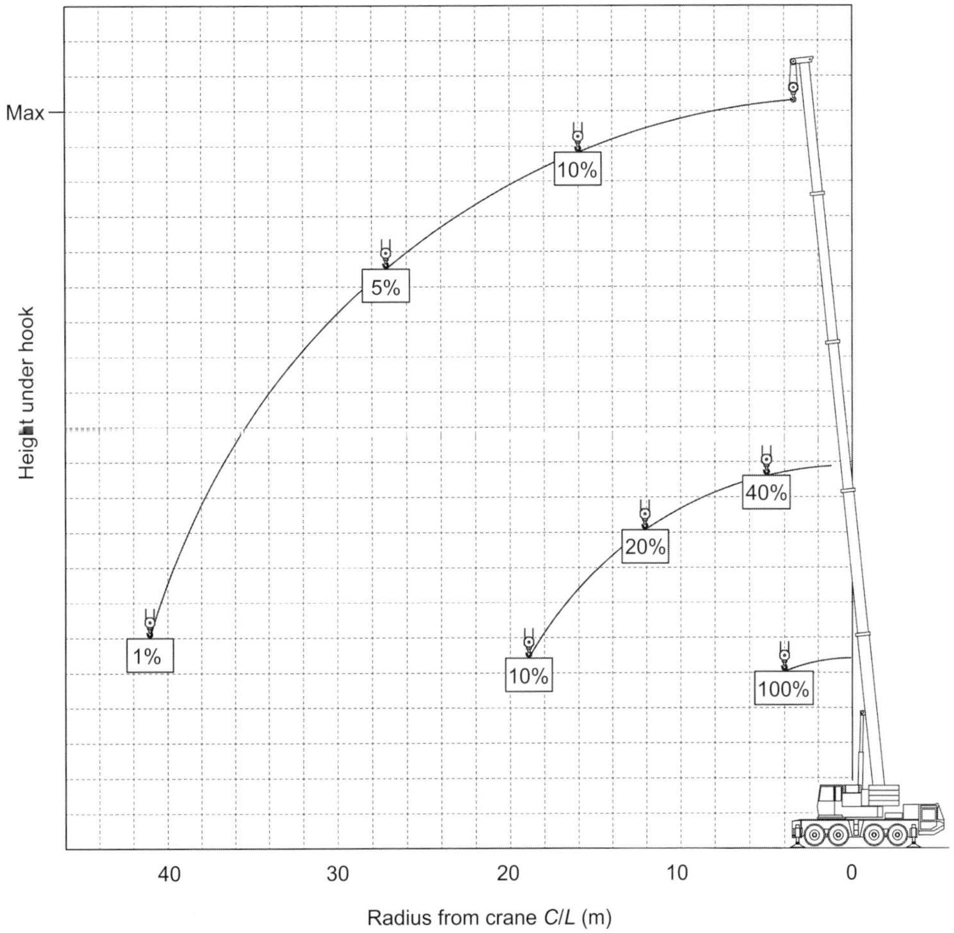

Figure 9.25 Crane chart, showing the maximum lift weight (as percentage of crane capacity) as a function of jib height and horizontal radius. For wind turbine construction, the maximum lift is typically 10% of the crane capacity at a horizontal radius of 15–30 m.

Winch erection is quite common for small wind turbines, but has been used for large ones too: the Howden HWP-300 in the Orkney Islands was lifted into position using a diesel powered winch in an exercise taking just 25 min (Brown, 1984). The disadvantages of winching large HAWTs are that a hinged and locally strengthened tower is required to take the high compressive loads of the initial lift, and an additional foundation plinth is needed to accommodate the winch.

9.4.5 Electrical Infrastructure

Electrical aspects of wind turbines are covered in detail in Chapter 5, but a summary of the physical infrastructure required to connect a wind energy plant to the grid can be made here. The details vary

Figure 9.26 Main rotor lift for a Vestas V52. A tailing crane (out of picture) is steadying the rotor. Single-crane rotor lift is also possible using a proprietary turning mechanism under the crane hook.

with project scale and location, but the schematic in Figure 9.28 contains the essential elements. The figure is representative of a single large wind turbine connected to a utility network. The generator (1) is located in the wind turbine nacelle and is commonly a low-voltage device with LV cables (2) carrying the power down inside the wind turbine; the contactor (3) is a switch to connect and disconnect the generator from the grid and is typically located in a cabinet in the tower base; the associated grid protection relay (R) detects abnormal conditions on the network and if necessary disconnects the generator automatically. The step-up transformer (4) raises the voltage from LV to HV for direct grid export: on many wind turbines the transformer is located in the tower base – see

Figure 9.27 Winch erection of a 50 kW wind turbine. Winch loading is highest at the start of the lift when the weight moment is greatest; once near the vertical, the winch load reduces almost to zero: the back stay seen on the right will shortly take up the load. (Photo courtesy Paul Pynn)

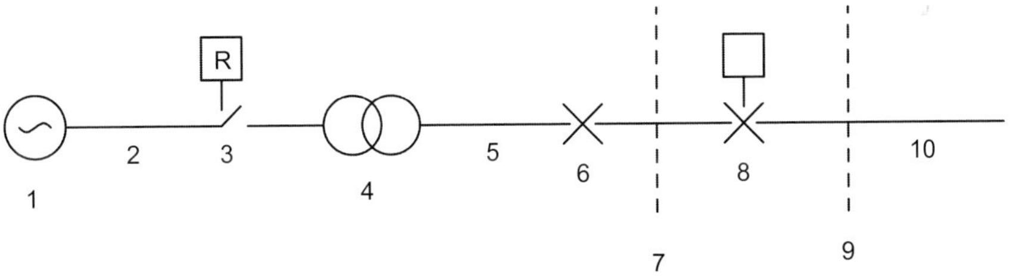

1. Generator
2. LV cabling
3. Contactor with grid protection relay
4. Transformer
5. HV cabling
6. HV circuit breaker
7. Boundary of private electrical network
8. HV circuit breaker with import/export meter
9. Physical boundary of wind turbine development
10. HV grid network

Figure 9.28 Schematic showing the electrical infrastructure between a wind turbine generator and the grid. The example shown is representative of a single large wind turbine.

Figure 9.29 Transformer for an 800 kW wind turbine. The lowest tower section will be lowered into place over the top of the transformer, with the control cabinets installed in a floor immediately above.

Figure 9.29 – though on very large machines it may be in the nacelle (to minimise LV cabling requirements) while small or medium WECs may have an external ground-mounted transformer in a separate cubicle. The transformer is connected by HV cabling (5) to a circuit breaker (6), which is essentially a protection switch that opens automatically on detection of current overload.

The boundary (7) marks the interface between the private and public electricity networks, which is usually within the physical curtilage of the development in a substation building. The network operator (DNO) may require an additional circuit breaker (8) to give them independent control of the windfarm connection: this breaker will open automatically on detection of a generator or transformer fault in order to protect the external network. The boundary (9) is the nominal extent of the windfarm site and beyond it the distribution network extends via (in most cases) overhead line conductors (10). Figure 9.28 is by no means exhaustive, and does not show any ancillary equipment that may be required, e.g. for voltage control (for more details, see Section 5.5). The schematic also represents just a single grid-connected wind turbine: in a large windfarm the grid protection relay (R) may be installed on a circuit breaker connecting more than one turbine. The HV network (5) within a windfarm is always in the form of underground cabling laid in trenches laid alongside the turbine WEC access tracks; in many cases the external HV network (10) is also cabled underground for environmental or safety reasons.

9.5 Exercises

9.5.1 Wind Class

Long-term measurements at a prospective windfarm site give the following results:

- average hub height wind speed 8.1 m s^{-1}
- characteristic turbulence intensity 17%
- extreme (50 year) gust 58.0 m s^{-1}
- wind shear exponent α 0.16

Which of the following IEC classes of wind turbine would be suitable for installation at the site: IA, IB, IIA, IIB, IIIA, IIIB?

9.5.2 MCP Analysis

The scatter plot in Figure 9.30 relates hourly wind speed measurements at a target site to a reference site over a 6 month period. The target site data were measured at 50 m above ground level, with a local wind shear index α of 0.139. The long-term mean wind speed at the reference site is 6.20 m s^{-1}. From these data estimate the long-term mean wind speed at the target site at heights of (a) 40 m, and (b) 64 m.

9.5.3 Wake Velocity Deficit

Two wind turbines with 80-m-diameter rotors are separated by a distance of 480 m on flat land. If one machine is directly downwind of the other, and the upwind machine is operating

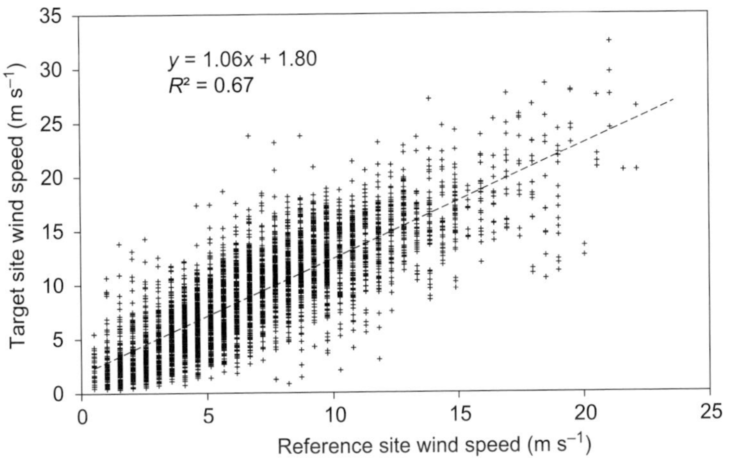

Figure 9.30 See Exercise 9.5.2.

at optimum efficiency (i.e. at the Betz limit), estimate the ratio of the hub height wind speed at the downwind machine to the freestream velocity. Repeat the calculation assuming the wind turbines are offshore.

9.5.4 Wake Extent

Calculate the horizontal width of the wake experienced by the downwind machine in Exercise 9.5.3 based on the Jensen–Katic wake deficit model, assuming an onshore installation. What range of incident wind directions (i.e. sector angle) will result in some degree of wake interference at the downwind machine?

9.5.5 Wake Turbulence

Using the formula from IEC 61400-1 calculate (a) the added wake turbulence and (b) the resulting net turbulence at hub height on the downwind machine in Exercise 9.5.3. Assume ambient turbulence intensity of 14% and a hub-height wind speed of 8.0 m s^{-1}. Repeat the calculation using the formula of Hassan assuming $C_T = 0.8$ and X_N equal to 2 rotor diameters.

9.5.6 Foundation Design

A wind turbine has a hub height of 49 m and an all-up weight (rotor, nacelle and tower) of 86 t. For the design of the foundation an extreme rotor thrust of 296 kN is assumed to act on the rotor centreline. The foundation comprises a square reinforced concrete plinth 0.90 m thick, with its upper surface buried 1.1 m below ground level. The density of reinforced concrete is 2500 kg m^{-3}, and the density of the soil and rock backfill (overburden) is 2200 kg m^{-3}. Calculate the minimum required horizontal dimension for the foundation slab (a) neglecting the weight of the wind turbine and (b) including its weight.

9.5.7 Crane Capacity

An all-terrain crane is to be selected for a wind turbine construction project. The heaviest lift is the fully assembled nacelle, weighing 23.5 t. What minimum capacity of crane (rounded to the nearest 100 t) would be required for this job assuming a horizontal lift radius of (a) 15 m, (b) 30 m, from the tower centreline?

Figure 9.31 See Exercise 9.5.8.

9.5.8 Crane Road Weight

Transport surveys must take account of the load-carrying capacity of bridges and culverts on the public roads leading to a windfarm site. Estimate the maximum road weight of the all-terrain crane shown in Figure 9.31, and explain your reasoning.

CHAPTER 10 PLANNING AND ENVIRONMENT

10.1 INTRODUCTION

Wind energy is a sustainable resource, so it follows that protection of the environment should be a major consideration when planning a wind turbine project, large or small. Planning regulations differ in scope and detail from one country to another, but the underlying principles of safeguarding the natural and human environment are the same. Wind energy developments will generally not be permitted where destruction of important plant or animal habitat might occur, and the planning process may require surveys of important species, assessment of likely impact, and agreed mitigation to protect or restore habitat. Impact on human activities is equally important. Wind turbines will not be permitted where they cause undue noise or visual impact (though more on this later), affect aviation, or disrupt radio frequency communications. This chapter contains a broad overview of these topics.

10.2 ECOLOGICAL IMPACTS

10.2.1 Birds

Wind turbines and birds can safely co-exist in many locations, but not all. Inappropriately sited projects can lead to bird mortality due to rotor collision, or reduced breeding success through disturbance or loss of habitat. A great deal of research has been conducted internationally into the impact of wind turbines on birds, and on the steps required to avoid unacceptable mortality or species displacement (see e.g. Langston et al., 2003). Some of the knowledge was hard won, and in the early days of the industry significant mortality occurred where large numbers of wind turbines were installed in areas with high density of vulnerable bird species. Well-documented examples include the Altamont Pass in California, and Tarifa in southern Spain: in these cases large birds such as eagles and vultures were found to be particularly at risk of rotor collision. Habitat loss, though less dramatic, is an equally important concern.

Leading ornithological bodies such as the US Audubon Society and the RSPB in the UK broadly support wind energy as a necessary response to the effects of climate change, but will oppose specific developments where they are believed to threaten local bird populations

(RSPB, 2018). The onus is on developers to prevent serious bird impact by judicious siting and/or habitat mitigation, and in the UK, site-specific bird surveys are mandatory before a development can obtain planning consent. The scope of such surveys typically includes

- compiling lists of all bird species in the immediate area around the proposed development, with map-based illustration of their distribution
- surveying breeding numbers of key species (e.g. with protected status); sampling typically covers 12 months with counts taken weekly throughout the breeding season
- recording flight lines: the commonest routes flown are often between roosting and feeding areas, and surveys can establish where a development will potentially obstruct or disrupt them
- carrying out post-construction bird counts, including mortality; searches for dead birds must be systematic, including frequent site visits to preclude carcases being removed by predators before being counted

Some examples may illustrate the process. Figure 10.1 is taken from a pre-construction survey that was carried out for a small wind energy development on the Island of Luing in Argyll, western Scotland. The barnacle goose (*Branta leucopsis*) is common here, and a significant percentage of the European population visits annually. The goose is a protected species, and some earlier

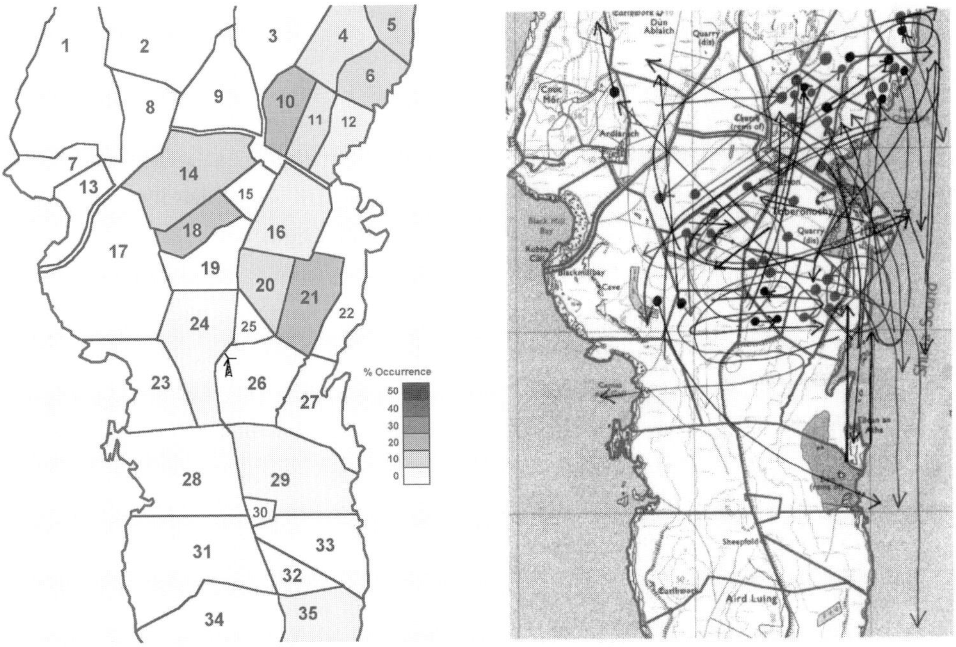

Figure 10.1 Pre-development species survey for a small wind turbine project on the Island of Luing, Argyll (Lamont et al., 2003). (*Left*) frequency distribution of barnacle geese in numbered fields; (*right*) observed flight lines from roost areas. (© Crown copyright and database rights 2019, OS licence number 100037385)

Figure 10.2 Barnacle geese in the vicinity of a 50 kW wind turbine on Luing. The site was monitored before and after construction to assess the likely impact on bird life; no goose mortality occurred in 10 years of turbine operation. (Photo courtesy Anja Lamont)

windfarm proposals in the same region had been rejected on a precautionary basis. Over the course of a year a qualified naturalist visited the Luing site at regular intervals and compiled a frequency map and flight lines for the goose population; the results indicated a low risk and the wind development was subsequently permitted subject to a programme of post-construction monitoring. The bird survey was extended for a year after the first turbine was built, and in the event the wind turbines were found to have little adverse impact, with no goose mortality (Lamont et al., 2003). In this instance the birds and wind turbines happily co-exist (see Figure 10.2).

For larger scale developments habitat mitigation may be possible, and a good example is the 30 MW Beinn an Tuirc windfarm, also in Argyll. This project was built in an area forming part of the territory for a pair of golden eagles, a species protected under UK and EU law. Pre-construction surveys revealed that the area had marginal food resources for the birds, whose low breeding rates were attributed to commercial forestry having replaced open moorland and displaced the eagles' natural prey (mainly rabbits). The windfarm developer agreed to fund a management plan to increase the availability of prey within the eagles' range, while at the same time reducing their risk of collision with wind turbines. This involved large-scale removal of immature commercial forestry and creation of open heather moorland at a suitable distance from the windfarm, where the eagles could hunt safely. Post-construction monitoring of eagle movements indicated that the measures were successful (Walker et al., 2005).

10.2.2 Mammals

Developments may also encroach on the habitat of mammals, reptiles, or other animals. In such cases pre-development surveys similar to those for birds are required. The main objective is again identification of sensitive sites, and, where necessary, measures to protect species during and after construction. At the planning stage it may be possible to call on the expertise of academic or local nature interest groups who have compiled species lists and are recognised as authoritative sources.[1] In the UK species such as badgers, deer, otters, and smaller mammals may be at risk of habitat disruption. In many cases the risk is greatest during construction, when appropriate measures should be taken to protect animals' pathways, burrows, or hides. Bats, however, share with birds the risk of rotor blade collision. Research in both the US (Horn et al., 2008) and UK (Mathews et al., 2016) links bat mortality to the presence of woodland within the vicinity of wind turbines; the current advice in the UK is to maintain a 50 m buffer around trees or hedges in the vicinity of wind turbines to minimise bat collision risk (Natural England, 2014).

10.3 Public Safety and Acceptance

10.3.1 Public Safety

Blade loss or other catastrophic damage to wind turbines is very rare, but not impossible, and public safety must be addressed during the planning process. A minimum set-back distance of 1.1 times the tip height of a wind turbine from any busy road or frequently occupied area was at one time proposed, the same 'toppling distance' as recommended by the UK electricity industry as safety clearance to overhead power lines (ENA, 2012). More detailed set-back recommendations from the US consider the possibility of total tower collapse, and include the observation that 'in cases where information is available, the majority of the major components (rotor, tower, and nacelle) have fallen to within one to two hub height distances from the base' (NYSERDA, 2017, p. 4), which implies a greater safety radius than the toppling distance criterion. In the UK there is no specified minimum set-back distance based purely on safety, but statutory noise limits generally result in an exclusion radius of 400 m or more round a large wind turbine, which effectively meets the criteria for public safety as well. Similarly, exclusion radii based on shadow flicker and visual impact (see below) are greater than those required on safety grounds.

In cold climates ice fragments may be shed from rotor blades, typically during start-up. Nowadays most wind turbines are equipped with temperature sensors and condition monitoring to detect ice build-up and curtail operation if necessary; on some machines blade de-icing systems are standard. Consideration of ice throw may nevertheless be required at the planning stage, and experimental studies (Tammelin et al., 2000) indicate an ice 'risk circle' around a turbine defined by

$$d = 1.5(D + H) \tag{10.1}$$

where d is the maximum ice throw distance, D the rotor diameter, and H the hub height. Ballistics analyses suggest that throw distance depends on tip speed, which is largely independent of turbine size, although greater tower height will effectively increase the throw radius. According to one

[1] A good example is NESBReC, which maintains comprehensive species databases and maps for northeast Scotland.

analysis, 'the risk of being struck by ice thrown from a turbine is diminishingly small at distances greater than approximately 250 m from the turbine in a climate where moderate icing occurs' (Morgan et al., 1998, p. 113).

10.3.2 Visual Impact

Visual impact of wind turbines is a highly subjective issue and whereas technical solutions exist for aspects such as noise, public safety, or ecological safeguarding, large wind turbines are difficult to hide. Nevertheless a systematic approach to landscape and visual impact assessment, based on independent guidelines, is attempted within the UK planning process. In assessments of new projects the initial step is to establish a zone of theoretical visibility (ZTV) using digital terrain mapping to identify locations from which all or part of the development may be visible. A typical example is shown in Figure 10.3, showing visibility of a three-turbine windfarm out to a radius of

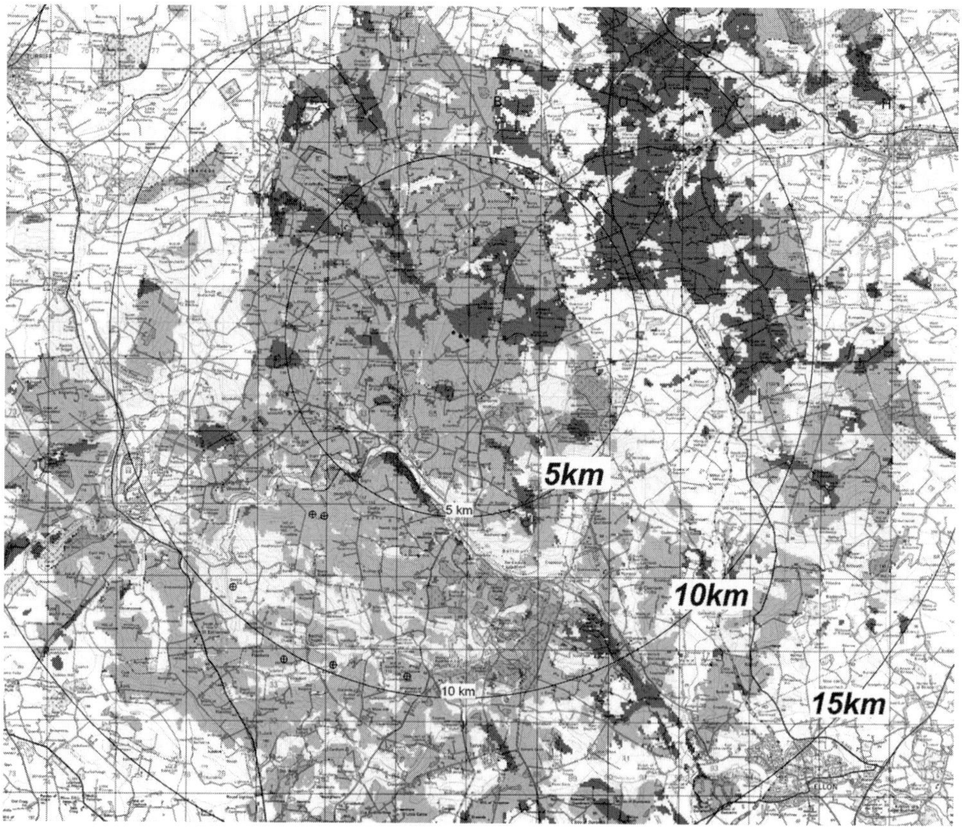

Figure 10.3 Zones of theoretical visibility for a small windfarm. Different shading indicates areas of cumulative impact assessment where other developments may also be seen. (Figure reproduced by permission of Greenspan Energy Ltd; © Crown copyright and database rights 2019, OS licence number 100037385)

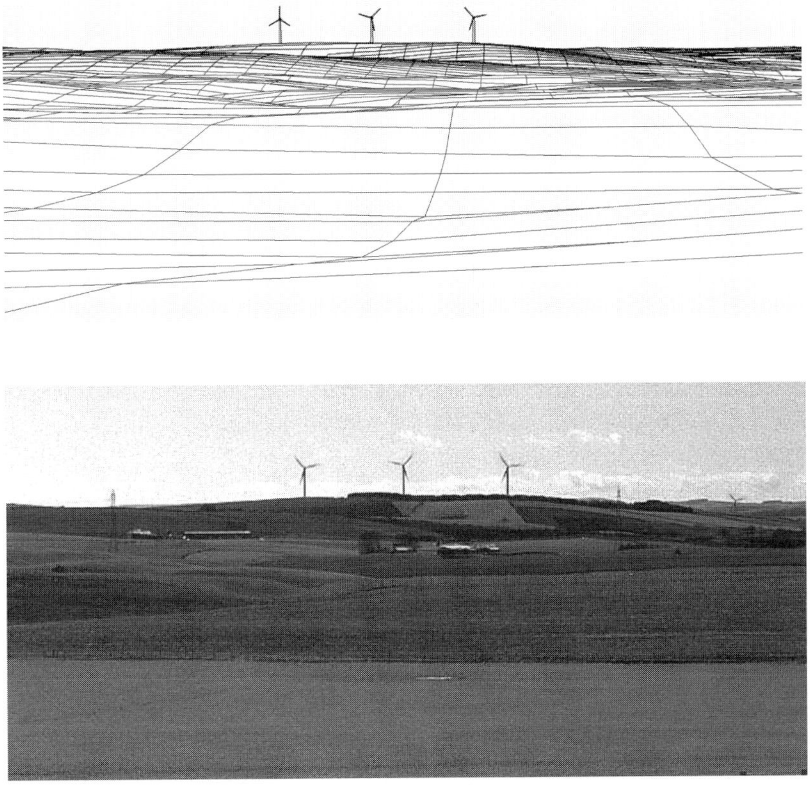

Figure 10.4 Visual representation of a proposed three-turbine windfarm: (*top*) wireframe plot based on digital terrain data; (*bottom*) photomontage. (Reproduced with kind permission of Greenspan Energy Ltd)

15 km. The ZTV is a fairly high-level tool that gives little indication of what the public will actually see, and does not account for the influence of competing landscape features such as buildings or vegetation (which are absent from the digital terrain database). It does, however, help identify important viewpoints from which the project may potentially be visible, and for which more detailed assessment can then be carried out.

The relative scale of the project and its perspective from an identified viewpoint are then assessed using 'wire frame' topographic images, again using digital terrain modelling. From such images accurately scaled photomontages can then be prepared, to provide a representative impression of the finished project; see Figure 10.4 for an example. Wire frame and photomontage work must follow strict procedures to ensure correct representation of scale, with camera focal length and viewing angles chosen to replicate the image seen by the human eye (SNH, 2002).[2] Similarly the background photographs used for photomontages must be taken under representative conditions of

[2] Original guidance recommended a focal length of 50 mm with 35 mm film-format camera; with modern digital cameras, the appropriate focal length must take account of the sensor size, which varies between types.

lighting and weather. While these techniques produce images that are optically accurate, subjective impact is harder to quantify. The UK planning process goes some way to address this issue via a ranking system for the perceived quality and value of a landscape and its capacity to accept wind turbines. Certain areas such as national parks may simply be off limits; in others the scale of the proposal may be the deciding factor.

In Scotland planning guideline NPPG18 describes the statutory protection for historic buildings, gardens, and other heritage sites (Scottish Government, 1999). Historic context is itself subjective, however; for example in 2007 a proposal for a small windfarm near a group of Neolithic sites on Orkney was rejected after a public enquiry, on the grounds that it would damage the 'authenticity' of the site. The debate that preceded the decision was nominally about visual impact, but expanded into topics ranging from aesthetics through to morality and the common good (McClanahan, 2013).[3] The importance of visual impact should not then be underestimated, and the issue played a significant part in the 2015 UK moratorium on support for onshore wind (see Section 11.6).

Public reaction to changes in a familiar landscape may, however, depend on many factors, and one that plays a significant role in the context of wind energy is the ownership or perceived benefit of a scheme. Projects owned by communities or local employers may be more positively regarded than those owned remotely by third parties (Warren et al., 2010). This point has been recognised in Denmark, where the early development of wind energy was characterised by widespread ownership of small-scale projects. Public resistance subsequently grew against windfarms owned by large energy companies, but wind power nevertheless remained popular, and in 2009 the Danish government legislated that developers must offer to sell at least 20% of any new project to citizens living within 4.5 km of the wind turbines (Hvelplund et al., 2017). This approach addressed the public's concern by offering them a greater involvement in onshore wind energy developments. The topic of ownership is discussed further in Section 11.5.

10.3.3 Shadow Flicker

At certain times of the year when the sun falls behind a wind turbine rotor it casts a moving shadow that may cause annoyance to nearby householders. The probability of shadow flicker occurrence depends on the relative locations of the turbine and affected property, the geographical latitude, time of day, and the degree of cloud cover. The maximum length of time for which the property will be affected can be accurately predicted and quantified in terms of annual hours of disturbance. In general, shadow flicker is unlikely to be a problem if

- the wind turbine is more than 10 rotor diameters from the property
- the property is (at UK latitudes) more than 130° either side of north of the wind turbine

[3] Arguably, the passage of time is an important factor too: in England, there are more than 1000 traditional windmills with the status of protected monuments (Historic England, 2018)!

The second of these rules must be modified for other latitudes but the recommended 10D separation is general, and is accepted in UK planning rules (Parsons Brinckerhoff, 2011). For closer distances detailed calculations may be needed and in some cases the developer is required to constrain turbine operation for the duration of the shadow occurrence. This can be achieved with sector management (see Section 6.5) in which the turbine controller is programmed to take account of the date and time of day; the resulting downtime and loss of energy output is generally low.

10.3.4 Pollution Risk

Pollution is normally a concern only during the construction phase of a wind energy project, and an environmental impact assessment detailing potential risks and appropriate mitigation strategies is often a statutory pre-construction requirement. Issues include temporary storage of oil, fuel, or other hazardous chemicals on site and these activities must be carefully planned, or potentially avoided, depending on the scale of the project. Construction work is ideally carried out avoiding periods of heavy rainfall when water run-off may occur, though the unpredictability of the weather and the tight schedules affecting large projects frequently conspire against this and work in bad weather may be inevitable. Spoil from foundation excavations must be responsibly managed, and contractors are encouraged to use recycled aggregate materials for roads and other civil work. Noise pollution during construction, though temporary, may have to be controlled within agreed limits and/or work restricted to certain hours of the day.

In the longer term there may be a risk of silting or pollution of natural watercourses due to surface water run-off from windfarm access tracks or crane hardstanding areas. This can be avoided by good design of compacted porous stone roads (see Figure 9.23), embankments, and drainage ditches. Windfarm track construction methods are generally similar to long-established forestry or farm practice (Forestry Commission, 2001); the use of tarmac or concrete site roads is unusual, at least in the UK. Planning authorities normally require a comprehensive method statement covering all aspects of site design, construction, and environmental management before work may proceed.

10.4 Noise

10.4.1 Origin of Wind Turbine Noise

Wind turbine noise arises from a combination of aerodynamic and mechanical sources. Aerodynamic noise is caused by shear stresses in the airflow around the rotor blades, particularly where relative velocities are high: trailing edge noise is generated by the shearing discontinuity between air leaving the upper and lower blade surfaces, and noise is also generated by the tip vortex (see Section 3.6.5) whose core contains high-velocity air. Mechanical noise sources include the generator and gearbox (unless direct drive) and miscellaneous cooling fans in the nacelle or tower base. Intermittent mechanical noise is also caused by yaw motors and pumps. In general aerodynamic noise is broadband in character, whereas mechanical noise may contain tones related to gear-meshing frequencies or motor rotation speeds. With good acoustic insulation mechanical noise can be largely suppressed but aerodynamic noise is harder to eliminate. The total acoustic energy emitted from all sources is the sound power L_W, given by

$$L_W = 10 \log_{10}(P/P_0) \text{ dB(A)} \tag{10.2}$$

where sound power P is in watts and P_0 is a reference level of 10^{-12} W (1 picowatt). The unit of decibels (dB) is here dimensionless and the suffix (A) indicates that the measured spectrum is acoustically weighted to match the response of the human ear, which is insensitive to very high or very low frequencies. Values of L_W for a range of commercial wind turbines are shown in Figure 10.5 as a function of rated output, ranging from 75 kW to 8 MW. The data are mainly from manufacturers' published data, and all relate to three-bladed upwind machines. The dashed line is a logarithmic fit of the form

$$L_W \cong 102.2 + 8.24 \log_{10}P \text{ dB(A)} \tag{10.3}$$

where P is rated power in megawatts. The absolute sound power in watts is shown on the right hand axis: its magnitude is somewhere between one ten-millionth and one hundred-millionth of the generated electrical power of the wind turbine. Equation (10.3) is empirical and makes no distinction between wind turbines on the basis of their power control method (stall or pitch-regulated), drivetrain configuration (geared or direct drive), or the degree to which noise optimisation has been addressed in their design. According to fundamental studies of rotor acoustics sound power is proportional to blade length, and increases as the fifth power of tip speed (de Wolf, 1987). Making this assumption for the machines represented in Figure 10.5 yields

$$L_W \cong 50\log_{10}V_{\text{tip}} + 10\log_{10}D - 6.7 \text{ dB(A)} \tag{10.4}$$

where V_{tip} is the tip speed in m s^{-1} and D is rotor diameter in metres.

Equation (10.4) predicts L_W within ±3 dB(A) for the range of wind turbine sizes shown. De Wolf derived a similar expression, but with slightly higher noise estimates:[4] this may be a sign of progress, as at the time of de Wolf's analysis the largest commercial wind turbines were rated below 500 kW, while larger machines were mainly one-off (and frequently noisy) prototypes. Since then advances in blade design, nacelle acoustic insulation, and anti-vibration mounting of gearboxes (or in some cases elimination of gearboxes altogether) have resulted in significantly quieter machines. This is reflected in the sound power values at the higher power ratings in Figure 10.5. More fundamental analyses of rotor aerodynamic noise have sought to quantify the contributions due to different mechanisms including trailing edge noise, separated flow, inflow turbulence, and tip vortex noise; examples of this work include Brooks et al. (1989), Lowson (1992), and Dunbabin (1994).

[4] De Wolf's constant term is −4 dB(A); this gives good results for wind turbines rated up to about 500 kW but overestimates the sound power of the larger (and newer) machines.

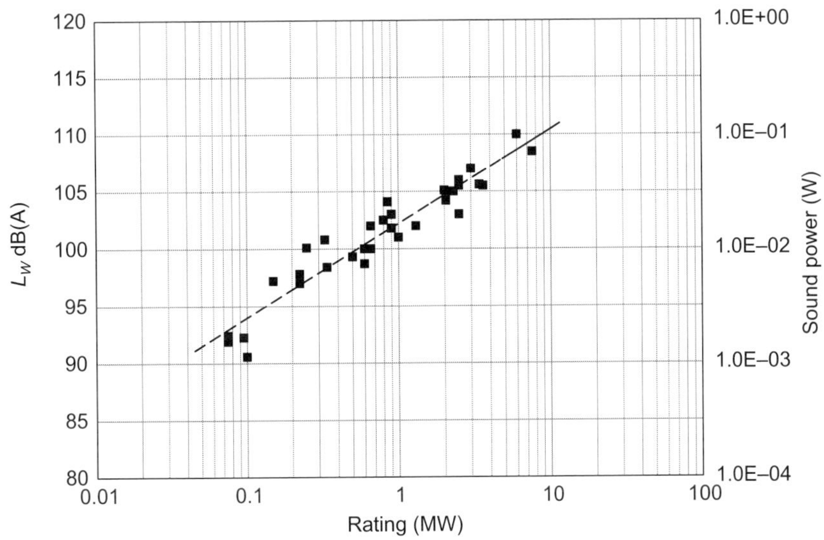

Figure 10.5 Sound power levels (at rated output) for a range of commercial wind turbines. The dashed line is a best fit of the form $L_W \cong 102.2 + 8.24\log_{10}(P)$, where P is rated power in MW.

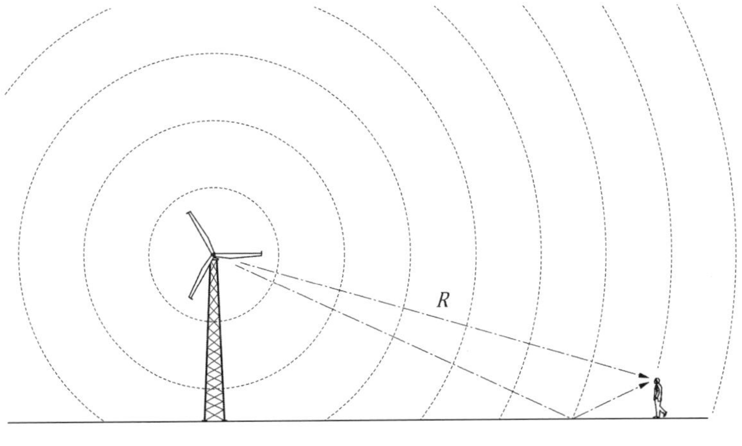

Figure 10.6 Spreading of sound energy from a point source at the rotor centre. Ground reflection doubles the energy received at the observer, equivalent to hemispherical spreading.

10.4.2 Noise Prediction

Predicting the magnitude of perceived noise at specific locations due to a wind turbine development is often mandatory at the planning stage. The general method can be explained with reference to Figure 10.6: the wind turbine is treated as a point source, radiating acoustic energy spherically outwards from the centre of the rotor. The strength of the source is the sound power W, and the sound intensity I is the power flux per unit area at a distance from the source; assuming spherical spreading the two are related according to

$$I = \frac{W}{4\pi R^2} \tag{10.5}$$

where R is the radial distance between source and observer. Beyond the immediate vicinity of the turbine base sound radiated downwards is assumed to reflect back off the ground so that the acoustic energy received at an observer is effectively doubled: the denominator in Equation (10.5) then becomes $2\pi R^2$, and the function is characterised as hemispherical spreading. In addition, some acoustic energy is lost due to atmospheric absorption, which causes an approximately linear drop in perceived noise with distance. The overall spreading function for broadband noise is then expressed in logarithmic form:

$$L_P = L_W - 10\log_{10}\left(2\pi R^2\right) - \alpha R \text{ dB(A)} \tag{10.6}$$

where

L_P: perceived noise or sound pressure level, dB(A)
L_W: sound power level of source according to Equation (10.2)
R: observer distance (m)
α: atmospheric absorption coefficient, dB(A) m^{-1}

Guaranteed values of sound power L_W are normally provided by the wind turbine manufacturer as functions of mean wind speed. The data are derived from IEC standardised measurements, usually made at a dedicated test site (IEC, 2012). The procedure for sound power measurement is essentially derived from the spreading function of Equation (10.6), with a number of detailed provisions regarding equipment calibration, measurement distance and position relative to the rotor, and correction for background noise. In addition certificated measurements must be taken at ground level on a hard reflecting board to eliminate variation due to ground conditions. Sound power values for a generic 2 MW–class wind turbine are shown in Figure 10.7. Note that for noise evaluation the reference wind speed is always referred to 10 m elevation, which is generally much lower than hub height (this was perhaps less the case when the standards were originally drawn up). Wind shear must then be considered, as the shear profile at a proposed site may differ from that obtaining during the reference noise measurements.

The atmospheric absorption coefficient α depends on frequency, with high-frequency sound absorbed more strongly than low. The applicable value in Equation (10.6) should therefore be a weighted average based on the acoustic noise spectrum of the wind turbine in question. Longer blades produce lower frequency sound, and consequently lower values of α must be assumed when evaluating noise due to large wind turbines. Indicative values of α are given in Table 10.1, based on octave-band analysis for a range of machine sizes. Assuming α is carefully chosen, the simple spreading function in Equation (10.6) is a good indicator of broadband noise levels, and predictions are generally conservative.

Table 10.1 Atmospheric Absorption Coefficients for Use in Broadband Noise Estimates

WEC rating	α(*dB(A)* m^{-1})
<500 kW	0.005
500 kW–1 MW	0.004
1–2 MW	0.003
>2 MW	0.002

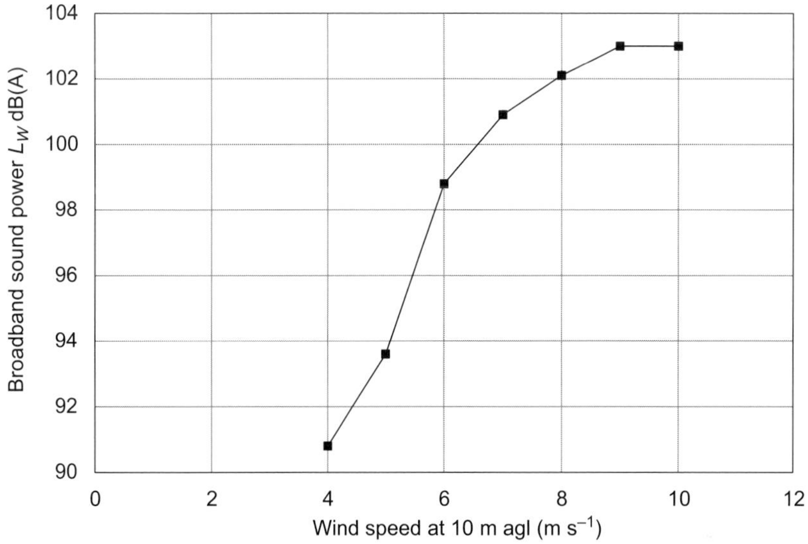

Figure 10.7 Certified sound power (L_W) for a 2 MW wind turbine as a function of mean wind speed. Above 10 m s^{-1}, the wind turbine operates at its rated power level, and L_W is assumed to remain constant.

Measured sound pressure levels for two early wind turbines are shown in Figure 10.8 based on octave-band measurements in (Dunbabin, 1994). The Vestas V17 was a commercial three-blade, stall-regulated machine rated at 75 kW; the WEG MS-1 was a two-bladed experimental wind turbine with 250 kW rating and variable blade pitch. The dashed lines correspond to hemispherical spreading according to Equation (10.6) with atmospheric absorption coefficient α of 0.005 dB(A) m^{-1}. The implied sound power levels are 89 dB(A) for the V17 and 98 dB(A) for the MS-1, with the difference largely explained by the much higher tip speed of the latter: 92 m s^{-1} compared to 41 m s^{-1} for the V17. In both cases measurements were made in the field, at hub height wind speeds of 8–10 m s^{-1}.

In some circumstances, e.g. where predicted noise levels may be close to allowable limits, the simple broadband approach described above is insufficient and more detailed analysis is required. Planning authorities may request noise calculations carried out to the ISO-9613 standard: this incorporates octave-band treatment of atmospheric absorption, and additional terms in the

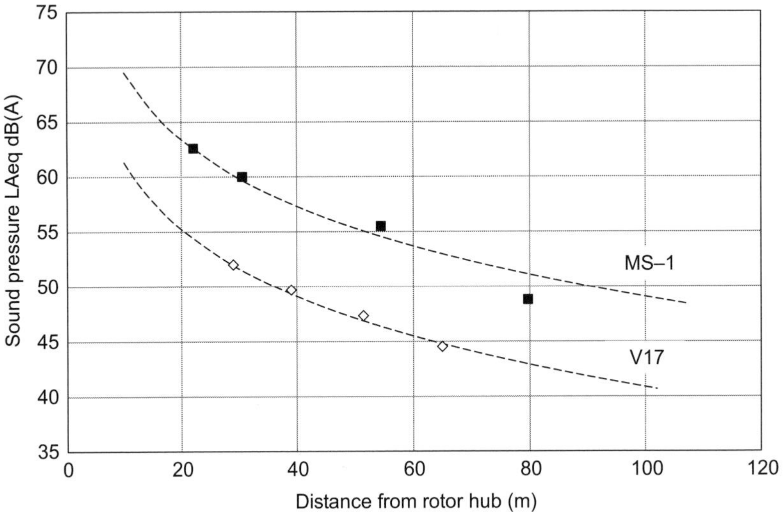

Figure 10.8 Broadband sound pressure (L_{Aeq}) measured downwind of two different wind turbines (Dunbabin, 1994); dashed lines correspond to hemispherical spherical spreading according to Equation (10.6).

spreading function to allow for attenuation of noise by the ground surface (rather than 100% reflection) and by terrain and/or vegetation barriers situated directly between the noise source and the observer. The influence of temperature and humidity are also included (ISO, 1993). In many cases these calculations result in lower noise estimates than the simple broadband procedure, which remains a useful and conservative approach.

When estimating the noise due to multiple wind turbines, individual values of L_P are first calculated using Equation (10.6); these must be converted to sound power (e.g. in picowatts) before being added to obtain cumulative sound power, which may then be converted back to decibels using Equation (10.2). Individual values of L_P in decibels must *not* be added directly. Windfarm design software makes such calculations relatively straightforward, even for large arrays, but for quick calculations two useful rules of thumb are as follows:

1. Doubling the number of wind turbines at a given distance adds 3 dB(A) to perceived noise, assuming an observer relatively far from the source, and wind turbines of equal sound power.
2. Perceived noise decreases by approximately 6 dB(A) per doubling of distance: this applies to a single wind turbine, or to multiple wind turbines at a significant distance from the observer where they can be treated as a point source.

10.4.3 Planning Limits

Permissible noise levels for wind turbines vary from one country to another, and different values generally apply by day and night, or depending whether an area is residential or industrial. The following guidelines for perceived noise are quoted from recommendations by the UK Department of Trade and Industry (ETSU, 1996) and are currently used by UK planning authorities:

- For single turbines or windfarms well separated from the nearest properties, a simplified condition applies with maximum allowable 35 dB(A) in wind speeds up to 10 m s^{-1}.
- In low-noise environments the daytime level of the windfarm noise should be limited to an absolute level within the range of 35–40 dB(A). This may be increased to 45 dB(A) in cases where the occupier of the affected property has a financial interest in the windfarm.
- Noise from a windfarm should be limited to 5 dB(A) above background for both daytime and night-time, with windfarm noise and background both measured as LA90, 10 min[5] (note that 'daytime' levels here refer to quiet periods based on evenings and weekends).

Based on these rules, if the predicted noise from a development is below 35 dB(A) under all conditions the project may be deemed acceptable. If the noise exceeds 35 dB(A) then background noise levels can be taken into account: this is because ambient noise, particularly due to vegetation, tends to increase with wind speed so that rotor noise is masked in high winds. Background noise measurements are then taken in the vicinity of a potentially affected property for an agreed period (which may be several weeks), with simultaneous measurements of wind speed, to cover all conditions. If the noise due to the wind turbine(s) is then predicted to be less than 5 dB(A) above average background at all times the development will be permitted.

The results of such an exercise are shown in Figure 10.9. The solid curve shows the predicted wind turbine noise at a property as a function of mean wind speed, based on the manufacturer's guaranteed sound power values, with spreading according to Equation (10.6). Also plotted are measured background noise data, with a best-fit polynomial indicating the average trend, and an offset curve denoting a level of 5 dB(A) above average. In most wind speeds the predicted wind turbine noise exceeds 35 dB(A), but it remains at all times within the permissible margin above background, and so would be acceptable. The plot also illustrates how the most noise-sensitive conditions are often in moderate winds, when the turbine is operating below rated power, but background noise is very low and the turbine is potentially more audible.

The above description of the UK noise recommendations is only an outline summary, and a detailed practical guide to the application of the ETSU noise assessment procedures has been published by the Institute of Acoustics (IOA, 2013). The best guarantee of preventing noise nuisance remains, however, to place enough distance between a wind turbine development and sensitive properties – this follows from the simple spreading function given in Equation (10.6). The choice of wind turbine also matters: modern variable-speed machines are inherently quieter than older fixed-speed types as they operate at lower rotor speed in light winds, when the masking effect of background noise is less. Variable-speed machines may also be operated on reduced-speed curves in high winds to further limit noise in sensitive conditions, with a small power output

[5] The level exceeded during 90% of a 10 min record (this parameter reduces the influence of intermittent noise).

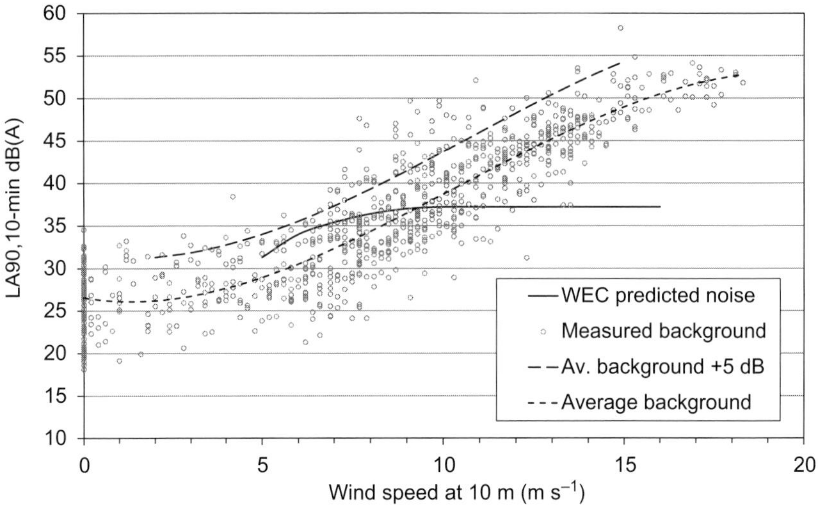

Figure 10.9 Comparison of predicted wind turbine noise with ambient background noise levels.

penalty. Sector management (see Section 6.5) enables speed control to be invoked for selected wind directions, to minimise the noise impact at specific locations.

10.5 Aviation

Wind turbines may conflict with civil or military aircraft activity due to (a) the risk of direct collision or (b) interference to critical radar systems. In both regards strict planning guidelines apply, although the detailed stipulations and degree of conservatism vary from country to country. The following observations are based on experience in the UK, where the Civil Aviation Authority has responsibility for all aspects of civil aircraft safeguarding, and has published detailed guidelines for wind energy development in CAP764 (CAA, 2011). The Ministry of Defence has equivalent responsibility for military aircraft safeguarding, but as the technical risks and remediation are essentially the same for all aircraft CAP764 is a useful starting point; it includes references to a range of subsidiary documents on detailed aspects of policy; most are freely available.

10.5.1 Collision Risk

Safeguarding of UK licensed aerodromes is covered by CAP168,[6] which contains guidelines for the maximum allowable height of structures near airfields (CAA, 2007). The rules define three-dimensional zones within which height limits are set by proximity to the airfield centre, or to runway take-off and approach paths. The inner horizontal surface (IHS) is a circle out to 4 km from the airfield centre, beneath which a height restriction of 45 m above runway elevation applies. From

[6] See in particular Chapter 4, 'The Assessment and Treatment of Obstacles'.

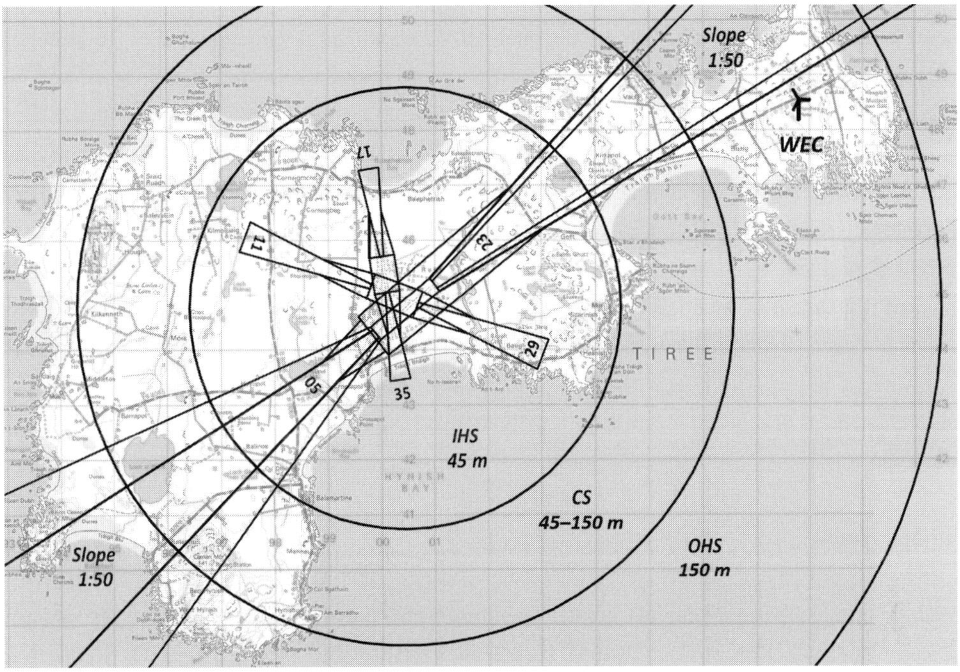

Figure 10.10 Obstacle height restrictions for Tiree Airport based on CAP168. A community wind turbine with 77 m tip height was successfully installed at the location shown in the upper right. (Obstacle map reproduced by kind permission of Highlands and Islands Airports Ltd; © Crown copyright and database rights 2019, OS licence number 100037385)

6 to 10 km radius an outer horizontal surface (OHS) applies with height limit of 150 m. Between the IHS and OHS the height is defined by a conical slope. Take-off and landing paths are protected by sloping surfaces of gradient between 1:20 and 1:50 depending on the runway type; if these slopes lie below the IHS or OHS the lower height restriction takes precedence. Restrictions become more severe at major airfields as runways with instrument landing systems (ILS) are subject to lower obstacle height limits than those for visual approach, and the take-off and landing gradients for larger aircraft are flatter.

The application of the CAP168 guidelines can be illustrated with an example. When the development of a community wind turbine was proposed for the island of Tiree in the Scottish Inner Hebrides, the local airport became an important factor in site selection. Tiree has a relatively large airfield, and Figure 10.10 shows the obstacle clearance map: the 1400 m long main runway is ILS-equipped, and clearance zones are relatively conservative as a result, restricting the available search areas for a wind turbine on the island. In addition the obstacle height limits are referred to runway elevation, which on Tiree is close to sea level, so elevated terrain further reduces the available headroom. The proposed wind turbine had 77 m tip height so the entire IHS and much of the conical surface were ruled out. The most promising locations were found within the OHS and away from the main runway flightpaths, but numerous other planning restrictions applied, including bird conservation areas, population centres, and (not least) a civil aviation radar located on a hill at the southwest of the island. The airport operator HIAL was helpful throughout the site selection process, which ultimately led to selection of the turbine position shown in Figure 10.10 and the wind turbine was successfully commissioned in 2010.

Note that the CAP168 guidelines apply only to licensed aerodromes. The CAA also provides guidance for unlicensed aerodromes, which are generally smaller and with lower traffic density, and the rules are accordingly less onerous: CAP793 states for instance that ‘anything that, because of its height or position, could be a hazard to an aircraft landing or taking off should be conspicuously marked if it cannot be practicably removed or minimised’. Wind turbines are often equipped with aviation warning lights, and in some cases (though not the UK) blades or towers may be painted with red warning stripes. The safeguarding rules for military aerodromes are similar to those described above, and areas designated for military low flying may also be deemed unsuitable for wind turbine developments, though in some cases turbines of modest tip height may be permitted.

10.5.2 Radar

Radar interference is a potentially greater impediment to wind energy than collision risk, as large areas of land and sea are covered by radar surveillance and are often excluded from development on a precautionary basis. At one time around 30% of otherwise consented UK wind energy projects were being held back by objections from either civil or military radar operators, but as a result of various mitigation strategies this situation has now improved. There are two types of radar in common use, namely primary and secondary surveillance radar: they are affected in different ways by wind turbines, and require different remedial solutions.

Primary surveillance radar (PSR) operates on the principle of echo-location, as illustrated in Figure 10.11. A rotating scanner sends out a pulsed beam that reflects back off an aircraft, whose position is determined by the rotation (azimuth) angle of the antenna and the time taken for the pulse to return. The PSR beam is vertically deep, however, and cannot easily discriminate an object’s height above ground: consequently, tall structures like wind turbines may be mistaken for aircraft flying high above them. Reflections from stationary objects can be removed by signal processing, using the frequency (Doppler) shift on the return pulse to indicate velocity. This technique cannot, however, be applied to eliminate wind turbines, whose rotor blades have similar velocity to low-speed aircraft or helicopters so cannot be safely filtered out. As a result wind turbines may appear on radar screens as intermittent reflections or ‘clutter’, which can cause genuine aircraft tracks (built up by successive antenna sweeps) to become distorted or lost altogether.

Several solutions to PSR interference exist, though none is yet comprehensive. The simplest technique is simply to ignore radar returns from certain locations,[7] similar to blanking out a single pixel on a TV screen. The smallest cell size is typically larger than a wind turbine and

[7] This technique is known as range-azimuth gating (RAG).

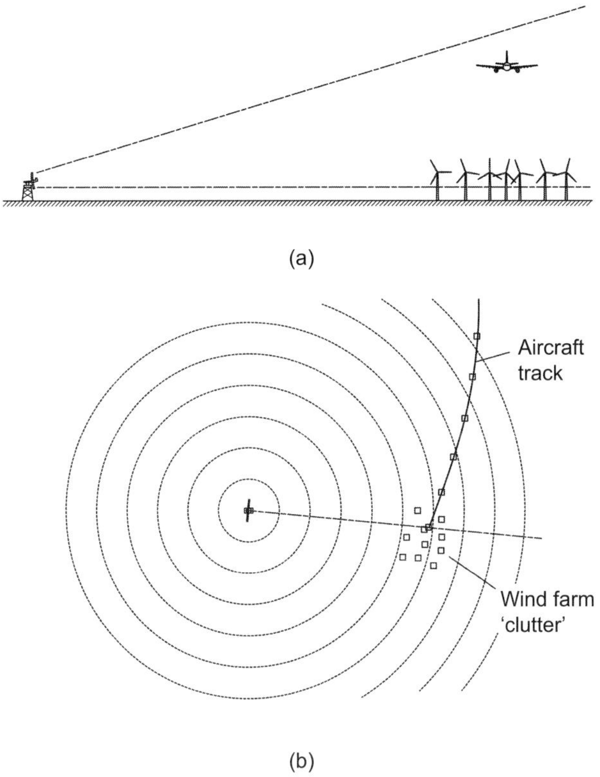

Figure 10.11 Interference to primary surveillance radar: (a) due to the depth of the scanning beam, it cannot discriminate height, and windfarm reflections may be mistaken for aircraft; (b) the track of an aircraft can be lost or 'seduced' due to windfarm clutter. Dimensions are not to scale.

the radar operator may agree (often for a fee) to blank the affected location. There is a penalty in terms of loss of radar coverage, so this technique cannot be applied indefinitely and is best suited to removing single turbines: in areas of high air traffic density only a few wind turbines may potentially be filtered out by blanking. More sophisticated improvements to signal processing have been explored, though the fundamental limitation of primary radar – inability to discriminate height – so far seems to have delayed any great progress. More promising has been the introduction of 3D or 'holographic' radar (Dodd, 2014). Once the preserve of the military this employs an electronically scanned antenna that enables distance, bearing, and (critically) height to be simultaneously determined. Wind turbine reflections can then be more effectively screened out.

The most comprehensive protection of PSR is via terrain shielding, where wind turbines are simply located out of the radar line of sight. This may be a case of selecting windfarm sites judiciously, using digital terrain maps to find screened locations. Often, however, this may not be convenient. The best wind sites are frequently on high ground where they are inherently visible to long-range radar; also, some projects may be limited to land owned or controlled by the developer, who has no alternative choice of site. In such cases a solution can be re-locating or providing alternative radar sets to achieve terrain shielding. A good example is Whitelee windfarm, near Glasgow Airport, whose developer Scottish Power paid for a new radar set to be built out of sight of the wind turbines but with visibility of the key airspace. Signals from the new 'infill' radar are integrated with those from existing sets to provide a complete and uncluttered picture. The cost of the new radar was estimated as £5M but as Whitelee was, at 322 MW capacity, one of the largest onshore windfarms in the world at the time this price could easily be justified (White, 2009).

A more radical solution is to incorporate radar-absorbing materials (RAM) in the structure of wind turbine blades. This again has military origins, in the development of 'stealth' technology for low-observable aircraft. Research by UK consultancy QinetiQ led to successful trials of blades with low radar cross section on a commercial 3 MW wind turbine, and subsequent commissioning of an 88 MW windfarm at Perpignan in France (Tovey, 2017). One strand of this technology exploits the principle of the Salisbury screen, illustrated in Figure 10.12. Two electrically conductive surfaces are separated by a distance equal to quarter of the wavelength of interest; the outer surface is a partial reflector, while the inner (the back plane) is fully reflective; radio waves reflected from the two surfaces are then one half-wavelength out of phase and destructive interference occurs.[8] The structure of a large HAWT blade is conveniently suited to this technique as the dimensions of a GFRP composite shell are comparable to the

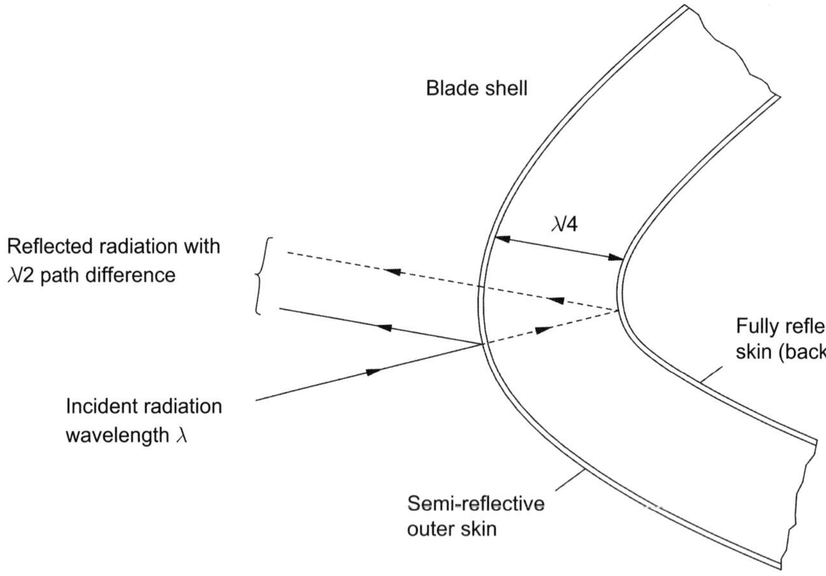

Figure 10.12 Principle of the Salisbury screen for 'stealth' proofing wind turbine blades.

[8] This explanation is somewhat simplified. In practice, RAM incorporates 'lossy' layers to dissipate energy as heat. Also the Salisbury screen is single-frequency: the multi-layer Jaumann absorber absorbs over a wider bandwidth.

quarter-wavelength of a typical aviation radar: for instance a 5 GHz radar signal has wavelength of 60 mm, and a blade skin sandwich panel with 15 mm spacing between conductive sheets is practicable. Electrical conductivity is achieved by 'doping' the glass fibre cloth with graphite prior to manufacture. For a more complete description of this topic, see Appleton (2005).

The other type of aviation radar in widespread use is secondary surveillance radar (SSR). In this semi-passive technique the rotating antenna sends out a continuous interrogating beam: when it impinges on an aircraft it triggers a signal from an onboard transponder broadcasting the aircraft identity and height. The bearing is determined (as with PSR) by the azimuthal angle of the transmitting antenna. Although unusual, interference to SSRs can be caused by wind turbines if they reflect the interrogating beam (uplink) so as to trigger responses from aircraft in unexpected positions. The situation is illustrated in Figure 10.13. False plots on the SSR at Copenhagen (Kastrup) Airport in Denmark were attributed to the Middelgrunden offshore windfarm, situated 5 km away: this large array (20 × 2 MW turbines) was found to be reflecting the SSR uplink signals and causing the same aircraft to appear simultaneously in two different places on the radar screen. Once aware of the situation, however, the Danish controllers were able to apply simple filtering in mitigation (Jago et al., 2002).

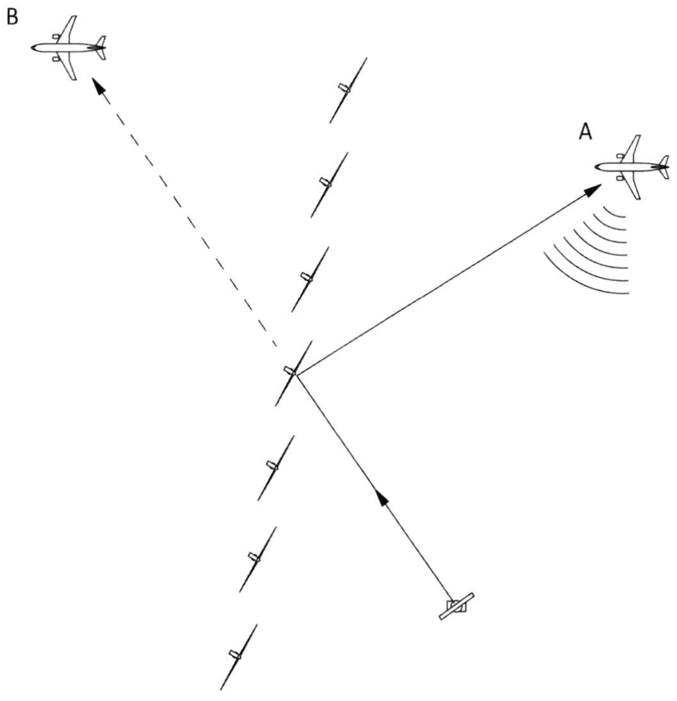

Figure 10.13 Interference to secondary surveillance radar. The aircraft at position A receives a reflected interrogating beam and transmits an identifying response; the radar receiver interprets the position as B (dimensions not to scale).

In the UK a conservative safeguarding policy was initially applied by national radar operator NATS, with a wind turbine exclusion zone of 10 km placed round any SSR. This caused delays to a number of planned wind energy projects, but the rules were subsequently relaxed following helicopter trials in the vicinity of the Arnish Moor windfarm, which is visible to Stornoway Airport SSR. The trials helped to establish the likely risk of interference and a number of wind projects in the area were subsequently permitted; interference to SSR is now considered on a case-by-case basis. Current UK policy is summarised in CAP764.

10.6 Radio and Microwave Communications

10.6.1 Microwaves

Wind turbines can potentially interfere with radio-frequency (RF) or microwave transmissions if sited too close to the beam path, transmitter, or receiver. The principal interference mechanisms are scattering and reflection. Scattering is the more significant concern and is caused when a wind turbine lies directly in the path of a point-to-point link, resulting in loss of onward signal strength. At the planning stage link operators may apply conservative beam avoidance rules, requesting clearance of 100–500 m around a beam centreline, and objecting to any wind turbine that intrudes within this zone. Such restrictions can exclude large areas from prospective wind turbine developments and are quite onerous, given that a GHz-band beam may be just a few metres wide. In such cases a detailed path analysis can be carried out to enable a closer separation.

The recommended criterion to avoid scattering is that all parts of the wind turbine should lie outside the second Fresnel radius (R_{f2}) of the microwave or RF link (Bacon, 2002). The geometry is illustrated in Figure 10.14: the beam is ellipsoidal in shape, with a circular cross section that expands to a maximum midway between transmitter and receiver. At an arbitrary distance from the transmitter the radius is found from

$$R_{f2} = \sqrt{\frac{600 d_1 d_2}{f(d_1 + d_2)}}\ (\mathrm{m}) \tag{10.7}$$

where d_1 and d_2 are the distances in kilometres from the two ends of the link and f is its frequency in GHz (care should be taken with the units). If the positions of transmitter and receiver are located with sufficient accuracy – fine scale mapping or direct GPS measurement can be used – then Equation (10.7) gives a good guarantee of beam avoidance. To illustrate, Figure 10.15 shows a site in Aberdeenshire where four medium-scale wind turbines were installed on a hilltop close to a group of microwave transmission masts, each hosting multiple links with a range of frequencies. The turbine positions were selected on the basis of the above Fresnel zone avoidance rule and laid down by accurate ground survey, taking sightings from the microwave transmission and receiving masts. The windfarm caused no interference in practice.

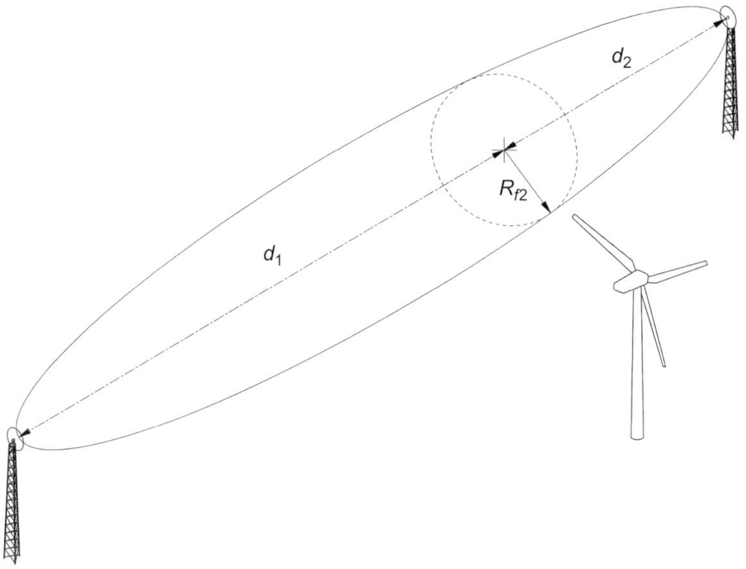

Figure 10.14 Microwave link avoidance. To prevent scattering interference, all parts of the wind turbine should lie outside the second Fresnel radius: Equation (10.7) applies. A GHz frequency beam may be only a few metres wide.

Less commonly, interference may be caused by reflection: in this case the wind turbine creates a secondary beam path, so that reflected and direct signals arrive at the receiver with a phase difference. The same phenomenon was responsible for the 'ghosting' often seen on analogue TV screens, and has in some instances caused interference to airport radar systems (see Section 10.5.2). To avoid reflection problems a minimum distance is prescribed between the wind turbine and either the transmitter or receiver, and the use of antennae with high directional discrimination is recommended. Reflection analysis is covered in detail in Bacon (2002).

10.6.2 Television

Terrestrial television signals at UHF frequencies may be vulnerable to reflection or scattering if wind turbines are located too close to the transmission path. The familiar 'ghosting' or double image seen on analogue TV pictures was the result of reflected signals reaching the receiver out of phase with the direct signal, having taken a longer path. Interference is most likely where the direct signal strength is low, but the reflected signal is relatively strong: these conditions typically arise where the reflecting structure is on a hilltop in view of both the TV transmitter and receiver, but the hill blocks the main signal path. The situation is shown in Figure 10.16. In such cases the ratio of carrier to interference signal strength (R_{ci}) may fall below a critical level; during the planning phase for a wind energy project a signal strength survey may then be required (see Figure 10.17). Remediation of TV interference can involve installing a local booster receiver, or providing an

Figure 10.15 Co-existing with microwave links. Accurate surveying of link paths (dashed lines) and turbine positions enabled this windfarm in Aberdeenshire to be constructed without causing interference, where conservative link avoidance criteria would have ruled it out. Equation (10.7) applies. (© Crown copyright and database rights 2019, OS licence number 100037385)

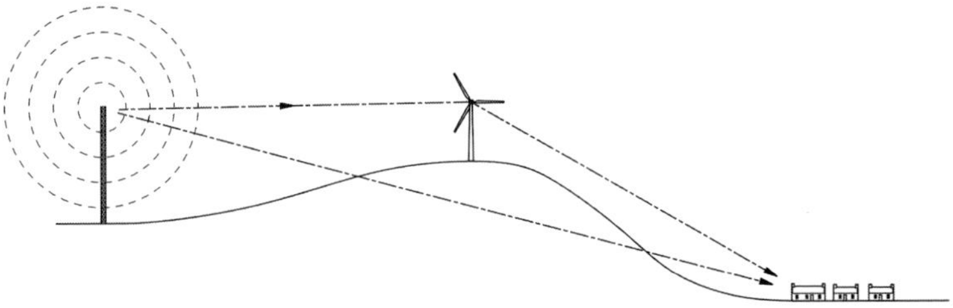

Figure 10.16 Conditions for interference to terrestrial TV. Due to terrain blockage, the reflected signal is relatively strong compared with the direct transmission. A double image can result.

alternative reception method, e.g. cable, satellite, or internet. With the spread of these digital platforms, however, TV interference due to wind turbines has become somewhat less of an issue.

10.7 Exercises

10.7.1 Ice Throw

What minimum set-back distance might be recommended to protect against ice throw for a wind turbine of tower height 80 m and rotor diameter 90 m?

Figure 10.17 Signal strength survey in area with poor television reception.

10.7.2 Shadow Flicker

In some countries the regulations regarding shadow flicker prevention are less conservative than the simple rules outlined in Section 10.3.3, and take account of additional environmental factors. Explain why the following factors may influence the severity of shadow flicker: (a) wind direction and (b) if the sun angle is very low, e.g. less than 3° above the horizon.

10.7.3 Sound Power

A wind turbine has a rotor diameter of 43 m, nominal rotational speed 31.0 rpm, and rated output 600 kW. Compare sound power (L_W) estimates for this machine using the empirical relationships in Chapter 10, Equations (10.3) and (10.4). Calculate the rotor speed that would be needed to effect a 2 dB(A) reduction in sound power output.

10.7.4 Perceived Noise

Two identical wind turbines are sited at a distance of 500 m from a house. The sound power of each wind turbine is 102 dB(A) and the applicable atmospheric absorption coefficient α is 4.0 dB(A) km^{-1}. Calculate (a) the perceived noise level at the house assuming only one wind turbine is operating and (b) the cumulative noise level with both turbines operating; (c) if the permissible noise level at the house is 38 dB(A), to what sound power level should the two turbines be restricted in order to comply?

10.7.5 Noise Measurement

As part of the environmental impact assessment for a planned windfarm, background noise measurements are required in the vicinity of a village in the countryside, over a range of representative conditions. Why might the measured background noise levels be significantly higher (a) by day rather than night and (b) in summer compared with winter?

10.7.6 Aircraft Collision Risk

A wind turbine of 99 m tip height is to be sited under the approach flightpath for a major airport. The airport runway is at 110 m elevation above sea level, and the flightpath has a 1:50 glide slope. If the wind turbine base is at 135 m above sea level, what is the closest horizontal distance to the airport that the wind turbine can be located without breaching the flightpath?

10.7.7 Radar Interference

Primary surveillance radars incorporate a moving target indicator (MTI) that exploits the Doppler frequency shift to filter out stationary objects. MTI is not entirely reliable for discriminating wind turbines, however, and its performance varies strongly with the ambient wind direction at the turbines' location. Suggest why.

10.7.8 Microwave Avoidance

A microwave transmitter and receiver are located 14 km apart, and the transmission frequency is 5 GHz. If a wind turbine with 70 m rotor diameter is sited midway between the two ends of the microwave link, how far to the side of the beam path must the turbine base be located to prevent interference?

CHAPTER 11 ECONOMIC AND POLITICAL CONSIDERATIONS

11.1 INTRODUCTION

The lifetime cost of a wind energy project is in principle straightforward to predict. Most of the capital cost is incurred up-front, there are no fuel costs, and O&M and labour costs vary more or less with normal inflation. Similarly, the lifetime energy yield of a windfarm can be accurately forecast using well-established techniques (see Section 9.2). As a result the long-term cost of wind energy is more predictable than, say, the price of oil. The question of how much a consumer should pay for wind energy is, however, less straightforward, as a range of economic and political factors come into play. Wind power is intermittent, so external balancing costs must be met; renewables capacity has historically been mandated by governments rather than direct consumer choice so wind programmes inhabit a grey area between free and state-controlled markets; and the technology lends itself to a range of ownership models with different outcomes at local and national scale. Some of these topics are reviewed in the present chapter and illustrated with examples.

11.2 THE COST OF WIND ENERGY

A standard economic measure is the levelised cost of energy, or LCoE, which is defined as the ratio of the cost to construct and operate a project to the amount of electricity generated over its lifetime. As project lifetimes are measured in decades the calculation must include a discount rate to allow for the change in the value of money over the term, and the following formula applies:

$$\text{LCoE} = \frac{\sum_{i=1}^{n} \frac{C_i + O_i}{(1+r)^i}}{\sum_{i=1}^{n} \frac{E_i}{(1+r)^i}} \tag{11.1}$$

where

C_i = capital expenditure in year i
O_i = operational and maintenance costs in year i
E_i = electricity generated in year i

r = discount rate

n = project lifetime in years

An allowance for intermittency must ultimately be included in the LCoE (the cost of balancing supplies is higher for wind generators than for conventional generating plant, and increases with the level of penetration into national supplies) and this topic is discussed in Section 11.2.3.

Equation (11.1) is conveniently solved by a spreadsheet calculation: capital expenditure and annual costs can be entered in arbitrary currency, and annual generation in MWh or GWh depending on project scale; the units of LCoE will then be self-consistent (e.g. £/MWh, €/GWh). Capital costs would normally include all planning and construction costs, including the grid connection and any civil infrastructure required to facilitate the project. In some countries the additional costs of grid reinforcement for wind projects do not fall to the developer but are met externally, and may then be excluded from the project LCoE calculation. The grid costs still have to be met by someone, however, so would normally be included if the calculation was e.g. a high-level comparison of different generation technologies.

Most of the capital expenditure for a wind power project is incurred before commissioning, and a simplified LCoE calculation can be made by assuming (a) all capex occurs in the first year and (b) energy production and operational costs are invariant from one year to the next. Making these assumptions,

$$\text{LCoE} = \frac{C}{S_n E} + O \tag{11.2}$$

where E is again annual energy production, but annual costs O are now expressed per unit output (£/MWh or similar). The factor S_n accounts for discounting over the project lifetime of n years, with

$$S_n = 1 + \frac{1}{r}\left[1 - (1+r)^{1-n}\right] \tag{11.3}$$

Equation (11.2) allows the LCoE calculation to be performed without a spreadsheet. Finally if a simple 'engineer's estimate' is required then discounting may be ignored, in which case the formula becomes

$$\text{LCoE} = \frac{C}{nE} + O \tag{11.4}$$

where C is capex, n is the lifetime in years, E is annual energy production, and operational costs O are again expressed per unit output. Estimates of levelised energy cost for a range of project cost and capacity factor are given in Section 11.2.4.

11.2.1 Installation Costs

The installed costs of a range of commercial wind energy developments in the UK are shown in Figure 11.1, illustrating trends for (a) onshore and (b) offshore projects since 1990. The data are taken from a number of published sources with all costs expressed in £M per MW corrected for inflation to 2017 values; adjustments for currency exchange rates were applied where necessary. The approximate rating

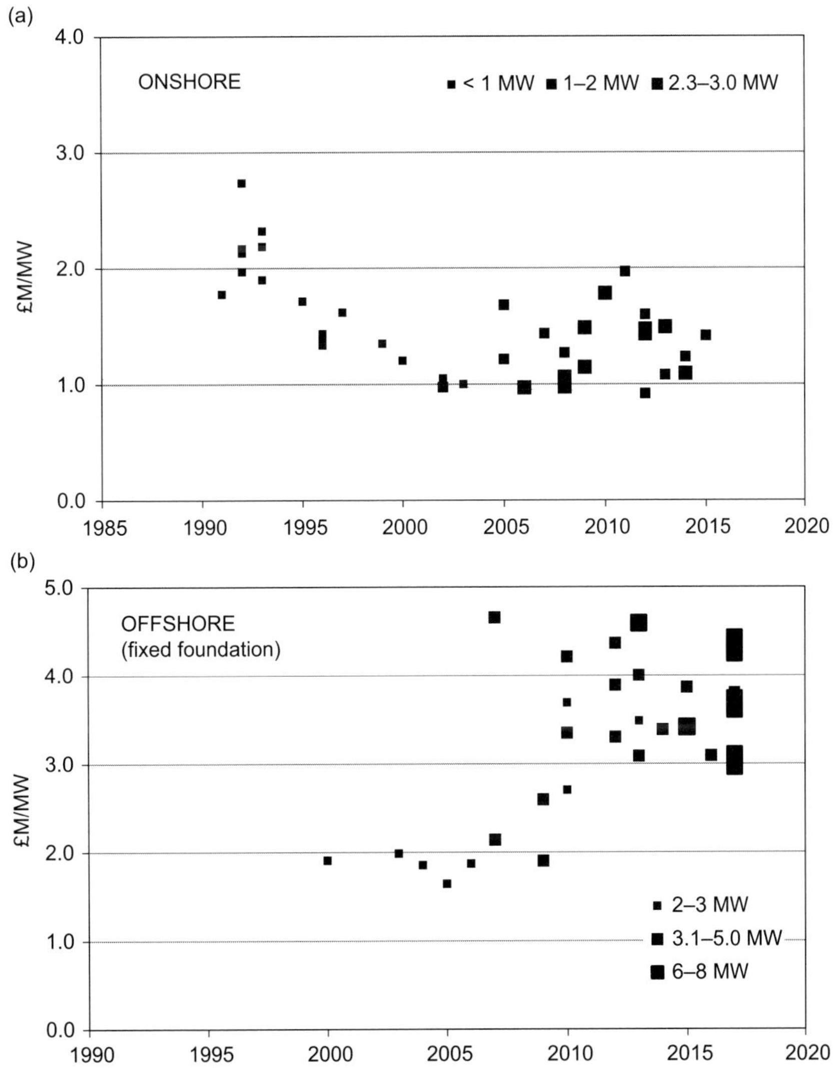

Figure 11.1 Installed costs of UK wind power. Historic trends for commercial projects (a) onshore and (b) offshore (excluding floating WECs) based on data published at the time of commissioning. All costs are in £M/MW adjusted to 2017 prices, with WEC rating indicated by marker size.

of the wind turbines is indicated by marker size and based on the data in Figure 11.1 the following observations are made:

- The cost of onshore projects roughly halved over the 20 year period following the first UK wind installations, from over £2M/MW in 1990 to around £1M/MW in 2010 (in 2017 prices). Thereafter the cost appears to have levelled out and may be rising again, although there is considerable scatter in the more recent data. Turbine sizes have progressively increased, with the largest onshore turbines currently in the 3 MW range.
- Offshore projects began with costs around £2M per MW in 2000, but after a decade these began to rise to current levels of £3.0–4.5M per MW. The size of offshore turbines is significantly greater now than onshore with the largest units in the 6–9 MW range; according to these figures offshore wind currently costs 2–3 times more than onshore to install.

In both cases it appears that installed costs are no longer falling, though whether this trend is temporary or not is difficult to predict. The initial decrease seen in the cost of onshore projects might be ascribed to the rapid expansion in the market for wind turbines at that time, with consequent economies of scale and efficiency in manufacturing; these factors may now be maturing. The recent increase in UK offshore costs may reflect the progression to deeper water and installation farther offshore with successive rounds of policy. In addition turbine sizes have risen significantly both onshore and offshore over the period shown, so it is possible that the 'square-cube law' that mitigates against scaling (see Section 1.5) is finally making itself felt – though it should be noted that some of the lowest installed costs are for projects using the largest wind turbines.

11.2.2 Operational Costs

Annual operational costs include maintenance (scheduled and unscheduled), warranty and insurance premiums, management costs, and miscellaneous business rates and taxes. For UK onshore wind projects total annual costs are currently of the order £15/MWh (2017 prices), about two-thirds of which may be accounted for by the manufacturer's 'availability warranty'. This is a catch-all agreement that guarantees a minimum level of availability (not to be confused with capacity factor) of typically 95%–97% with the manufacturer responsible to rectify any fault in order to achieve this figure; in the event of under-performance compensation is paid at an agreed rate. Availability warranties involve a measure of risk-sharing by the equipment manufacturer and have largely superseded traditional maintenance agreements; they reduce the need for equipment breakdown insurance and warranty fees are charged on the basis of generated output, subject to a minimum premium.[1] Older equipment that is out of warranty may still be covered by third-party insurance or a reserve fund to allow for unscheduled breakdown or repairs; in such cases the net costs can be similar to those under an availability warranty. A 'ballpark' figure for the annual costs of an onshore wind energy

[1] An incentive to reward the manufacturer for availability above the guarantee threshold may be included.

project is 3% of the installed capital cost; the figure for offshore costs is broadly comparable to this, bearing in mind the higher capex involved.

11.2.3 Intermittency

As noted earlier wind energy is intermittent, and other plant must be available to supply the network demand when wind conditions do not allow. In the UK balancing is primarily achieved using gas-fired power stations due to their fast response and relatively low emissions; some large hydro schemes are also used. The requirement for backup generation is not, however, confined to renewable generators: due to the inherent variability of national demand there must always be a percentage of spare capacity or 'spinning reserve' available for rapid despatch. The difference between UK daytime and night-time electricity demand is almost a factor of 2, so 'baseload' generation represents less than half of maximum demand. In addition no large power station is 100% reliable, and significant backup capacity is necessary to cover for generation lost through unscheduled faults or trips.[2]

Nevertheless intermittent generation does lead to increased balancing costs compared with conventional non-renewable plant whose output is more controllable. According to a 2009 EWEA report, at wind penetration of up to 20% of gross demand balancing costs increase the wholesale cost of wind energy by 5%–10% (Krohn, 2009). A similar study in the UK examined the implications of adding 29.5 GW of wind capacity to the grid, concluding that Short-Term Operating Reserves (STOR) would have to increase by 6.5 GW to cover intermittency, with unit cost of around £5.4/MWh (National Grid, 2009); these figures fell to 4 GW of additional STOR and £3/MWh assuming improvements in wind forecasting, bringing them roughly into line with the EWEA analysis. Based on the above the increase in the final consumer price of electricity due to intermittency is about 5% assuming 20% renewables penetration.

11.2.4 Total Generation Costs

Figure 11.2 shows levelised costs (LCoE) for wind energy as a function of capacity factor and installed cost, based on Equation (11.3). In all cases a 20 year project lifetime is assumed, with annual costs estimated at £20/MWh (based on 2017 assumptions) including a £5/MWh allowance for balancing. Discount rates of 5% and 7.5% are assumed, and the following outcomes are noted:

- Assuming a discount rate of 7.5% an onshore project with installed cost of £1 M/MW and 30% capacity factor achieves LCoE of £55/MWh. This is in reasonable agreement with published estimates (e.g. Lemming, 2008) and is similar to the marginal generation price (i.e. the unsubsidised cost of electricity) currently paid in the UK.

[2] From the *Edinburgh Evening News*, 5 February 2010: 'Nuclear reactor shut down at Torness power station: The reactor supplies more than half a million households but the National Grid said there is spare capacity to maintain the supply when such events occur'. There were no power cuts, which implies several hundred MW of spinning reserve was available to cover this event.

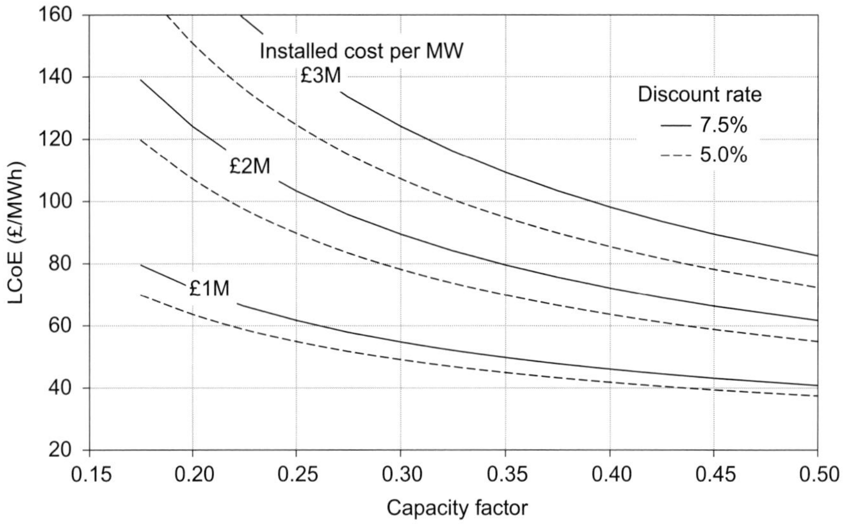

Figure 11.2 Levelised cost of wind energy (LCoE) as a function of capacity factor and installed cost, for discount rates of 7.5% and 5%. *Assumptions*: annual costs £20/MWh including balancing, project lifetime 20 years.

- For an offshore project costing £3 M/MW and achieving 40% capacity factor the LCoE is £98/MWh assuming 7.5% discount rate, and £85/MWh at 5% discount rate. This sector is seeing rapid economic change at the time of writing, and the above figures are already higher than the strike price paid for the latest European offshore projects (see below).

The appropriate discount rate to use is a matter for debate. Prior to 1989 a figure of 5% was applied in the UK's nationalised electricity sector, with 11% anticipated post-privatisation (Dimson, 1989). Discount rates for private sector energy projects are often quoted in the range 8%–12% but these figures generally pre-date the 2008 financial crisis. Since then the cost of borrowing has been historically low (making projects cheaper to finance) while return on general investments has been lower (meaning investors have had to lower their expectations): these trends are illustrated in Figure 11.3. The discount rates for wind assumed above (5.0%–7.5%) may therefore be more representative of projects currently under development, and this appears to be reflected in recent strike prices.

11.3 The Price of Wind Energy

When wind power began to take off seriously in the 1980s the cost of the technology was significantly higher than traditional power generation, but there was a strong political drive for its introduction (see Chapter 1). As a result a feature of the early wind energy markets was government intervention, and this largely remains the case today: as yet there are few examples of wind power being sold directly to consumers without some form of legislated support. Usually this takes the form of a guaranteed tariff for a power purchase contract sufficiently long to ensure that developers see a return on their

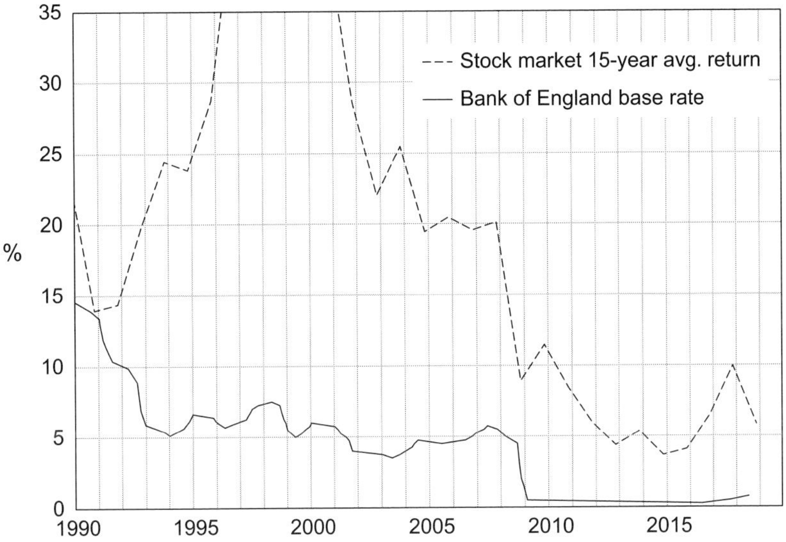

Figure 11.3 The value of money. Bank base rates have been historically low since the financial crash of 2008, making wind energy projects cheaper to finance. The dashed line shows the annual average stock market return over a 15 year period. (*Source*: Bank of England, Dow Jones)

investment. Alternatively the purchase or construction of wind power plant may be directly subsidised: although this approach is less favoured, it still plays a part in some important markets.

The evolution of the Danish wind industry is instructive. In the late 1970s prospective owners were offered 30% grants for purchase of wind turbines, with the capital subsidy directed to the purchaser rather than the manufacturer in order to stimulate industrial competition. In 1984 an energy tax of 15.5 øre/kWh was introduced, again payable to the wind turbine owner and representing an early example of a 'feed-in tariff'; at the same time the purchase subsidy was progressively withdrawn (Madsen, 1985). The principle of subsidising electricity production rather than the manufacture or purchase of equipment was recognised as key to long-term growth of the technology, and became the model widely adopted elsewhere in Europe. The German Feed-In Law of 1991 subsequently stimulated one of the largest European wind energy markets, with a guaranteed tariff set at 90% of the average consumer price of electricity (using UK figures as a very rough guide the consumer price is 2–3 times the raw generation cost). In addition the German policy was formulated to ring-fence the cost of wind energy within the electricity market, rather than cross-subsidise it from general taxation.

In the US in the early boom years of the Californian 'wind rush' development was initially driven by investment tax credits, effectively a subsidy based on installed capacity rather than output. This led to rapid growth but also included some rather dubious installations, and the policy was fairly quickly superseded by one based on production tax credits (PTCs), a form of guaranteed energy tariff. US policy has since then proceeded via a series of PTC rounds, each guaranteeing a premium payment for the first 10 years of operation of a wind project, and including a mix of federal and state-level incentives. The US policy is somewhat complex – the PTC also enables elements of investment tax credit and other incentives, and its implementation varies from state to state – and has been criticised for its short-term uncertainty; the US nevertheless remains one of the countries with the highest installed capacity of wind power. The current value of the production tax credit is around $23 per MWh, but including other incentives the net subsidy for US wind is higher; in 2012 the Wall Street Journal estimated an applicable figure of just over $52 per MWh (Gramm, 2012).[3]

In the UK the principal support mechanism for onshore wind energy is the Renewables Obligation (UK energy policy is discussed in Section 11.6) and in 2018 a typical project would sell energy at a net price of around £100/MWh, of which roughly half was the value of the subsidy (aka ROC certificate) and half the marginal export price. Offshore wind projects installed under the same scheme receive two ROC certificates with the overall tariff then worth around £150/MWh. Under the more recent CFD legislation (see Section 11.6) offshore projects must win contracts by competitive bidding, and strike prices (the guaranteed price including subsidy) as low as £67/MWh[4] have been achieved (Weston et al., 2018). Some even more dramatic reductions have been signalled elsewhere in Europe, with strike prices of €54.5/MWh for the 700 MW Dutch Borssele 3&4 project, and €49.9/MWh for the Danish 600 MW Krieger's Flak project (Appleyard, 2017). Some care is needed in making comparisons, however, as UK offshore projects must pay the full cost of their grid connections, whereas elsewhere in Europe transmission system costs are borne by the public network operator. This may mean that the European figures are more representative of the immediate economics of generation, but the UK data more reflective of the total costs; either way the full costs eventually come back to the consumer.

The above figures can be compared with the levelised cost estimates shown in Figure 11.2. What is notable is that the latest offshore generation costs are now apparently as low as onshore (or in some cases lower) and not much above the marginal cost of electricity, despite the significantly higher capital costs offshore. Whether this is due to lower expectations from investors, high anticipated capacity factors, or the availability of very low interest finance, will presumably become apparent in time.

11.4 Matching Supply and Demand

11.4.1 Background

Many of the challenges faced by wind energy at a national level are replicated at small scale, for instance where a business installs a wind turbine in order to reduce its reliance on imported electricity. Ideally the owners' desire may be energy self-sufficiency, but this runs up against the overarching problems of intermittency and balancing, and the need to match supply to demand at all times. There are nevertheless good arguments for installing wind turbines at a local level, not least that power is generated close to the

---

[3] The $52 figure was based on information from the US Energy Information Administration. The *WSJ* article was unsympathetic to wind energy and carried the strapline 'Producers get so much from the government that they can pay utilities to take their power and still make a profit'.

[4] The strike price for the Moray East 950MW development, adjusted for inflation (£57.50/MWh in 2012).

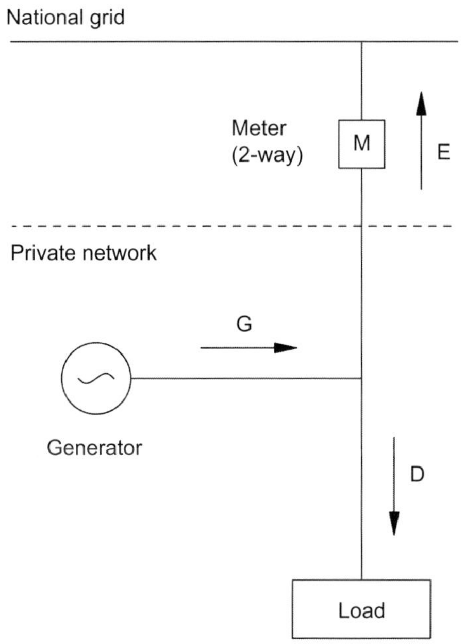

Figure 11.4 Schematic of a renewables generator serving an on-site load. Power flows shown are generation (G), demand (D), and export (E). Imported power corresponds to negative E.

point of consumption, avoiding network losses. In an 'on-site' configuration the generator is connected in parallel with the grid but on the customer's side of the meter, as shown in Figure 11.4. When wind energy is available it serves the customer load first, with any excess power exported to the grid. The instantaneous export (E) is the difference between generation (G) and demand (D), i.e.

$$E = G - D \tag{11.5}$$

When generation exceeds demand, E is positive and excess power is exported. If demand exceeds generation (e.g. in low winds) then E becomes negative; the business then imports power from the grid as previously. No switching or control systems are necessary to achieve this arrangement – electricity inherently flows to the point of lowest potential – and a two-way meter records the direction of power flows and cumulative energy totals (import and export). The only physical difference between an on-site generator and one connected directly to the grid is in the metering arrangements: the power flows would be the same in the absence of the meter.

On-site generation brings several potential advantages to a business, not least of which is the avoided cost of electricity and as the import tariff is generally higher than the export, an on-site wind turbine is a better investment than one selling power directly to the grid. The generator also provides a hedge against electricity price increases, improving the economic stability of the owner's business; there are wider benefits too, in becoming become more environmentally sustainable. Some of these points are illustrated with the example in the following section.

11.4.2 On-Site Generation Example

An example of a business that made an early investment in on-site wind generation is Mackies, a family farming business based in Aberdeenshire, north-east Scotland. In 1983 Mackies were among the first UK companies to install a grid-connected wind turbine: this was a 60 kW machine connected on the private side of the meter to offset farm electricity usage. At that time their electricity demand was relatively small, and the wind turbine output was a modest 50 MWh per annum. Although its impact on the business was not dramatic, it did provide useful data, and a report on its performance (Saluja, 1990) concluded that projects of this kind could be economically justified under the following conditions:

1. A reasonable wind regime is present.
2. A very high proportion of wind-generated electricity is consumed by the customer.
3. Local authority rates (taxes) are low.
4. The performance of the WTG matches the performance curve supplied by the manufacturer.

Mackies' first wind turbine did not meet all of these conditions, but the analysis of energy cost savings in the study was nevertheless fundamentally positive; moreover the project provided a valuable early introduction to wind energy at the level of owner and user at a time when it was still a relative novelty in the UK: the first utility-owned wind project on the UK mainland was not commissioned until 1991. Mackies' business meanwhile had begun to grow rapidly as they diversified from traditional farming into ice cream manufacture. Growth was accompanied by a significant rise in electricity demand: ice cream production involves high electricity usage for blast freezing and refrigeration. More power was also needed for production and packaging machines, and robotic milking stalls for the cattle.

By 1993 Mackies' annual electricity consumption had risen to 1.8 GWh, with a cost in excess of £100 000. Commercial wind turbine technology had by now progressed, however, so the prospect for larger scale on-site generation began to be visited in earnest. A wind measurement campaign was carried out at a hilltop site, around 1 km from the dairy. The average wind speed was found to exceed 7.5 m s^{-1} at 49 m height, with favourable long-term wind statistics found from correlation with Aberdeen Airport (see Section 9.2.3). Although planning consent was initially obtained for a single 225 kW wind turbine in 1996 the project did not proceed then due to unfavourable economics. Renewable incentives were available, but only via a competitive bidding process which did not favour small projects.[5] In addition, the fixed cost of the electrical connection to the dairy was relatively high, favouring a much larger wind turbine.

Following the introduction of the 2001 UK Renewables Obligation, which offered a fixed incentive for all wind generation, Mackies revisited the project on the basis of an 850 kW wind turbine, for which planning consent was sought and obtained. The project was delivered via

[5] The Scottish Renewables Obligation (SRO); see Section 11.6.

Figure 11.5 Construction of Mackies 850 kW on-site wind turbine. (Photo reproduced with kind permission of Mackies Ltd)

a turnkey contract covering purchase and installation of the turbine and all associated civil and electrical works, including a 1 km HV underground cable connection. Civil works comprising access track and foundations were carried out in autumn 2004, with delivery and erection of a Vestas V52 in March 2005 (see Figure 11.5). The total cost of the project was then £680 k (£800/kW), which was typical for onshore wind at that time. The single most expensive item was the wind turbine itself at around 66% of project cost; civil works accounted for 16%, with electrical works making up the remaining 18%.

During its first 2 years of operation Mackies' V52 achieved a capacity factor of 32% with around two-thirds of its output used on-site and the remainder exported to the grid; the on-site component supplied slightly over half of Mackies' electricity needs. Following the success of the installation two further wind turbines were added in 2007 bringing the installed capacity to 2.55 MW. The additional turbines were 'daisy chained' to the first by underground cable, so the entire windfarm remained within the private 11 kV network (Figure 11.6). With the expanded windfarm annual generation rose to around 7 GWh, now significantly greater than the average consumption of the business. Figure 11.7 shows monthly energy figures for export and on-site use as a function of total generation; the site average electricity demand is indicated by the dashed line. On-site supply and export both increase with total generation, the former tending asymptotically towards the site demand, and export rising roughly in proportion to generation.

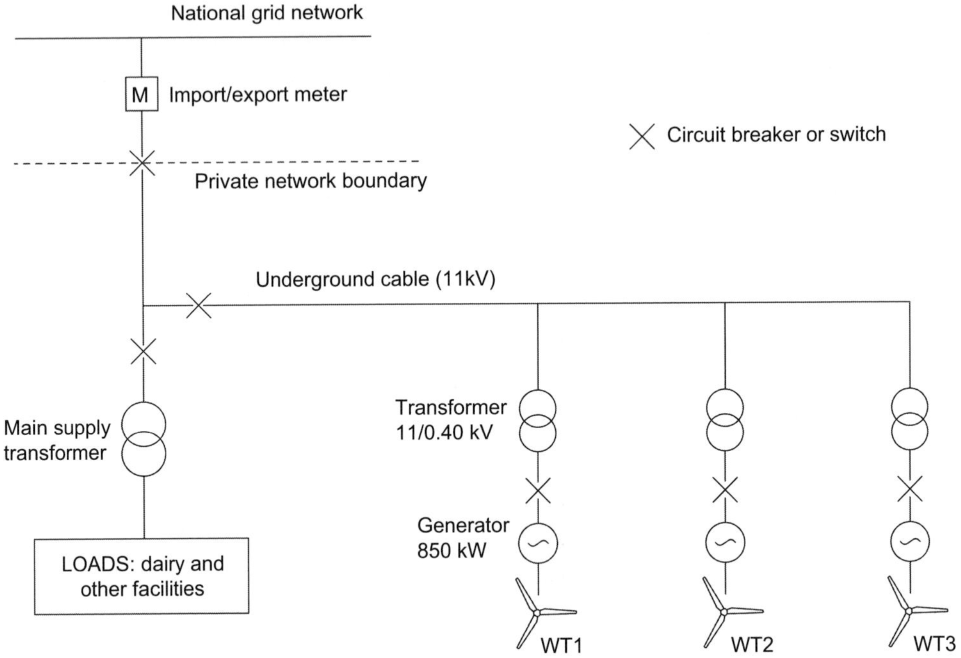

Figure 11.6 Single line diagram of Mackies private electrical network with on-site wind generators. The original single-turbine project was extended by 'daisy-chaining' two additional machines.

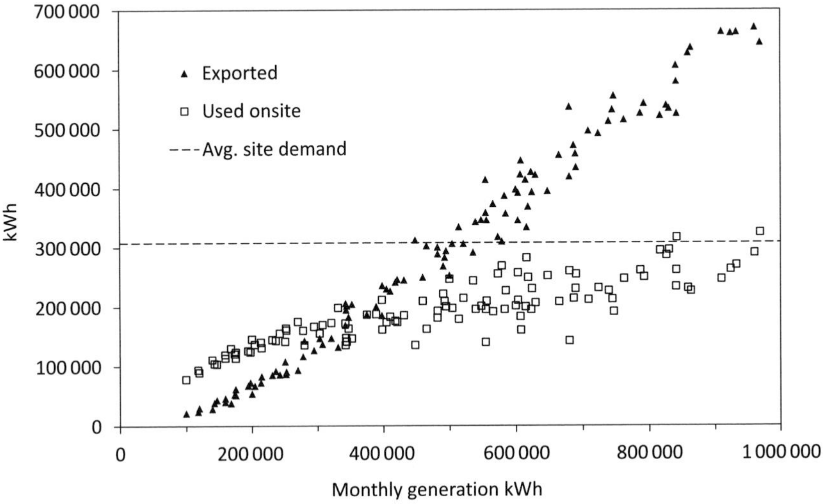

Figure 11.7 Monthly energy statistics for Mackies with on-site wind generation. On average, the windfarm supplies 70% of the business electricity demand; excess generation is sold to the grid.

Figure 11.7 illustrates the benefit of on-site wind generation, but also the underlying challenges of intermittency and balancing. On the one hand the windfarm supplies over two-thirds of the annual electricity demand of the business (and in a windy month the figure can rise to 90%), representing a significant economic benefit. On the other hand there is a law of diminishing returns, and no matter how much generation capacity is installed, wind power alone will not supply the total annual demand. This is simply a reflection of the statistical nature of the resource: there will always be periods when the wind speed is too low. In the present case an estimated 12% of the year is spent in winds below the turbine cut-in, when wind energy cannot provide any local energy supply.

11.4.3 Extrapolating to Wider Scale

The relationships between energy export and onsite use are illustrated more generally in Figure 11.8. The theoretical curves show the proportion of wind energy exported, and the proportion of load demand met by import (top up), as functions of total wind generation (normalised with respect to demand). The calculations assume a flat load profile, 32% wind capacity factor, and a Rayleigh wind distribution. Monthly export figures from Mackies windfarm are overlaid and show good agreement. It can be seen that even a dramatic over-installation of wind capacity (note the log scale) cannot completely remove the need for imported power – there will always be calm days. Based on the above some general rules are:

- To avoid export the installed wind turbine capacity (rated power) must be below the site minimum demand.
- The maximum proportion of the onsite demand that can be supplied by wind energy without incurring export is then equal to the WEC capacity factor.

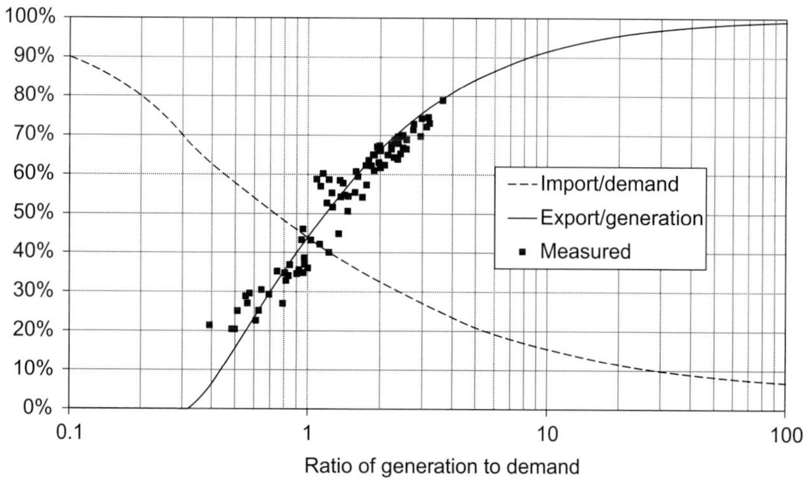

Figure 11.8 Theoretical relationships between on-site wind generation, export, and import (*assumptions*: Rayleigh wind distribution, 32% capacity factor, flat on-site demand). Monthly export data from the Mackies example are overlaid.

- If the average wind generation output equals the average site demand, then approximately half the generation will be exported; at the same time around half the site demand will have to be met by imported power.

These rules are somewhat approximate (the figures for export and import in Figure 11.8 are closer to 45% when generation equals demand) and they depend on the generator capacity factor, and to some extent on the on-site load profile. They can, however, be treated as useful rules of thumb, and as such can help to illustrate the challenge of implementing renewable energy at national scale. The power flow statistics from Mackies can be extrapolated to those of a whole country, in this case Scotland: 'local' load becomes national electricity demand, installed wind energy capacity is likewise on a national scale, and exported generation flows out of the country via an interconnector.

The Scottish government has set a target of meeting the equivalent of 100% of national electricity demand from renewables by 2020 (Scottish Government, 2017). Assuming wind power is the main source, the trends in Figure 11.8 suggest that only around half of the energy generated would meet the local (i.e. Scottish) demand, with the remainder exported to the rest of the UK. Similarly, half the demand would have to be met by other (presumably non-renewable) generation. This scenario is technically feasible as the Scottish grid is strongly interconnected with the rest of the UK, so the necessary balancing is easily accomplished. At the time of writing the policy appears to be on target: by the end of 2015 national renewables capacity was around 8 GW, with annual output equivalent to 59% of equivalent national demand. Approximately 30% of generation was exported, as seen in see Figure 11.9 (based on BEIS figures for the period 2000–15). The export cannot be directly compared with Figure 11.8 as it includes legacy non-renewable generation, but

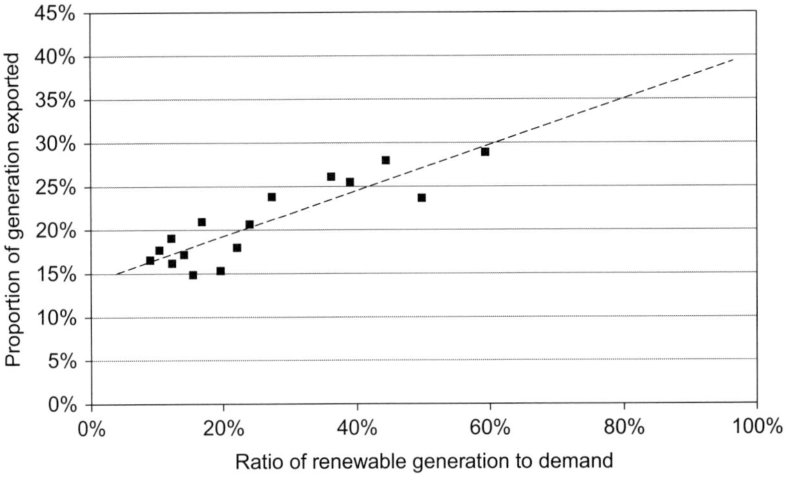

Figure 11.9 The proportion of Scottish electricity generation exported as a function of equivalent renewables output (ratio of generation to demand). According to this trend, when 100% equivalence is achieved, 40% of generation will be exported. (Based on BEIS data for the period 2000–15)

the rising trend is solely due to renewables as no conventional power stations were commissioned over the period shown. The trend in Figure 11.9 suggests that the Scottish policy objective of 100% equivalent renewables generation will coincide with an export proportion of around 40%, which is in good agreement with the small-scale results from Mackies.

The Scottish government policy of 100% renewable electricity is of course not the same as energy autonomy and a policy to achieve supply solely from renewables will be much harder to implement without major progress in large-scale energy storage, load management, and/or despatchable renewable generation (e.g. thermal generators burning biomass, or hydroelectric schemes with storage capacity). This is the wider challenge currently facing renewables.

11.5 Ownership

11.5.1 Overview

Unlike conventional forms of generation, wind energy lends itself to distributed ownership. This is seen in most countries pursuing wind energy policies, and in the UK in addition to the large-scale projects owned by utilities and corporations, there can be found projects owned by farmers, private investors, small businesses, communities, and various non-profit organisations. The ownership of wind power is a topical issue, invoking arguments about economics and social benefit that are not often heard in the context of conventional power stations. This may have something to do with the historic roots of wind power: only a generation ago most wind turbines were small, with individual machines installed by farmers or rural communities in a spirit of self-reliance or economic independence. The trend was most notable in Denmark, where wind power has been pursued since the beginning of the twentieth century (see Section 1.3), and where in the more recent era 'wind guilds' were formed to enable co-operative ownership of wind turbines. In 2002 an estimated 80% of wind turbines in Denmark were still owned by small-scale cooperatives (Krohn, 2002) and despite the growth of large projects a more recent estimate puts the proportion of Danish turbines in co-operative ownership still at around 50% (Vindenergi Danmark, 2017).

As the technology has grown in size and national economic importance it has, however, increasingly been viewed in the same light as conventional forms of generation, with utilities and institutional investors taking a major interest. This trajectory is common in emerging industries, where corporate ownership (it is argued) leads to economies of scale, better access to finance, and in the case of wind power cheaper electricity for the consumer. Yet even at large scale the idea of local or widespread ownership persists, as shown in Denmark where the concept of community participation is extended to very large projects including offshore windfarms such as Middelgrunden (Larsen et al., 2005). Danish legislation in 2009 mandated that 20% of any large-scale wind development should be offered for sale to those living within a 4.5 km radius (Hvelplund, 2017). In Germany, which has the highest installed wind energy capacity in Europe, over half is in private ownership (Renewable Energies Agency, 2018).

There is, too, a political dimension. This was seen in the grass-roots wind power movement in the 1970s which rejected nuclear power, with the two forms of generation seen by many as opposing ideologies both environmentally and in terms of citizen empowerment; the rejection of nuclear power by the Danish parliament in 1985 was a major spur for wind energy. Sustainability is also perceived by some in a wider context in which distributed energy production offers an opportunity for similarly distributed wealth. A counter-argument is that wind energy is an increasingly mature technology and as its importance grows the more it will be bound by the laws of market economics including large-scale corporate ownership and control. The debate is too large a subject for this book, but the following observations are offered in support of a distributed ownership model:

- The countries which gave most encouragement for participation in wind energy at the individual or co-operative level became, and remain, the major centres for the design and manufacture of the technology. Denmark and Germany are the prime examples (see above).
- Wind power is different from other 'big business' in that markets are driven by government policies rather than discretionary consumer choice: the price of wind energy is effectively set by law, and paid by everyone. On the basis that 'he who pays the piper calls the tune' governments have some legitimate power to determine where the economic benefit will fall.

The UK has pursued a largely free-market approach to electricity production since privatisation in the 1980s. There is little indigenous wind turbine manufacture (see Section 11.6) and the majority of wind energy capacity is owned by utilities or corporations. Local or community level ownership represents only a small proportion of UK installed capacity, although it does now account for a significant number of projects. Ownership data for Scotland are shown in Figure 11.10, based on 2015 figures from the Energy Saving Trust (Young, 2015): of a total installed capacity of 5587 MW

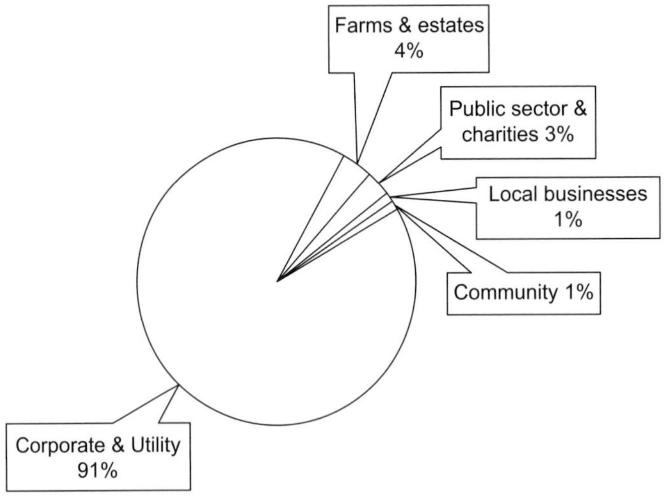

Figure 11.10 Ownership of wind energy capacity in Scotland in 2015. (Based on data in Young, 2015)

around 9% (508 MW) was in local ownership. This total includes a range of private, farm, or co-operative ownership models similar to those seen in other countries. There are a variety of definitions of 'community' ownership, ranging from investment cooperatives who collectively own a project and distribute shareholder profits, through to non-profit energy companies, which are effectively a form of local public ownership; somewhere in-between lie 'Industrial and Provident Societies' whose shareholders receive a fixed rate of interest, with the remainder of profit distributed to local public causes.

A good example of a community-owned project is the Isle of Gigha windfarm. Built in 2004, it has two notable points of interest: it was Scotland's first community-owned grid-connected windfarm; it was built on an island that, a few years earlier, had been purchased outright by the community and taken into public ownership. The Gigha project also gives an insight into the economic and technical challenges of wind energy development at small scale, but with evidence that local ownership is not a disadvantage in these respects but in some cases the reverse. The following is a short history of the project.

11.5.2 The Gigha Community Windfarm

The island of Gigha[6] lies off the west coast of Scotland in the county of Argyll; its land area is 14 km^2, and current population around 165. During the late twentieth century the population was in decline: in the early 1970s there were 200 inhabitants, but by 2002 fewer than 100. The island had been owned by a succession of private landlords, and when it was put on the market in 2001 the inhabitants proposed a community buyout. They were supported by the Scottish government and via a combination of grant and debt finance successfully met the asking price of £4 M. In 2002 the island was taken into public community ownership and a charitable trust established to manage and run it. One of the first priorities for the Isle of Gigha Heritage Trust was to identify viable economic activities to provide a long-term future for the island, and help reverse the declining population trend. A feasibility study was commissioned to examine the prospects for renewable energy (Bonnar, 2002); this identified wind power as an attractive possibility, and indicated suitable areas of the island for siting a medium-sized turbine.

The island and local electricity grid network are shown in Figure 5.15. The best location for a wind project was found at the south end of the island, 30 m above sea level with good all-round exposure. Further studies assessed the wind resource, grid capacity, transport access, and environmental suitability. The preferred scheme was for a wind turbine directly connected to grid, with energy sales via the recently introduced UK Renewables Obligation, which offered an attractive long-term tariff. A stand-alone energy company, Gigha Renewable Energy Ltd (GREL), was set up to progress the project, and a business plan was written on the basis of projected costs and revenue. Originally the plan was based on a single 250 kW wind turbine, but in order to maximise the

[6] Old Norse: *Gud-Øy*, 'God's Island' or 'the good island'.

potential yield from the available grid capacity (based on system studies by the network operator) this was changed to two 225 kW machines; accordingly a site layout plan was drawn up and access and infrastructure costs assessed.

Aside from the wind turbines themselves one of the highest-cost items was the 500 m site access track needed to connect the public road to the windfarm site; GREL calculated that its fixed cost could be better justified if the project capacity were increased to cover three wind turbines with combined rating of 675 kW. This would, however, exceed the grid export limit based on local voltage rise. The way in which this was resolved is described in detail in Section 5.6.1, but in summary, the network operator agreed to the higher export capacity on condition that the windfarm should incorporate dynamic power factor control. This was achieved at low cost by including a static VAr compensator (SVC) in the private electrical network, and the project was then progressed as a development of three Vestas V27 wind turbines.

A full planning application was submitted to the local authority in September 2003. It won a high level (99%) of public support with no formal objections from any inhabitants of the island; at one meeting a resident rose to say that although he personally disliked wind turbines he would support the project for the wider benefit of the community in which he lived. The project was granted full planning consent in June 2004, 8 months after submission. By this time, however, Vestas had ceased V27 production, and with limited alternatives the Gigha Trust explored the possibility of purchasing second-hand wind turbines. By chance an early windfarm at Haverigg in the north of England was then being re-powered, and five V27s were up for sale: the Trust negotiated the purchase of three with the owners, who were helpful and supportive. The machines were inspected while still operational, and their service records were made available. An agreement was then signed to purchase three of the V27s 'as seen'.

The turbines were dismantled and transported to Gigha in late 2004. Project delivery was achieved via two contracts covering (a) turbine procurement and (b) civil and electrical balance of plant. The latter included dismantling the turbines at Haverigg, transportation, and re-erection on Gigha. Because the V27s were second-hand a single turnkey contract was not available, as it would have exposed the principal contractor to risks associated with the turbines, which were 12 years old and long out of warranty. Although the wind turbines were a bargain the overall project costs had still to be carefully controlled. The cost of the site access track and civil infrastructure remained a concern, and to address them the Trust re-opened a dormant quarry on the island to provide stone for aggregate; the same quarry was used to supply a new community housing development, so the fixed costs of re-opening it were shared between two projects.

The island location presented logistical challenges and a transport survey identified pinch points on the local single track roads (helicopter delivery was considered for the main turbine components, but quickly ruled out on cost grounds). Happily the V27 towers were three part, so all delivery vehicle lengths were within normal road limits and no modifications to the public road were needed; the scheduled Gigha car ferry was also deemed suitable for all deliveries with the proviso that the crane was stripped down to meet the ramp weight limit; Figure 11.11 shows ferry

Figure 11.11 Ferry delivery of V27 tower section to Gigha in 2004. (Photo courtesy Donald Mackay)

delivery of a V27 tower section. Construction and commissioning proceeded without major difficulty, and in December 2004 the Gigha community windfarm was formally commissioned. At the time of writing the windfarm has been operational for over 14 years, and despite the overall age of the wind turbines – now 26 – their availability remains high. The average capacity factor of 32% is among the highest for windfarms in the region, and performance has not noticeably degraded since commissioning (in 2015 the capacity factor was 36%).

The capital cost of the project at the time of construction was £440 k or £652/kW, which was then comparable to the cost of large-scale utility windfarms.[7] The cost of electricity from the project is also comparable as all electricity sold under the UK Renewables Obligation receives roughly the same unit subsidy irrespective of project scale. Gigha's operating costs are similar to new projects, although the age of the V27s has implications for the extent of insurance cover available. Overall, the technical and economic performance of the project is comparable to that observed with significantly larger scale developments, the principal difference being that the economic benefit is wholly retained within the community. One key factor in the success of the project is the sound design of the wind turbines: although less electrically sophisticated than modern machines, the V27 is a match in terms of output and reliability. Some of the wider lessons learned from the Gigha project are the following:

[7] The nearby 30 MW Beinn an Tuirc windfarm on Kintyre cost £21M to build in 2002, or £690/kW.

- Projects in community ownership can demonstrate output and availability as good as, and often better than, large-scale utility-owned projects.
- Local support can help keep planning and development costs low and proportionate to project scale. This has been noted more widely (Warren, 2010).
- Where a renewable energy project is important to the local economy, technical problems are dealt with as a matter of priority. Communities can tackle difficult technical challenges.

Following their initial success the Isle of Gigha Heritage Trust invested in a fourth turbine, this time a new Enercon E33, in 2013 (Figure 11.12). In the years since the first project a number of other community groups around Scotland have developed projects, many on remote or island sites including the Orkneys and Outer Hebrides. Among the most ambitious to date are the 6.9 MW Lochcarnan Community Windfarm on South Uist, and the 9 MW Beinn Ghrideag windfarm on Lewis with three Enercon E-82s; commissioned in 2015, the latter is currently the largest community-owned wind project in the UK. All these projects sell power directly to the grid with revenues retained locally for non-profit purposes. In one estimate the employment impact of re-investing income from community-owned wind projects is up to eight times that arising from conventional, i.e. non-owned, developments (Okkonen, 2016).

Figure 11.12 The Isle of Gigha Community windfarm today. The original project of three second-hand V27s was commissioned in 2004, with a fourth turbine (Enercon E33 in foreground) added in 2013.

11.6 UK Renewables Policy

The three main drivers for the development of renewable energy have historically been energy security, economics, and the environment. The oil crisis of the 1970s brought the first two of these sharply into focus in the UK. Renewable energy policy here followed a broadly similar path to that in the US and Europe, though progress was initially slow. This can be attributed to a number of factors but an important one was undoubtedly the country's historic fuel reserves: Britain was always energy-rich. Wood from extensive native forests was the primary fuel source until the eighteenth century, when it gave way to abundant deep coal reserves; nuclear power was embraced in the mid twentieth century, and in the 1970s came the unexpected bonus of offshore oil and gas. Ironically the British Isles also has one of the best wind regimes in the world, but it was always going to be harder to sell wind energy politically in the UK than in other countries less well endowed with fossil fuel reserves. By contrast the development of wind power in Denmark began before the First World War and was driven by the need for rural electrification in a country with limited energy resources. Despite its massive coal reserves, however, the 1970s oil crisis seriously affected the UK and marked the beginning for a progressive, though at times tentative, renewable energy policy.

The first renewables legislation also had a political dimension. At the time of the energy crisis UK electricity supply was 100% publicly owned; power generation was dominated by large coal-fired stations plus several first-generation (Magnox) nuclear plants. In late 1973 the country was suffering widespread industrial unrest, including a major dispute within the coal mining industry, which became a full-blown strike in early 1974 with a knock-on effect on electricity generation. Power cuts and a 3 day working week were enforced to conserve coal stocks and, coming on top of the oil price shock, the UK's energy security suddenly appeared fragile. The Department of Energy was at the time overseeing the development of second-generation nuclear reactors, but in response to the crisis began a modest R&D programme for renewables including wind, wave, geothermal, and tidal power.

The industrial unrest eventually brought down the ruling Conservative government, but an incoming Labour administration had scarcely better luck. Strikes persisted, as did the high oil price. The Department of Energy funded some MW-scale wind turbine prototypes, but the overall budget was small and the programme lacked urgency. Energy security had become as much a political issue as one of resources, and in 1979, under the leadership of Margaret Thatcher, a new Conservative government came into office promising liberal economic policies and a free-market approach to energy production. Another coal strike in 1984 escalated into a major dispute that again threatened electricity supplies, but this time the government succeeded in 'keeping the lights on' by judiciously managing coal stocks and relying on baseload output from nuclear power stations (McSmith, 2011). There were no power cuts, and an emboldened government prepared plans for the complete privatisation of the electricity industry. Nuclear power was seen as a key component of future energy strategy, immune to the influence of coal miners or foreign oil supplies, and the North Sea was by now becoming a major source of oil and gas.

The head of the CEGB (the nationalised power company) Sir Walter Marshall was tasked with preparing the way for the sale of all the UK's generating assets, including the nuclear stations, under the 1989 Electricity Act. The Department of Energy was abolished on the grounds that the free market, rather than the state, should henceforth dictate which forms of energy production were economic. There would be no more subsidies: the price of electricity, and the means of generating it, would be decided by market forces without political intervention. Or so it was intended – at this point, however, it became clear that free-market principles would be difficult to reconcile with energy security. Market investors were unwilling to underwrite the (largely unknown) costs associated with nuclear power, and the nuclear stations had to be withdrawn from the privatisation programme. Sir Walter, who had been knighted for his role in maintaining electricity supplies during the 1984 miner's strike, lost his job. The government conceded that some form of price subsidy to support nuclear power was unavoidable: the alternative of allowing the nuclear stations simply to be switched off was not an option. Accordingly in 1990 the first Non-Fossil-Fuel Obligation (NFFO) was introduced. This legislation obliged the privatised electricity suppliers to include a minimum component of non-fossil output in their mix, but allowed them to charge customers a higher tariff to cover the cost.

Although the NFFO was primarily introduced to rescue the nuclear industry, it opened the door for renewables, whose costs had by now significantly dropped. The legislation was then extended to provide a rolling programme of renewable energy installation, with periodic capacity auctions over an 8 year period against a target of 3% of UK electricity supply. This policy led directly to the building of the UK's first commercial windfarms. The NFFO applied in England and Wales only, and subsequent legislation was introduced for Scotland via the Scottish Renewables Obligation (SRO) and Northern Ireland (NI-NFFO). The NFFO legislation had mixed success. On the one hand, it began the widespread introduction of wind power in the UK, and its competitive nature was credited with reducing the cost of generation; on the other hand it led to modest installed capacity, and failed to promote either widespread project ownership, or a domestic manufacturing industry (Mitchell, 2000).

A major failing of NFFO was that although competitive bidding led to very low generation prices there were no penalties for non-delivery, and by 2003 around two-thirds of contracted capacity remained unbuilt (Hartnell, 2003). The limitations of the policy were finally recognised when it was replaced with the 2001 UK Renewables Obligation (RO), which placed a legal obligation on electricity suppliers and imposed penalties for non-compliance. One attractive aspect of the RO legislation was that the tariff paid to renewable generators was effectively set by the penalty paid by non-compliant suppliers; in this way 'brown' generators subsidised 'green'. At the same time the capacity targets were set high, with initially 10% of electricity to come from renewables by 2010, and subsequently 15% by 2015. The RO legislation led to a rapid and widespread expansion of wind energy and, without the competitive bidding element of the NFFO, also encouraged projects over a wide range of scale and ownership. The relative success of the two policies may be judged by Figure 11.13, which shows the annual average power output of UK wind projects (onshore and offshore) since 1990.

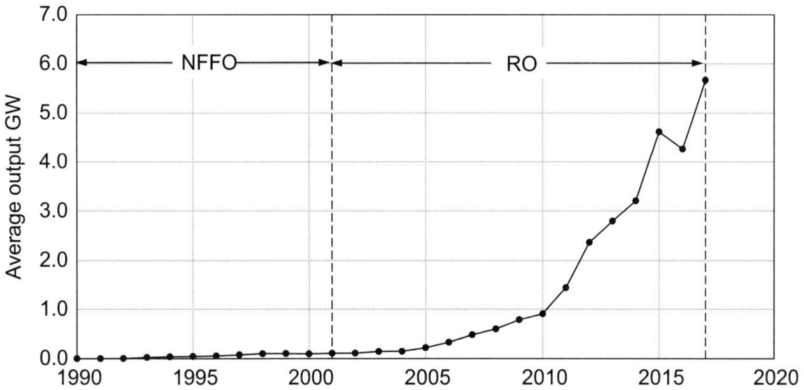

Figure 11.13 Annual average output of wind power in the UK since 1990. The NFFO legislation brought down costs but resulted in limited capacity; installation finally took off with the 2001 Renewables Obligation. (*Source*: DUKES)

Arguably the RO policy should have been implemented a decade earlier, but in 1990 the political climate was set firmly against a widespread subsidy for renewables, or anything that might prejudice the sale of the public electricity assets. The NFFO was also somewhat grudging in the level of capacity it issued and the policy lacked commitment. The legislation did not, for instance, include Scotland until 1994: the reasons for this were related to the privatisation agenda but the result was that the majority of the UK onshore wind resource remained untapped for another half decade. The introduction of the Renewables Obligation in 2001 finally ensured an ambitious market for wind energy across the entire UK, and the resulting distribution of capacity today can be seen in Figure 11.14. As a footnote, the privatisation of UK electricity had mixed results. Although the objective was to remove state control from the UK energy sector the majority of regional electricity companies were subsequently sold overseas, in many cases to state-controlled corporations, in what seems a peculiarly British interpretation of a free market.

In 2010 further legislation was introduced to favour smaller-scale renewables via long-term fixed price contracts. The UK Feed-In Tariff (FiT) targeted projects up to 5 MW, with a range of capacity and technology bands. The FiT legislation ran into difficulties due to its complexity, however, and on account of the high prices paid in some technology bands. The regressive effect of banding thresholds was also controversial: tariffs were set by generator nameplate rating rather than energy output, which led to under-utilisation of sites (with effectively higher prices paid for less output). The scheme was amended in a series of rear-guard actions but eventually closed to new entrants in 2017.

In 2015 new legislation in the form of 'Contracts for Difference' (CFD) was brought in to replace the Renewables Obligation; this reintroduced a bidding process with periodic allocations of capacity, similar to the NFFO. The first allocations included a number of onshore windfarms offering strike prices (guaranteed tariffs) of around £82.5/MWh and offshore projects with prices

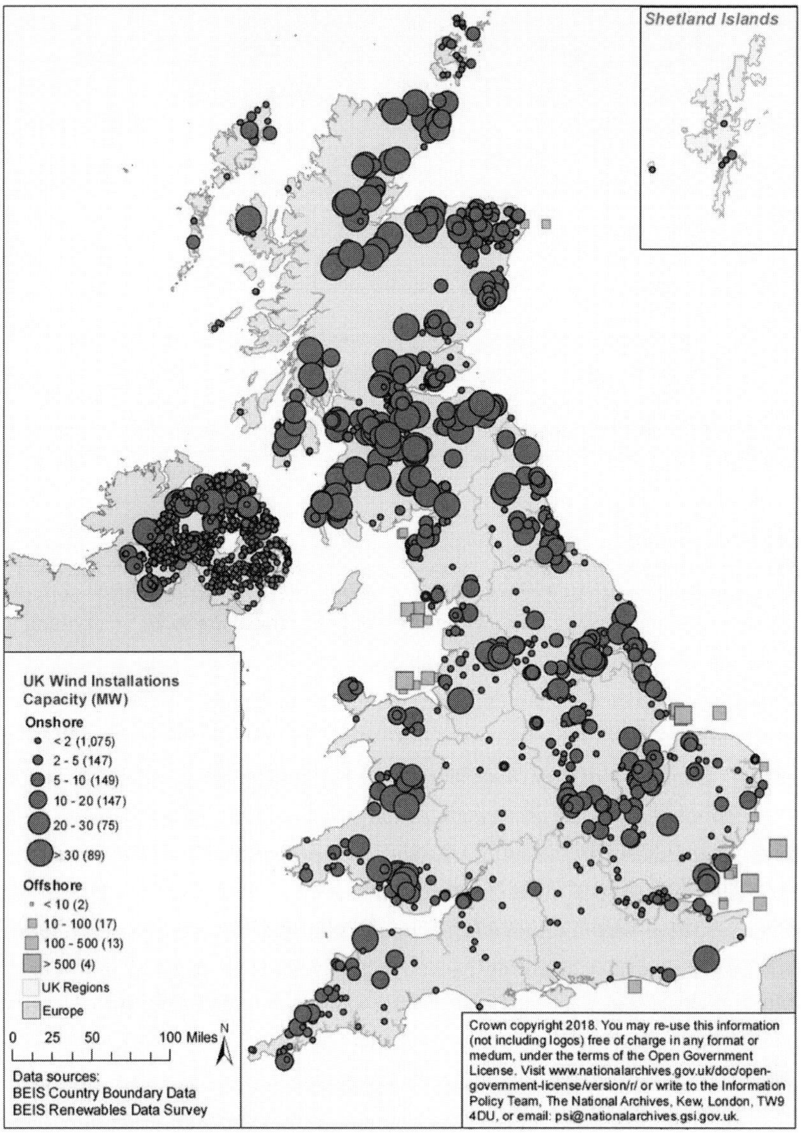

Figure 11.14 Distribution of UK onshore and offshore wind capacity in 2017. Wind energy then provided 15% of UK national demand. (*Source*: DUKES 2018; contains public sector information licensed under the Open Government Licence v2.0)

down to £115/MWh (DECC, 2015). The new policy initially co-existed with the Renewables Obligation, but the latter was closed to new entrants from 2017. In a surprise move onshore wind and solar PV were subsequently excluded from bidding for CFD contracts, which in tandem with the closing of the RO and FiT schemes effectively ended the development of new onshore wind in the UK. The government's decision was partly a reaction to perceived public attitudes, with the

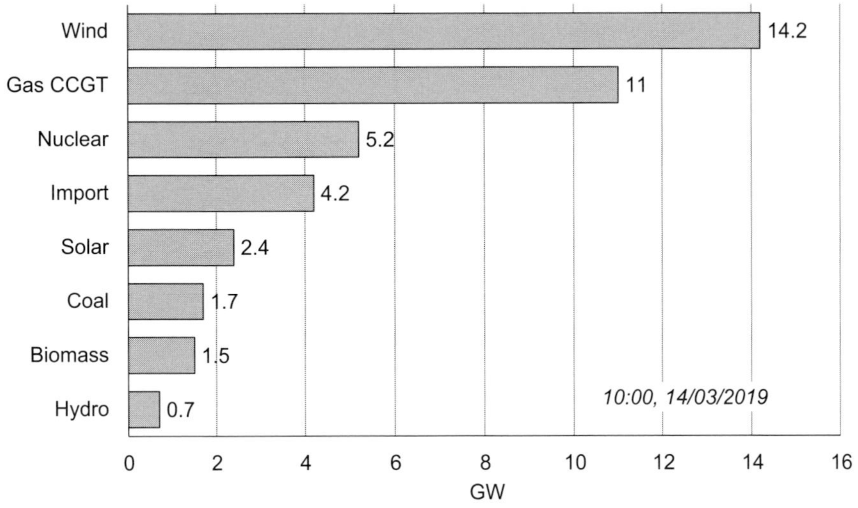

Figure 11.15 Powering ahead. On the morning shown, wind power was making the largest contribution to UK electricity supply, at over one-third of total generation. Collectively renewables (wind, solar, biomass, and hydro) were contributing 46% of total supply. (*Source*: http://energynumbers.info, from data supplied by Elexon, National Grid, and the University of Sheffield)

responsible Minister stating, 'Onshore wind is an important part of our current and future low-carbon energy mix. But we are reaching the limits of what is affordable, and what the public is prepared to accept' (Rudd, 2015).

Ending support for onshore wind was controversial, coming at a time when the technology was widely acknowledged to be cheaper than either offshore wind or nuclear power, both of which were supported under the CFD legislation; the affordability argument seemed weak. Meanwhile the government's own attitude surveys continued to indicate a sizeable majority in the UK in favour of onshore wind (BEIS, 2018). At the time of writing a future policy reversal seems likely, whether through a change of government or because the economic case for renewable energy is increasingly hard to refute. The growth of wind power meantime continues apace, and on some days wind is now the largest contributor to UK electricity supply: Figure 11.15 shows a snapshot of the generation mix on a windy day in March 2019.

11.7 Exercises

11.7.1 Project Appraisal

The table below contains a breakdown of the construction and operational costs of a 5 MW onshore windfarm. Calculate the levelised cost of energy for the project over a 20 year lifetime, assuming a target discount rate of 10%, and annual capacity factors of (a) 25% and (b) 30%.

Capital costs	£
Wind turbines	4 800 000
Civil and electrical work	1 840 000
Planning and development costs	200 000
Annual costs (fixed)	
Insurance and taxes	38 000
Electrical O&M	3000
Management	8000
Annual costs (variable)	
Availability warranty	£15 per MWh

11.7.2 Offshore Installed Costs

The table below gives cost and capacity figures for a number of offshore wind projects that achieved final investment approval in 2018 (*Source*: Wind Europe, 'Offshore Wind in Europe: Key Trends and Statistics 2018'.) Calculate the installed cost in €M/MW for each country's developments, then compare with the data shown in Figure 11.1(b) assuming an exchange rate of £0.87/€1.

Sector	Investment cost €M	Capacity MW
UK	5400	1858
Netherlands	1400	732
Denmark	1100	605
Belgium	1800	706
Germany	400	258

11.7.3 Levelised Cost of Offshore Wind

For each of the offshore projects listed in the table above calculate the levelised cost of energy (LCoE). In each case assume an output capacity factor of 40%, annual costs of €23/MWh, 20 year project lifetime, and a discount rate of 7.5%.

11.7.4 Network Restriction

A community owns an 800 kW wind turbine. They wish to install a second, identical, machine but the local network capacity is limited to a peak output of 1.4 MW. The community has a choice of either restricting the second turbine to 600 kW maximum output, or de-rating both turbines to 700 kW. Which option should they choose, and why?

11.7.5 De-rating Windfarms

An 800 MW offshore windfarm comprises 100 wind turbines with individual rating of 8 MW. There are strategic network advantages in restricting the nominal array output to 600 MW by lowering the power set-point of all the wind turbines to 6 MW, but allowing them to operate back at full power at certain times. Explain what the advantages of such a policy might be, compared with the case where the turbines are operated at their nominal rating.

11.7.6 Grid Penetration

A small country has an average electrical demand of 5 GW, and wishes to be self-sufficient on the basis of renewable energy. If the average capacity factor of wind plant in the country is 28%, what installed wind power capacity would be needed to generate the equivalent of the country's average electricity needs? And if excess wind generation can be exported to a neighbouring state, what would be the required power rating of the interconnector?

11.7.7 Battery Storage

A recent US Department of Energy report gave the installed cost of long-term battery storage systems as $400 per kWh capacity (EIA, 2018). Estimate the capacity and cost of a battery suitable for storing the full output of a 300 kW wind turbine for 8 hours. If the battery has a round trip efficiency of 90% and the stored energy is sold back to the grid at a premium of $50 per MWh calculate the number of full discharge cycles necessary to pay back the cost of the battery. Ignore battery degradation, interest rates, and cash flow discounting.

11.7.8 Feed-In Tariffs

The UK Feed-In Tariff legislation was introduced to stimulate the development and ownership of smaller-scale renewables. The tariffs paid for energy varied according to generator nameplate rating, so that a wind turbine rated below 500 kW was paid a unit price 40% more than one rated at 500 kW or above. Explain (a) why this policy could lead to market distortion and (b) why an alternative strategy with tariff bands based instead on energy production might be preferable.

REFERENCES

Abiven, C, Palma, J M L M, and Brady, O. 2009. Time dependent CFD analyses of wind quality in complex terrain. Paper presented at EWEC 2009, Marseille, France.

2011. High-frequency field measurements and time-dependent computational modelling for wind turbine siting. *Journal of Wind Engineering and Industrial Aerodynamics*. Vol. 99, pp. 123–29.

Ackermann, T (ed.). 2012. *Wind Power in Power Systems*. Wiley.

Albers. 2009. Comparison of lidars, German test station for remote wind sensing device. Paper presented at EWEC 2009, Marseille, France.

Anderson, C G, et al. 1993. Yaw System Loads of HAWTs. ETSU W/42/00195/REP.

Anderson, C G, Heerkes, H, and Yemm, R. 1999. Use of blade mounted dampers to eliminate edgewise stall vibration. *1999 European Wind Energy Conference*. James & James, pp. 207–11.

Anderson, C, and Jamieson, P. 1988. Mean load measurements on the Howden 33 m wind turbine. *Procs 10th BWEA Conference, London, 1988*. Mechanical Engineering Publications Ltd.

Anderson, C, Niven, A J, Jamieson, P, Knight, R R, and Milborrow, D. 1987. Flow visualisation studies on rotating blades. *Procs 9th BWEA Conference, Edinburgh, 1987*. Mechanical Engineering Publications Ltd.

Anderson, C G, Richon, J-B, and Campbell, T J. 1998. An aerodynamic moment-controlled surface for gust load alleviation on wind turbine rotors. *IEEE Transactions Control Systems Technology*. Vol. 6, pp. 577–95.

Anderson, J D. 1985. *Fundamentals of Aerodynamics*. McGraw-Hill.

1997. *A History of Aerodynamics*. Cambridge University Press.

Anderson, M B. 1981. An experimental and theoretical study of horizontal axis wind turbines. PhD thesis, University of Cambridge.

Appleton, S. 2005. Design and manufacture of radar absorbing wind turbine blades, DTI Report W4400636, QinetiQ, February.

Appleyard, D. 2017. Dutch offshore consortium win signals industry-wide efficiency gains. http://newenergyupdate.com/, 13 February.

Archive, Scottish Screen. 1955. http://movingimage.nls.uk/film/6387.

ASTM. 2017. Standard practices for cycle counting in fatigue analysis, ASTM Designation: E-1049 85 (Reapproved 2017), ASTM International.

Bacon, D. 2002. A proposed method for establishing an exclusion zone around a terrestrial fixed radio link outside of which a wind turbine will cause negligible degradation of the radio link performance. *Fixed-Link Wind-Turbine Exclusion Zone Method*. Ofcom.

Bak, C, et al. 1998. Double stall. *Riso-R-1043*. Riso National Laboratory.

Bathurst, G. 2009. A simplified method for estimating voltage dips due to transformer inrush. *20th International Conference on Electricity Distribution, Prague*. CIRED.

Bedford, L A W, Simpson, P B, and Stevenson, W G. 1985. The Orkney 60 m horizontal axis wind turbine generator – a progress report. *Procs. 7th BWEA Conference*. Mechanical Engineering Publications Ltd, pp. 241–51.

BEIS. 2018. *Energy and Climate Change Public Attitude Tracker – Wave 25*. Department for Business, Energy, and Industrial Strategy.

Björck, A. 1990. Coordinates and calculations for the FFA-W1-xxx, FFA-W2-xxx, and FFA-W3-xxx series of airfoils for horizontal-axis wind turbines, FFA TN 1990–15, Stockholm.

Bonnar, J. 2002. Energy usage and options for renewable energy on the island of Gigha. MSc dissertation, Open University.

Bramwell, A R S. 1976. *Helicopter Dynamics*. Edward Arnold.

Brooks, T, Pope, D, and Marcolini, M. 1989. Airfoil self-noise and prediction, NASA Reference Publication 1218, NASA Langley Research Center.

Brown, A. 1984. Operating experience with the Howden HWP-300 generator. *Procs. 6th BWEA Conference*. Cambridge University Press, pp. 59–69.

BS6399-2. 1997. Loading for buildings. Part 2: Code of practice for wind loads.

Burch, S F, et al. 1988. Estimation of the UK wind energy resource using computer modeling techniques, ETSU WN 7053, DTI.

Burton, T, et al. (eds.). 2011. *Wind Energy Handbook*. Wiley.

BWEA. 1982. *Wind Energy for the Eighties*. Peter Peregrinus.

CAA. 2007. *Licensing of Aerodromes*. Civil Aviation Authority.

2011. *CAA Policy and Guidelines on Wind Turbines*. Civil Aviation Authority.

Carter, R A. 2006. Boat remains and maritime trade in the Persian Gulf during sixth and fifth millennia BC. *Antiquity*. Vol. 80, pp. 52–63.

Cleijne, H, and Scott, J. 2012. *Smart Grid Strategic Review: The Orkney Islands Active Network Management Scheme*. KEMA report for SHEPD.

Cook, N J. 1985. *The Designer's Guide to Wind Loading of Building Structures, Part 1*. Butterworths.

Corbel, G, et al. 2007. Wind trends in the highlands and islands of Scotland and their relation to the North Atlantic Oscillation. Paper presented at 19th Conference on Climate Variability and Change, San Antonio, TX.

Corten, G P. 1999. The tip commands – a hypothetical explanation of double stall. *Procs 21st BWEA Conference*. Professional Engineering Publishing, pp. 383–89.

Crandall, R A, Dahl, N C, and Lardner, T J. 1978. *An introduction to the mechanics of solids*. McGraw-Hill.

Dalpane, E, and Dutting, S. 1988. Status of M30 single bladed WECS. *Procs. European Community Wind Energy Conference*. H S Stephens, pp. 85–89.

de Vries, E. 2011. Vestas builds on earlier innovations. *Windpower Monthly*, 11 October.

2013. Close up – Aerodyn's 6MW offshore turbine design. http://windpoweroffshore.com, 16 August.

de Wolf, W B. 1987. Aerodynamisch geluid van windturbines, NLR-MP87004 U, January.

DECC. 2015. *Contracts for Difference (CFD) Allocation Round One Outcome*. Department for Business, Energy, and Industrial Strategy.

Derrick, A. 1992. Development of the measure-correlate-predict strategy for site assessment. *Procs. 14th BWEA Conference*. Mechanical Engineering Publications.

Dimson, E. 1989. The discount rate for a power station. *Energy Economics*. Vol. 11, pp. 175–80.

DNV-GL. 2019. Bladed (aero-elastic wind turbine design tool), DNV-GL, Det Norske Veritas Group. www.dnvgl.com/energy/generation/software/bladed/index.html.

Dodd, C W, McCalla, T, and Smith, J G. 1983. How to protect a wind turbine from lightning. NASA-CR-168229, September.

Dodd, Jan. 2014. Wind and radar can learn to co-exist. *Windpower Monthly*, 27 November.

Dodge, D. 2014. Illustrated history of wind power development. www.telosnet.com/wind/early.html.

Dunbabin, P A. 1994. Noise from wind turbines. PhD thesis, University of Edinburgh.

Eggleston, D, and Stoddard, F. 1987. *Wind Turbine Engineering Design*. Van Nostrand Reinhold.

EIA. 2018. *U.S. Battery Storage Market Trends*. US Energy Information Administration, May.

ENA. 1989. Planning limits for voltage fluctuations caused by industrial, commercial, and domestic equipment in the United Kingdom. *Engineering Recommendation P28*. Energy Networks Association.

2001. Planning levels for harmonic voltage distortion and the connection of non-linear equipment to transmission systems and distribution networks in the UK. *Engineering Recommendation G5/4*. The Electricity Association.

2012. Separation between wind turbines and overhead lines. *Engineering Recommendation L44*. Energy Networks Association.

2013. Engineering Recommendation G59/3. *Recommendations for the Connection of Generating Plant to the Distribution Systems of Licensed Distribution Network Operators*. Electricity Networks Association.

ESQCR. 2002. The electricity safety, quality and continuity regulations 2002. *Statutory Instrument 2665*. HMSO.

ETSU. 1996. ETSU-R-97. *The Assessment and Rating of Noise from Wind Farms*. DTI.

Fletcher, J, and Yang, J. 2010. Introduction to doubly-fed induction generator for wind power applications. *Paths to Sustainable Energy*. Ng, A (ed.). Intech.

Flight. 1921. Pierre Levasseur. *Flight International*, 17 November.

Forestry Commission. 2001. *Road Specification with Reference to the DfT Design Manual for Roads and Bridges (DMRB)*. Forestry Commission.

Fraunhofer Institute. 2013. *Levelized Cost of Electricity Renewable Energy Technologies*. Fraunhofer Institute for Solar Energy Systems.

Freris, L. 1990. *Wind Energy Conversion Systems*. Prentice Hall.

Freris, L, and Infield, D. 2008. *Renewable Energy in Power Systems*. Wiley.

Germanischer Lloyd. 1993. Regulations for Certification of Wind Energy Conversion System (Rev. 1998). Germanischer Lloyd.

Gipe, P. 1993. *Wind Power for Home and Business*. Chelsea Green.

2017. Wind-Works. www.wind-works.org.

Goodfellow, D, Smith, G A, and Gardner, G. 1986. Control strategies for variable-speed wind energy recovery. *Procs. 8th BWEA Conference*. Mechanical Engineering Publications Ltd, pp. 219–28.

Gougeon Brothers. 2018. GBI history. www.westsystem.com/gbi-history/.

Gramm, P. 2012. The multiple distortions of wind subsidies. *Wall Street Journal*, 26 December.

Gunneskov, O, et al. 2002. Wind turbine blade. US Patent 7198471B2, July.

Hansen, M H, Gaunna, M, and Madsen, H A. 2004. A Beddoes–Leishman type dynamic stall model in state-space and indicial formulations, Riso-R-1354, Riso National Laboratory.

Harris, B. 1999. *Engineering Composite Materials*. Institute of Metals.

Hartnell, G. 2003. Powering the renewable future (focus on renewable energy). *Energy & Environmental Management*, March–April.

Hassan, U. 1992. A wind tunnel investigation of the wake structure within small wind turbine farms, E/5A/CON/5113/1890, UK Department of Energy.

Henderson, G, Bossanyi, E A, Haines, R S, and Sauven, R H. 1990. Synchronous wind power generation by means of a torque-limiting gearbox. *Procs. 12th BWEA Conference*. Mechanical Engineering Publications Ltd, pp. 41–46.

Himmelskamp, H. 1945. Profile investigations on a rotating airscrew. PhD thesis, University of Gottingen.

Historic England. 2018. Listed buildings, scheduled monuments, protected wrecks, registered parks and gardens, and battlefields. https://historicengland.org.uk.

Hock, S M, et al. 1987. Preliminary results from the dynamic response testing of the Howden 330kW HAWT, SERI/TP-217-3243, Solar Energy Research Institute.

Horn, J W, Arnett, E B, and Kunz, T H. 2008. Behavioral responses of bats to operating wind turbines. *Journal of Wildlife Management*. Vol. 72, pp. 123–32.

Hoskin, R E, Warren, J G, and Henderson, G M. 1988. Testing of the MS2 stall-regulated rotor. *Procs. 10th British Wind Energy Asociation Conference*. MEP Ltd, pp. 57–62.

Hughes, P, Childs, S, and Facchetti, A. 1993. Test results from the AOC 15–50 wind turbine development programme: Tip brake life test. *Procs. 15th BWEA Wind Energy Conference*. MEP Ltd, pp. 347–52.

Hughes, P, and Sherwin, R. 1994. Advanced wind turbine design studies – advanced conceptual study final report, NREL TP-442–4740, August.

Hvelplund, F, Østergaard, P A, and Meyer, N I. 2017. Incentives and barriers for wind power expansion and system integration in Denmark. *Energy Policy*. Vol. 107, pp. 573–84.

IEA. 1994. Acoustic measurements of noise emission from wind turbines. *Recommended Practices for Wind Turbine Testing*. International Energy Agency.

IEC. 1990. Protection of structures against lightning – Part 1: General principles, IEC 61024–1, International Electrotechnical Commission.

2002. Wind turbine generator systems – Part 24: Lightning protection, IEC 61400–24, International Electrotechnical Commission.

2005. Wind turbines – Part 1: Design requirements. *International Standard IEC 61400 Part 1 (2005–08)*. International Electrotechnical Commission, p. 3.

2008. Windturbine generator systems, Part 21: Measurement and assessment of power quality characteristics of grid connected wind turbines, IEC 61400–21, International Electrotechnical Commission.

2012. Wind turbines – Part 11: Acoustic noise measurement techniques Edition 3.0, IEC 61400–11, International Electrotechnical Commission.

IOA. 2013. *A Good Practice Guide to the Application of ETSU-97 for the Assessment and Rating of Wind Turbine Noise*. Institute of Acoustics.

IPCC. 2007. *Climate Change 2007: Mitigation of Climate Change*. Cambridge University Press.

ISO. 1993. Acoustics – Attenuation of sound during propagation outdoors, Parts 1 and 2, ISO 9613–1:1993(E) and ISO 9613–2:1996(E).

Jago, P, and Taylor, N. 2002. Wind turbines and aviation interests: European experience and practice, ETSU Report W/14/00624/REP, DTI.

Jamieson, P. 2011. *Innovation in Wind Turbine Design*. Wiley.

Jamieson, P, and Anderson, C. 1988. Power performance measurement on the Howden 33 m wind turbine. *Procs. 10th BWEA Conference*. Mechanical Engineering Publications Ltd, pp. 75–79.

Jamieson, P, and Brown, C J. 1992. The optimisation of stall regulated rotor design. *Procs. 14th BWEA Conference, Nottingham*. Mechanical Engineering Publications Ltd.

Jamieson, P, and McLeish, D. 1983. The HWP-300 wind turbine. *IEE Procs*. Vol. 130.

Jenkins, N, et al. 2000. *Embedded Generation*. The Institution of Electrical Engineers.

Jensen, N O. 1983. A note on wind generator interaction, RISO-M-2411, Riso.

Kane, L, and Ault, G. 2014. A review and analysis of renewable energy curtailment schemes and Principles of Access: Transitioning towards business as usual. *Energy Policy*. Vol. 72, pp. 67–77.

Katic, I, Hojstrup, J, and Jensen, N O. 1986. A simple model for cluster efficiency. *Procs. EWEC '86*. Rome, October.

Krohn, S. 2002. Danish wind turbines: An industrial success story. Danish Wind Industry Association, February. www.windpower.org.

(ed.). 2009. *The Economics of Wind Energy*. European Wind Energy Association.

Lamont, A, and Anderson, C. 2003. The impact of a coastal wind farm on bird populations, ETSU W/13/00612/00/RE, ETSU for DTI.

Langston, R H W, and Pullan, J D. 2003. *Windfarms and Birds: An Analysis of the Effects of Windfarms on Birds, and Guidance on Environmental Assessment Criteria and Site Selection Issues*. BirdLife International on behalf of the Bern Convention.

Larsen, J H M, Soerensen, H C, Christiansen, E, Naef, S, and Vølund, P. 2005. Experiences from Middelgrunden 40MW Offshore Wind Farm, Copenhagen Offshore Wind, 26–28 October.

Leithead, W E, Wilkie, J, and Anderson, C G. 1989. Simulation of wind turbines by simple models. *Procs. EWEC 89*. Peter Peregrinus, pp. 336–40.

Leithead, W, et al. 1990. Classical control of a pitch control system. *Procs. 12th BWEA Conference*. MEP Ltd, pp. 85–93.

Lemming, J, et al. 2008. *Contribution to the Chapter on Wind Power Energy Technology Perspectives 2008*. Risø National Laboratory.

Link, H, et al. 1995. The US Department of Energy Wind Turbine Development Program. Paper NREL/TP-441–7390 presented at International Solar Energy Conference, Hawai'i.

Linscott, B S, Perkins, P, and Dennett, J T. 1984. Large, horizontal-axis wind turbines, NASA TM-83546, March.

Lowson, M V. 1992. Applications of aero-acoustic analysis to wind turbine noise control. *Procs. 14th BWEA Conference*. MEP Ltd, pp. 91–99.

Madsen, B T. 1985. The industrial development of the windpower industry in Denmark. *Procs. 7th BWEA Conference*. Mechanical Engineering Publications Ltd, pp. 37–43.

Madsen, H A. 1990. Measured airfoil characteristics of three blade segments on a 19 m HAWT rotor, Riso-M-2826, Riso National Laboratory, January.

Madsen, P H, Hock, S M, and Hausfeld, T E. 1987. Turbulence loads on the Howden 26 m-diameter wind turbine, SERI/TP–217–3269, Solar Energy Tresearch Institute.

Maegaard, P, Krenz, A, and Palz, W. 2013. *The Rise of Modern Wind Energy (Wind Power for the World)*. Pan Stanford.

Martinez, V. 2011. Numerical and experimental analysis of stresses and failure in T-bolt joints. *Composite Structures*. Vol. 93, pp. 2636–45.

Massey, B S. 1979. *Mechanics of Fluids*. 4th edn. Van Nostrand Reinhold.

Mathews, F, et al. 2016. Understanding the risk to European protected species (bats) at onshore wind turbine sites to inform risk management, WC0753, University of Exeter.

Matsuishi, M, and Endo, T. 1968. Fatigue of metals subjected to varying stress, Japan Society of Mechanical Engineering.

McClanahan, A. 2013. Curating 'northernness' in Neolithic Orkney: A contemporary monumental biography. *Visual Studies*. Vol. 28, pp. 262–70.

McSmith, A. 2011. *No Such Thing as Society*. Constable.

Megson, T H G. 1972. *Aircraft Structures for Engineering Students*. Edward Arnold.

Milborrow, D. 1985. Changes in aerofoil characteristics due to radial flow on rotating blades. *Procs. 7th BWEA Conference, Oxford*. Mechanical Engineering Publications Ltd.

Milborrow, D J, Bedford, L A W, and Passey, D J. 1988. Development of an advanced low cost 1MW wind turbine for general utility application. *Procs. European Community Wind Energy Conference*. H Stephens, pp. 60–65.

Miller, D R. 1986. Summary of NASA/DOE Aileron-Control Development Program for wind turbines, NASA TM-88811, NASA.

Mitchell, C. 2000. The England and Wales Non-Fossil Fuel Obligation: History and lessons. *Annual Review of Energy & Environment*. Vol. 25, pp. 285–312.

Moller, T. 1997. Blade cracks signal new stress problem. *Windpower Monthly*, May.

Moores, W H. 1988. Laymans' guide to wind. *Procs. 10th BWEA Conference*. Mechanical Engineering Publications Ltd, pp. 81–88.

Morecroft, J, and Hehre, F. 1924. *Electrical Circuits and Machinery, Vol. II: Alternating Currents*. Wiley.

Morgan, C, and Bossanyi, E. 1998. Assessment of safety risks arising from wind turbine icing. Proceedings of Boreas IV Conference, Hetta, Finland, 31 March - 2 April 1998.

Mosley, W H, Bungey, J H, and Hulse, R. 1999. *Reinforced Concrete Design*. 5th edn. Palgrave.

Musial, W, et al. 2001. Static testing of the Aerpac UK AOC 15/50 FRP blade – first article, NWTC-ST-AOC-STA–01–0700, NREL.

National Grid. 2009. Operating the electricity transmission networks in 2020, initial consultation.

Natural England. 2014. Bats and onshore wind turbines Interim guidance, Technical Information Note TIN051.

Nielsen, E-G. 2000. Winds of change – 25 years of wind power development. www.windsofchange.dk.

Nikuradse, J. 1950. Laws of flow in rough pipes, NACA TM-1292 (translated from the 1933 original), Washington, DC.

NREL. 2019. NWTC Information Portal. *FAST – An aeroelastic computer-aided engineering (CAE) tool for horizontal axis wind turbines*. https://nwtc.nrel.gov/FAST.

NYSERDA. 2017. Public safety and setbacks. *New York Wind Energy Guide for Local Decision Makers*, LSR-LSW-guidesec8-1–v1. New York State Energy Research and Development Authority.

Okkonen, L, and Lehtonen, O. 2016. Socio-economic impacts of community wind power projects in Northern Scotland. *Renewable Energy*. Vol. 85, pp. 826–33.

Øye, S. 1986. Unsteady wake effects caused by pitch angle changes. Paper presented at IEA Joint Action on Aerodynamics of Wind Turbines – First Symposium, London, October.

Parsons Brinckerhoff. 2011. *Update of UK Shadow Flicker Evidence Base*. DECC, March.

Pedersen, J T, et al. 1998. Prediction of dynamic loads and induced vibrations in stall, R-1045(EN), Risoe National Laboratory.

Pedersen, T F, and Madsen, H A. 1988. Location of flow separation of an 11 m wind turbine blade by means of flow vizualisation and a two-dimensional airfoil code. *Procs. 10th BWEA Conference, London*. Mechanical Engineering Publications Ltd.

Pena, A, et al. 2013. Results of wake simulations at the Horns Rev I and Lillgrund wind farms using the modified Park model, DTU Wind Energy E-Report-0026(EN), Technical University of Denmark, October.

Poul la Cour Foundation. 2005. Poul la Cour Museum. www.poullacour.dk/engelsk/menu.htm.

Price, T J. 2005. James Blyth – Britain's first modern wind power pioneer. *Wind Engineering*. Vol. 29, pp. 191–200.

Pryor, S C, and Barthelmie, R J. 2011. Assessing climate change impacts on the near-term stability of the wind energy resource over the United States. *Proceedings of the National Academy of Sciences of the United States of America*. Vol. 108, pp. 8167–71.

Renewable Energies Agency. 2018. Ownership structure of onshore wind power installations. www.unendlich-viel-energie.de/english.

Robb, C. 2003. Low cost infrastructure solutions for small embedded wind generators, W/25/00583/REP, ETSU for DTI.

Ronsten, G. 1991. Static pressure measurements on a rotating and non-rotating 2.375 m wind turbine blade. *Procs. Amsterdam EWEC '91*. Elsevier.

RSPB. 2018. Wind farms – publications. www.rspb.org.uk.

Rudd, A. 2015. Statement on ending subsidies for onshore wind. Oral statement to Parliament, 22 June. https://www.gov.uk/government/speeches/statement-on-ending-subsidies-for-onshore-wind.

Saad-Saoud, Z, Craig, L M, and Jenkins, N. 1995. Static VAr compensators for wind energy applications. *Procs. 17th BWEA Conference, 1995*. Mechanical Engineering Publications Ltd, pp. 347–52.

Saluja, G S, and MacMillan, S. 1990. Operating experience of a 60kW wind turbine generator on an agricultural farm. *Procs. 12th BWEA Conference, Norwich*. Mechanical Engineering Publications Ltd, pp. 251–57.

Scottish Government. 1999. Planning and the historic environment, NPPG 18, 29 April.

2017. Renewable energy. www.gov.scot/Topics/Business-Industry/Energy/Energy-sources/19185.

Shearer, D L, and Brown, A. 1986. The construction of a 26MW wind park in the Altamont Pass region of California. *Procs. 8th BWEA Conference*. Mechanical Engineering Publications Ltd, pp. 23–26.

Shigley, J E. 1981. *Mechanical Engineering Design*. 3rd edn. McGraw-Hill.

Smeaton, J. 1759. An experimental enquiry concerning the natural powers of water and wind to turn mills and other machines, depending on circular motion. *Philosophical Transactions*. Vol. 51, pp. 100–74.

Snel, H, et al. 1999. Progress in the JOULE project: Multiple stall levels. *1999 European Wind Energy Conference*. James & James, pp. 142–45.

SNH. 2002. Visual assessment of windfarms: Best practice, SNH Report F01AA303A, University of Newcastle.

Sørensen, T, Nielsen, P, and Thøgersen, M L. 2006. Recalibrating wind turbine wake model parameters – validating the wake model performance for large offshore wind farms. *Procs. EWEC 2006*. Athens.

Spera. 1990. Structural properties of Laminated Douglas Fir-Epoxy Composite Material, NASA RP-1236, NASA.

SSP Technology. 2018. The SSP M36 root joint, Technical Data Sheet DK-5771, SSP Technology A/S.

Stevens, R J A M, and Meneveau, C. 2017. Flow structure and turbulence in wind farms. *Annual Review of Fluid Mechanics*. Vol. 49, pp. 311–39.

Stiesdal, H. 1994. Extreme wind loads on stall regulated wind turbines. *Procs. 16th BWEA Conference*. MEP Ltd, pp. 101–6.

Stiesdal, H, et al. 2006. Method for Manufacturing Windmill Blades. European Patent 1,310,351 (A1), 19 April.

Tammelin, B, et al. 2000. Wind energy production in cold climates, WECO Report JOR3-CT95 -0014, Finnish Meteorological Institute.

Tangler, J L, and Somers, D M. 1995. NREL Airfoil families for HAWTs, NREL/TP-442–7109, US Department of Energy, January.

Thomson, W. 1993. *Theory of Vibration with Applications*. 4th edn. Chapman & Hall.

Timmer, W A, and van Rooy, R P O J M. 1991. Thick Airfoils for HAWTs. *Procs. EWEC'91*. Amsterdam, October.

Tovey, A. 2017. Blown over: QinetiQ baffles radar with 'stealth' wind turbines. *The Telegraph*, 1 January.

Veldkamp, H. 1998. Wind turbine blade with u-shaped oscillation damping means, US Patent 6626642 B1, July.

Vermeer, N J, Sorensen, J N, and Crespo, A. 2003. Wind turbine wake aerodynamics. *Progress in Aerospace Sciences*. Vol. 39, pp. 467–510.

Vindenergi Danmark. 2017. *Annual Report 2016*.

von Mises, R. 1959. *Theory of Flight*. Dover.

Walker, D, et al. 2005. Resident golden eagle ranging behaviour before and after construction of a windfarm in Argyll. *Scottish Birds*. Vol. 25, pp. 24–40.

Warren, C R, and McFadyen, M. 2010. Does community ownership affect public attitudes to wind energy? *Land Use Policy*. Vol. 27, pp. 204–13.

Wehrey, M, et al. 1988. Dynamic response of a 330kW horizontal-axis wind turbine generator, US DoE Report SERI/STR-217-3203, Solar Energy Research Institute, February.

Westerhellweg. 2010. One year of LIDAR measurements at Fin01-Platform. Paper presented at DEWEK 2010, Germany.

Weston, D, and Richard, C. 2018. Updated: Moray East reaches financial close. *Windpower Monthly*, 29 November.

Whale, J. 1996. A study of the near wake of a model wind turbine using particle image velocimetry. PhD thesis, University of Edinburgh.

Whale, J, Anderson, C G, Bareiss, R, and Wagner, S. 2000. An experimental and numerical study of the vortex structure in the wake of a wind turbine. *Journal of Wind Engineering and Industrial Aerodynamics*. Vol. 84, pp. 1–21.

Whale, J, Papadopoulos, K H, Anderson, C G, Helmis, C G, and Skyner, D J. 1996. A study of the near wake structure of a wind turbine comparing measurements from laboratory and full-scale experiments. *Solar Energy*. Vol. 56, pp. 621–33.

White, G. 2009. Wind farm off the radar and powering ahead. *Daily Telegraph*, 17 May.

Wieringa, J. 1973. Gust factors over open water and built-up country. *Boundary-Layer Meteorology*. Vol. 3, pp. 424–41.

Williamson, K. 2012. Push and pull – testing wind turbine blades. *Reinforced Plastics*, 9 February.

Wright, A D, and Thresher, R W. 1988. Prediction of stochastic blade response using measured wind speed data as input to the FLAP code, SERI/TP–217–3394, Solar Energy Research Institute.

Yang, Z F, Sarkar, P P, and Hu, H. 2012. Visualisation of the tip vortices in a wind turbine wake. *Journal of Visualisation*. Vol. 15, pp. 39–44.

Young, F, and Georgieva, K. 2015. *Community and Locally Owned Renewable Energy in Scotland at September 2015*. Energy Saving Trust.

INDEX